HEAT TRANSMISSION IN BUILDINGS

HEAT TRANSMISSION IN BUILDINGS

A. W. Pratt
Department of Construction and Environmental Health,
University of Aston in Birmingham

A Wiley–Interscience Publication

JOHN WILEY AND SONS
Chichester · New York · Brisbane · Toronto

***British Library Cataloguing in Publication Data*:**

Pratt, A. W.
Heat transmission in buildings.
1. Buildings—Thermal properties
2. Heat—Transmission
I. Title
697 TM6025 80-42021
ISBN 0 471 27971 4

Filmset and printed in Northern Ireland at The Universities Press (Belfast) Ltd.,
and bound at the Pitman Press, Bath, Avon.

Contents

Preface

The book is a monograph of research intended especially for those seeking an interpretation of the thermal performance of buildings through a rigorous examination of heat transmission in the building fabric. Although mainly theoretical it includes an account of experiments on heat transmission through walls. The initial introduction of the fundamentals of steady and dynamic heat flow through structures is developed further in later chapters with applications to an examination of the energy balance of buildings for a variety of circumstances, including ventilation.

Many different combinations of boundary conditions may be devised in formulating mathematical models of heat exchange in buildings. The more important of these have been selected for their practical interest and the solutions evaluated to measure the interplay and relative importance of the factors considered. It has been the aim to express the solutions where possible in alternative forms, albeit with a calculable reduction in accuracy, that are more suitable for practical design purposes, and to identify parameters that characterize the thermal performance of buildings as demonstrated by the analysis and the results of experiments.

The experimental studies recounted were carried out at the Building Research Station, England. They are the most comprehensive investigation of their kind undertaken in any country and, in Britain, are the major source of data recommended in official regulations for calculating the space heating requirements of buildings. The test walls in these experiments were exposed to the weather as in a normal installation (a unique and valuable feature of the research) and the dates therefore of each period of measurement, usually a whole winter, are included in the record of results so that they may be related to the prevailing weather conditions and other factors of relevant interest.

The experimental work at the BRS involved several laboratory staff. Some were my colleagues during the later years when I was directly responsible for this research and acknowledgement is due to them for the benefit especially of their experimental ingenuity.

The book is composed largely of the contents of research papers that I have published. Thanks are due to the copyright holders and I am pleased to make the following acknowledgements:

Chapter 1. Figs. 1 to 6, Tables 1, 2, 3, and portions of text. Copyright holder The Institution of Mechanical Engineers.

Chapter 2. Figs. 1 to 15, Tables 1, 2, 3, 4, and portions of text with permission from *Thermal Conductivity* Tye, R. P., Vol. 1, 1969. Copyright by Academic Press Inc. (London) Ltd.

Chapter 3. Figs. 1, 3 to 11, Tables 1 and 2, and portion of text. Copyright holder Chartered Institution of Building Services.

Chapter 4. Figs. 1 to 5 and portions of text. Source of material and copyright The Institute of Physics.

Figs. 6, 7, 8, and portions of text. Copyright holder Pergamon Press Ltd., Oxford.

Chapter 5. Figs. 1 to 5, Tables 1, 2, 3, 4, 5, 6, and portions of text. Copyright holder Pergamon Press Ltd., Oxford.

Chapter 6. Fig. 1 and portions of text. Copyright holder Pergamon Press Ltd., Oxford.

Chapter 7. Fig. 1. Copyright holder Pilkington Flat Glass Ltd.

Figs. 2 to 5 and portions of text. Copyright holder Pergamon Press Ltd. Oxford.

Chapter 8. Figs. 1 to 7, reproduced by courtesy of the Director, Building Research Establishment, Crown Copyright Controller, HMSO.

Table 1 and data sheets, copyright holder Pergamon Press Ltd., Oxford.

Chapter 9. Figs. 1 to 14 reproduced from BRS Current Paper, Research Series 30/1964 by courtesy of the Director, Building Research Establishment. Crown Copyright Controller, HMSO.

Figs. 22, 23, 24, 25. Copyright holder Pergamon Press Ltd., Oxford.

Introduction

This analysis of the thermal behaviour of buildings begins with a theoretical examination of steady heat flow through a cavity wall, the most common form of building construction, exposed on one side to the weather elements and on the other facing an enclosure at a constant and uniform temperature. It is shown how the heat flux may be calculated to take account of orientation, different degrees of natural exposure, and of airspace ventilation in open cavity structures. The calculations are aided by introducing an equivalent outdoor temperature; this identifies a 'shade' transmittance for the wall and leads to a common form of expression for the flow of heat through structures generally, including glass. When the ambient temperature is unsteady the flow of heat is determined additionally by the thermal capacity of the structure. This is considered in the analysis by calculating solutions for the temperature in a homogeneous wall exposed to transient and steady periodic variations in the ambient temperature at either surface. By superimposition of these solutions expressions are obtained that give the resultant temperature in an enclosure for patterns of space heating that are typical of the normal usage of buildings.

A quantity that figures prominently in the design of buildings is thermal conductivity. This specific property of a material is discussed in Chapter 2 under the following divisions

1. Classification of building materials as heat insulators.
2. Heat transfer in porous material.
3. Mechanism of heat transfer in fibrous material examined by the methods of the Kinetic Theory of Gases.
4. Heat transmission in damp material.
5. Measurement of thermal conductivity using the steady state method, and dynamic methods with line, and plane, heat sources.

The control of heat flow in buildings relies basically on the low thermal conductivity of air. The air may be contained within small spaces or pores, or between the separate layers in cavity walls and roofs. In Chapter 3 an analysis is given of the thermal resistance of airspaces in naturally exposed cavity structures, in particular of the manner in which the resistance is affected by ventilation. The analysis is illustrated with a detailed calculation of steady heat flow through a cavity roof exposed to typical weather conditions in winter.

In practice it is often difficult to arrange for the flow of heat to conform with design requirements. This is particularly so with buildings where, for simplicity, it is assumed that heat passes through all elements of the structure in one direction only, normal to the surfaces. Edges and corners distort this simple pattern of heat flow and thereby introduce error in the statutory calculations recommended for determining the thermal requirements of buildings. These errors are considered in Chapter 4 by analysing two characteristic problems: (a) edge heat loss in the measurement of thermal conductivity by the hot plate method, with or without a guard ring, and (b) heat flow through a right-angled corner in which the adjoining slabs are isotropic but differ from each other in thickness and thermal conductivity.

The daily variation in the outdoor temperature is roughly sinusoidal superimposed with random fluctuations of shorter duration. Occasionally this diurnal temperature pattern is disturbed by a rapid change to a substantially different mean level which, if maintained long enough, will affect the mean indoor temperature by a similar amount. Such a change is a cold spell in winter when the outdoor temperature falls at an appreciable rate through several degrees and the lower temperature level is maintained for a prolonged period. The thermal response of a building to such an event is considered in Chapter 5 with a theoretical examination of the transient change in the air temperature inside an enclosure, in which heat is produced at a constant rate, following a step-function change in the outdoor temperature. The effects of ventilation, internal heat storage, and the thermal properties of the external fabric are analysed and compared by numerical calculation of the rates of cooling of various types of buildings representing heavyweight and lightweight structures.

The results of the analysis of dynamic heat flow given in Chapter 1 apply only to a solid wall of uniform material. In practice the majority of walls consist of two or more layers of different thermal properties and usually include a cavity. In Chapter 6 the analysis is extended to two problems of non-steady heat flow in a cavity wall. The first considers periodic heat flow applied as a regular sinusoidal variation in the ambient temperature at the outside surface. The second examines transient flow in which the applied temperature condition is a single step function of time. The solutions obtained may readily be adjusted to define dynamic heat flow in various forms of construction and to compare, for example, the effect of placing a layer of highly insulating material at various positions in a wall.

Also described in Chapter 6 is a method for calculating the passage of regular and random wave forms of temperature through structures. It is suitable for practical use and applies to any form of construction for which the appropriate transient response curve of temperature is available. The computation is reducible to a simple drill and the results of several worked examples are given.

Windows are an important consideration in the thermal design of a building owing to their high thermal conductance and the distinctive property that glass possesses of transmitting solar radiation. The effect of these properties of windows upon the indoor temperature is examined in Chapter 7 with a theoretical analysis of the heat gain to an enclosure due to the direct transmission of solar radiation through glass and subsequent absorption at opaque interior surfaces. Solutions are obtained that define the resultant heat load in terms of the ratio of glass to opaque wall area, the thermal properties of the walls, absorption of short wave radiation by the glass, and the ventilation rate of the enclosure. Transient and steady periodic wave forms of the incident solar radiation are considered.

The remaining Chapters, 8 and 9, are devoted entirely to an account of experimental studies of heat transmission through structures, notably the work carried out in Britain over a period of twenty-five years at the Building Research Station (BRS) of the former Department of Scientific and Industrial Research (DSIR). The building specially constructed for this purpose at the BRS was named the Wall Laboratory. The laboratory had facilities for making simultaneous measurements of temperature and heat flow on six coplanar test walls each measuring 2.4 m square and corresponding therefore to storey height in buildings. The test walls were exposed on one side to the weather, as in a practical installation, and faced a sealed space on the other which was maintained at a constant temperature of 18.3 degC throughout each test period of a whole winter. Experimental data were recorded continuously and used to calculate values of thermal transmittance and other thermal properties. An account is given in Chapter 8 of the details of construction of the Wall Laboratory and experimental facilities together with a summary of the results obtained for each of the seventy-three walls that were investigated. Many of the walls tested were of established construction and others were specially designed for research purposes which included investigations, described in Chapter 9, of

1. The moisture dependence of heat transmission through masonry walls.
2. Surface resistance.
3. Cavity resistance: ventilated cavity.
4. Thermal conductance of perforated bricks.
5. Thermal transmittance of windows, single- and double-glazed.
6. Diurnal variation of thermal transmittance.
7. Measurement of the thermal diffusivity of walls.

The theoretical studies in the early chapters relate to the experiments in the BRS Wall Laboratory.

CHAPTER 1

Fundamentals of Heat Transmission through Walls

The determination of the transfer of heat through walls is treated for normal design purposes as an example of the linear flow of heat in a slab of finite thickness bounded by a pair of parallel planes of infinite extent. Each of the layers of material of different thermal properties that compose the wall is assumed to be a homogeneous isotropic solid, and the temperature changes in natural exposure sufficiently small for the temperature dependence of thermal conductivity, *k*, to be neglected. The equation for the conduction of heat in each layer of the wall therefore is

$$K\left(\frac{\partial^2\theta}{\partial x^2}+\frac{\partial^2\theta}{\partial y^2}+\frac{\partial^2\theta}{\partial z^2}\right)=\frac{\partial\theta}{\partial t} \tag{1}$$

where θ is temperature, t is time, and x, y, z are orthogonal space co-ordinates.

The quantity K is the thermal diffusivity of the material; it has a significance in non-steady heat flow equivalent to that of thermal conductivity in steady flow. Briefly, conductivity is a coefficient of heat flow whereas diffusivity is a coefficient of temperature flow, a property that has been aptly termed ‘thermometric conductivity’.

A standard specification test is available in many countries for determining the thermal conductivity of heat insulating materials and in all cases is based on the steady state method of measurement. A description of the apparatus for use with materials at normal temperatures is given in Chapter 2. Thermal diffusivity is determined experimentally with methods that entail variable heat flow. Experimental methods suitable for use with small test specimens, i.e. homogeneous slabs about 25 mm thick and 150–300 mm square, have been reported by several authors. Certain of these are described in detail also in Chapter 2. An account is given in Chapter 9 of experiments for making *in-situ* determinations of the diffusivities of solid walls measuring 2.4 m square corresponding to single-storey height in rooms.

1. STEADY HEAT FLOW

In steady heat flow, temperature is a function of position only and the equation of heat conduction becomes Laplace's equation

$$\frac{\partial^2\theta}{\partial x^2}+\frac{\partial^2\theta}{\partial y^2}+\frac{\partial^2\theta}{\partial z^2}=0 \tag{1.1}$$

The assumption that the wall may be represented formally as a large, finite slab implies that the flow of heat at all positions is unidirectional. The error in heat conduction calculations due to ignoring the greater heat flow through corners in a practical installation is examined in detail as a steady state problem in Chapter 4.

Assuming the flow in the slab therefore as being entirely along the x-axis normal to the surfaces, the conduction equation (1.1) reduces to

$$\frac{d^2\theta}{dx^2}=0 \tag{1.2}$$

Integration of (1.2) gives

$$q=\frac{k}{l}(\theta_1-\theta_2)$$

where q is the heat flux at any point in a homogeneous slab, thickness l, of an isotropic material of thermal conductivity k, and θ_1, θ_2 are the temperatures of the hot and cold surfaces of the slab. In practical calculations of heat transmission through structures, usually the only temperatures known are those of the air inside and outside the building denoted by θ_i and θ_o, respectively. These temperatures are introduced via the surface thermal resistances to give, for steady heat flow

$$q=\frac{\theta_i-\theta_{si}}{R_{si}}=\frac{\theta_{si}-\theta_{so}}{\sum(l/k)}=\frac{\theta_{so}-\theta_o}{R_{so}}$$

or

$$q=\frac{\theta_i-\theta_o}{R_{si}+\sum(l/k)+R_{so}} \tag{1.3}$$

where R_{si} and R_{so} are the respective resistances at the iside and outside surfaces, and $\sum(l/k)$ denotes the surface-to-surface resistance of the structure. By definition the reciprocal of the sum of the thermal resistances from air to air, $\{R_{si}+\sum(l/k)+R_{so}\}$, is the thermal transmittance or U-value of the structure.

It will be appreciated that thermal transmittance is not a characteristic of the structure alone but varies according to the prevailing conditions of radiation and air movement. The complex nature of the variation is indicated by the set of equations (1.4) below, defining heat transfer in the steady state through a wall facing a heated enclosure on one side and exposed at

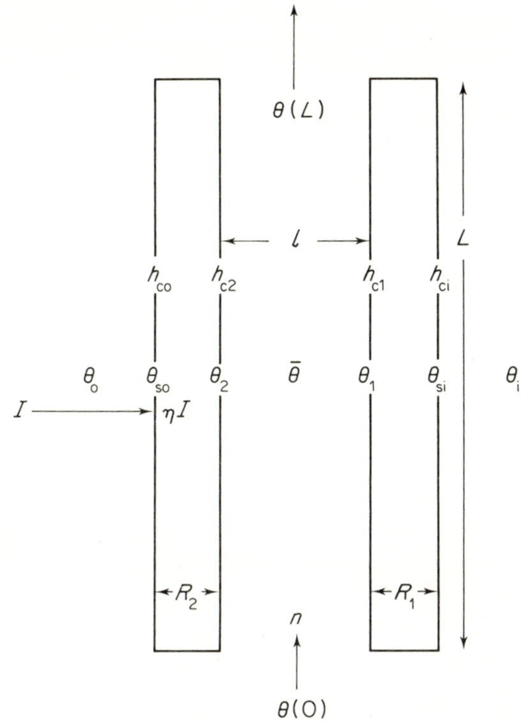

Fig. 1. Representation of cavity structure

the other to the weather. The formulation is generalized to include many forms of construction by incorporating an air space (cavity) that is ventilated to the outside air. The rate of ventilation is denoted as n air changes per hour.

Referring to Fig. 1 it will be seen that the following system of heat balance equations applies

$$
\begin{aligned}
&h_{ci}(\theta_i - \theta_{si}) + \sigma \sum_j (F_a)_{si,pj}(A_{pj}/A_{si})(T_{pj}^4 - T_{si}^4) = (\theta_{si} - \theta_1)/R_1 \\
&(\theta_{si} - \theta_1)/R_1 = \sigma(F_a F_e)_{2,1}(T_1^4 - T_2^4) + h_{c1}(\theta_1 - \bar{\theta}) \\
&h_{c1}(\theta_1 - \bar{\theta}) = nlc\{\theta(L) - \theta(0)\} + h_{c2}(\bar{\theta} - \theta_2) \qquad (1.4) \\
&h_{c2}(\bar{\theta} - \theta_2) + \sigma(F_a F_e)_{2,1}(T_1^4 - T_2^4) = (\theta_2 - \theta_{so})/R_2 \\
&(\theta_2 - \theta_{so})/R_2 = h_{co}(\theta_{so} - \theta_o) + \sigma(F_a F_e)_{s.o,so}(T_{so}^4 - T_{s.o}^4) - \eta I
\end{aligned}
$$

Values for the convective transfer coefficients h_{ci}, h_{c1}, and h_{c2} may be obtained from a comprehensive summary given by Hutchinson (1962) of formulae defining heat flow in systems corresponding to different structural arrangements. A result due to Jurges (1924) gives the convection coefficient at the outside surface as

$$h_{co} = 5.7 + 4.1v, \quad \mathrm{W/m^2\,degC} \tag{1.5}$$

where v is the prevailing wind velocity in m/s.

The quantity $\sigma \sum (F_a)_{si,pj}(A_{pj}/A_{si})(T_{pj}^4 - T_{si}^4)$ expresses the net energy received by radiation, after multiple reflection, at unit area of the wall, total area A_{si}, from the surfaces of the room, area A_{pj} ($j = 1, 2, \ldots$) that completely enclose it. The term F_a is a configuration factor representing the fraction of the energy received directly at a surface by radiation from another surface.

The corresponding expressions for the radiation heat exchange in the cavity and at the outside surface include a term F_e; this allows for the effect of the emissivities of the surfaces on the energy exchanged between them. At the outside the area of the surround is very large compared with that of the surface of the building. All the energy, therefore, radiating from the building will be received by the surround and, in turn, a negligible fraction of the energy radiated by the surround will fall on the relatively small surface of the building. It follows that at this position $(F_aF_e)_{s,o,so}$ reduces simply to the emissivity of the building surface.

Calculation of the heat transfer by radiation between two surfaces at temperatures θ_1 and θ_2 is facilitated by using a radiation transfer coefficient defined as

$$h_r = \sigma \frac{(T_1^4 - T_2^4)}{\theta_1 - \theta_2} \tag{1.6}$$

Values of the coefficients for the range of temperature likely to be encountered in buildings are shown plotted in Fig. 2 against the average temperature of the heat exchanging surfaces. To obtain a value for h_r it is normally sufficiently accurate to enter the diagram at a temperature equal to the air temperature.

The cavity in a wall may be sealed or ventilated. If ventilated, the mean temperature of the air in the cavity, $\bar{\theta}$, may be calculated from a consideration of the heat balance for an air stream bounded by a pair of plane isothermal surfaces of length L in the direction of the air flow and forming a channel of uniform width that is small compared with the edge dimensions. It can then be shown (see Chapter 3) that

$$\bar{\theta} = \frac{1}{L}\int_0^L [\psi + \{\theta(0) - \psi\} \exp\{-x(h_{c1} + h_{c2})/ncl\}]\,\mathrm{d}x$$

where $\psi = (h_{c1}\theta_1 + h_{c2}\theta_2)/(h_{c1} + h_{c2})$ and c is the volumetric specific heat of air.

By substituting appropriate formulae for the convection transfer coefficients, the system of equations (1.4) may be solved to give the temperature distribution through the structure for any combination of values assigned to

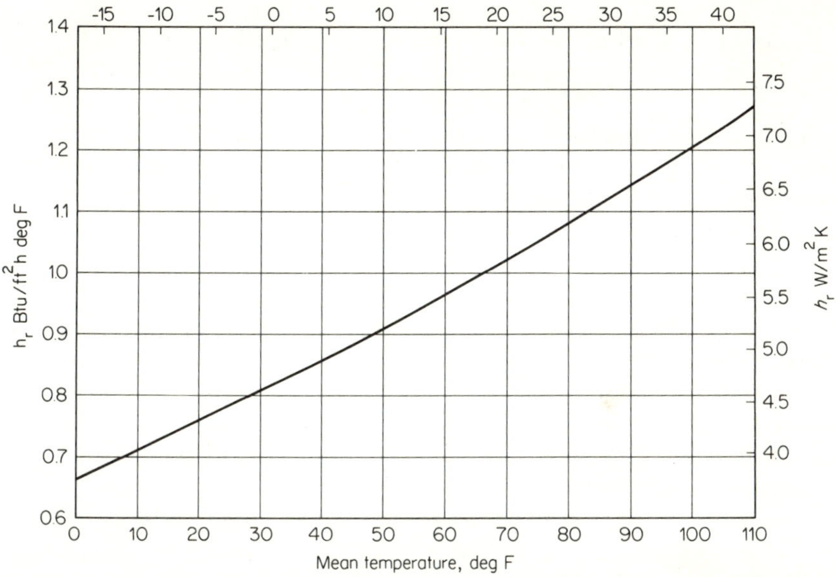

Fig. 2. Radiation transfer across an air layer: variation of radiation coefficient with mean temperature (Robinson, Powlitch, and Dill, 1954)

the several variables. With these temperatures the values of resistance are then obtained using the defining equations below

$$
\begin{aligned}
R_{si} &= R_1(\theta_i - \theta_{si})/(\theta_{si} - \theta_1) \\
R_{so} &= R_2(\theta_{so} - \theta_o)/(\theta_2 - \theta_{so}) \\
R_{a.a} &= R_1(\theta_i - \theta_o)/(\theta_{si} - \theta_1) \qquad (1.7) \\
R_c &= R_1(\theta_1 - \theta_2)/(\theta_{si} - \theta_1) \\
R_c' &= R_{a.a} - (R_1 + R_2 + R_{si} + R_{so}), \qquad n \neq 0
\end{aligned}
$$

where R_c is the thermal resistance of the cavity when sealed and R_1 and R_2 denote the resistances of the inner and outer leafs of the structure, rspectively. The primed quantity R_c' represents the resistance of the cavity when ventilated and is calculated as the value that combines linearly with the component resistances of the structure to conform with the conventional method for calculating $R_{a.a}$, the air-to-air resistance of the structure. Cavity ventilation and associated thermal effects in naturally exposed structures are examined in detail in Chapter 3.

In practice the heat flux at the inside surface of the structure has the greatest interest since this is the quantity that determines the space heating, and cooling, load of the building due to heat exchange with the outdoor environment by conduction through the external fabric. Standards relating to the computation of the heat requirements for buildings define this amount by equation (1.3) which is written more conveniently as

$$q = U(\theta_i - \theta_o) \tag{1.8}$$

The conventional method for calculating the U-value of a structure relies on agreed values of cavity and surface resistances for which the IHVE *Guide* Book A (1970) is the authoritative source in Britain. The values of outside surface resistance tabulated in the *Guide* represent average conditions for three degrees of natural exposure, classified as sheltered, normal, and severe, and are given for a wall and a roof separately with provision for the emissivity of the surface. These values of R_{so} are accepted for design purposes, they are few in number and insufficient to include the many possible combinations of meteorological factors. Likewise, at the inside surface a single value of R_{si} is recommended irrespective of the method of space heating as represented by characteristic values of the air temperature and radiation temperature distribution indoors. Furthermore, the air space in a cavity structure is asumed, almost without exception, to be sealed.

It will be evident that the system of steady state equations (1.4) provides a method for calculating heat transmission through structures that overcomes the restrictions of the present conventional system. By substituting appropriate values to represent orientation, different degrees of exposure, and indoor temperature requirements, the equations may be solved to give the temperature distribution, and hence the heat flow, through homogeneous and composite structures represented by an extensive range in the values of the individual-layer resistances and climatic data. This method of calculation may be applied also to fenestration using single- and multi-glazed sealed window units, also double windows. In such cases the equations (1.4) are extended to include the short-wave solar radiation absorbed increasingly with distance in the glass by representing this as a plane heat source of uniform strength situated mid-way through each separate sheet of glass. The method is described in Chapter 7 in an analysis of the thermal load on a building due to solar heat gain through glass.

As an example of the use of the steady state equations (1.4) these have been solved and the results evaluated to illustrate the effects of construction, air space ventilation, and outdoor radiation conditions on the thermal properties of structures that incorporate a cavity. The results are given in Table 1 and apply to a sheeted-roof type of construction consisting of an external cladding lined internally with insulation board to enclose an air space. In cases 1–4 the greater part of the air-to-air thermal resistance is on the warm side of the air space. In cases 5–8 the air space is sandwiched between roughly equal amounts of resistance. The interplay and relative importance of the factors in the second and third columns of the Table will be apparent from an inspection of the results, although a more detailed interpretation is given in Chapter 3.

The solution of the system of equations (1.4) for computing steady heat transfer through building structures involves a considerable amount of

Table 1 Numerical results for steady heat flow through a structure incorporating a cavity

Cavity	Cavity	Period	Cavity resistance (m^2K/W) Cavity sealed	Cavity ventilated	Thermal transmittance (W/m^2K)
1	Sealed	Night	0.174	0.174	2.16
2	Sealed	Day	0.171	0.171	1.48
3	Ventilated	Night	0.143	0.144	2.27
4	Ventilated	Day	0.122	0.072	1.65
5	Sealed	Night	0.167	0.167	1.36
6	Sealed	Day	0.176	0.176	0.91
7	Ventilated	Night	0.118	−0.014	1.93
8	Ventilated	Day	0.099	−0.317	1.56

numerical work that clearly would not be acceptable for practical purposes. The difficulty may be overcome by tabulating results for a variety of structure types exposed to a comprehensive range of environmental conditions. If, however, the structure is solid, of single- or multi-layer composition, or contains a cavity that is sealed, then the heat flow is expressed by (1.3) which may be modified slightly by replacing θ_0, the outside air temperature, with an equivalent outdoor temperature, θ_e, as described below.

Equivalent Outdoor Temperature (eot)

Consider the heat exchange at the outside or naturally exposed surface of a structure. The heat balance equation at this position is

$$h_{co}(\theta_{so}-\theta_o)+\sigma\varepsilon(T_{so}^4-T_{s.o}^4)-\eta I=-k\frac{\partial\theta}{\partial x} \tag{1.9}$$

Assume the emissivity of the surface, ε, to be unity and rewrite (1.9) as

$$h'_{co}(\theta_{so}-\theta_e)+\sigma(T_{so}^4-T_e^4)=-k\frac{\partial\theta}{\partial x} \tag{1.10}$$

Equation (1.10) defines an ambient temperature θ_e (hypothetical) at the surface that gives the same rate of heat flow through the surface as exists with the actual temperature and radiation environment. It is termed the 'equivalent outdoor temperature'. In this and the equations that immediately follow the primed quantities denote coefficients that relate to this temperature.

Introducing coefficients of radiation heat exchange defined as

$$h_{ro}=\sigma(T_{so}^4-T_{s.o}^4)/(\theta_{so}-\theta_{s.o})$$
$$h'_{ro}=\sigma(T_{so}^4-T_e^4)/(\theta_{so}-\theta_e)$$

the equations (1.9) and (1.10) are written more conveniently as

$$\left.\begin{array}{l} h_{co}(\theta_{so}-\theta_o)+h_{ro}(\theta_{so}-\theta_{s.o})-\eta I \\ h'_{co}(\theta_{so}-\theta_e)+h'_{ro}(\theta_{so}-\theta_e) \end{array}\right\}=-k\frac{\partial\theta}{\partial x} \qquad (1.11)$$

to give the following expression for the equivalent outdoor temperature

$$\theta_e=\theta_{so}+\{h_{co}\theta_o+h_{ro}\theta_{s.o}-(h_{co}+h_{ro})\theta_{so}+\eta I\}/(h'_{co}+h'_{ro}\} \qquad (1.12)$$

In natural exposure the ratio $(h_{co}+h_{ro})/(h'_{co}+h'_{ro})$ has a probable mean value of unity. If this value is accepted then θ_{so} conveniently disappears from (1.12) to give

$$\theta_e=\theta_o+\{h_{ro}(\theta_{s.o}-\theta_o)+\eta I\}/(h_{co}+h_{ro}) \qquad (1.13)$$

Now, in steady state

$$\frac{\theta_i-\theta_{so}}{R_{si}+\sum(l/k)}=(h'_{co}+h'_{ro})(\theta_{so}-\theta_e) \qquad (1.14)$$

Putting $h'_{co}+h'_{ro}=h_{co}+h_{ro}=R_{so}^{-1}$ and representing the steady heat flux in the structure by q, then from equation (1.14)

$$q=\frac{\theta_i-\theta_e}{R_{si}+\sum(l/k)+R_{so}} \qquad (1.15)$$

which, by introducing the expression (1.13) for θ_e, becomes

$$q=\frac{1}{R_{si}+\sum(l/k)+R_{so}}[\theta_i-\theta_o-\{h_{ro}(\theta_{s.o}-\theta_o)+\eta I\}R_{so}] \qquad (1.16)$$

By definition an expression for the prevailing transmittance or U-value follows from (1.16) by dividing the steady heat flux by the air-to-air temperature difference $(\theta_i-\theta_o)$ giving

$$U=\bar{U}\left[1-\frac{\{h_{ro}(\theta_{s.o}-\theta_o)+\eta I\}}{\theta_i-\theta_o}R_{so}\right] \qquad (1.17)$$

where $\bar{U}=\{R_{si}+\sum(l/k)+R_{so}\}^{-1}$ represents the thermal transmittance of the wall when shaded from solar radiation and exposed at the outside surface to surroundings at a mean radiant temperature $\theta_{s.o}$ equal to the air temperature θ_o. It would seem appropriate to describe thermal transmittance defining the unit rate of heat flow through a structure under these particular ambient conditions as the 'shade transmittance'.

An extension of the foregoing analysis to heat transmission through glass (see Fig. 4 and equations (1.4) Chapter 7) gives the following results, corresponding to (1.16) derived for an opaque structure,

$$q=\bar{U}\left|\begin{array}{l} \text{single-glazing:} \\ \theta_i-\theta_o-h_{ro}(\theta_{s.o}-\theta_o)R_{so}-\beta(R_{so}+R/2) \\ \\ \text{double-glazing:} \\ \theta_i-\theta_o-h_{ro}(\theta_{s.o}-\theta_o)R_{so}-\beta_2(R_{so}+R_2/2)-\beta_1(R_{so}+R_2+R_c+R_1/2) \end{array}\right. \qquad (1.18)$$

where R_c denotes the thermal resistance of the sealed cavity, β the amount of short-wave radiation absorbed by single-glazing of resistance R, and β_1, R_1; β_2, R_2 are the corresponding quantities for the inner and outer sheets of glass, respectively, in double-glazing. Formulae for calculating the total amount of heat absorbed by each sheet of irradiated glass due to direct solar energy and radiation from the sky are summarized by Hutchinson (1962).

The similarity in form between the corresponding equations (1.16) and (1.18) for heat transmission through opaque structures and glass, respectively, will be noted. In these expressions $\bar{U}$ denotes the shade transmittance, introduced above, and the terms in the brackets provide corrections to this quantity that take account of variations in radiation heat exchange and wind speed, the principal climatic factors that determine the heat exchange at the outside surface with a natural environment.

With certain types of space heating systems the mean radiant temperature viewed by the inside surface of the wall may be significantly different from that of the indoor air. In such cases an equivalent indoor temperature may be defined, as at the outside surface, by using the first expression in the system of equations (1.4) and following the analysis in equations (1.9)–(1.13) to obtain a similar adjustment in the value of θ_i.

2. NON-STEADY HEAT FLOW

Calculations, as specified in *Standards*, of the heat requirements for buildings assume for simplicity that the flow of heat in the building fabric takes place in the steady state. Clearly this is contrary to reality. Because of the diurnal variation in the ambient thermal conditions, as determined by the weather on one side and the space heating requirements on the other, the temperature of the structure is changing all the while. A statement of the heat equation involving time as a variable and more appropriate, therefore, than equation (1) for defining heat flow in a wall under practical conditions is

$$\frac{\partial^2 \theta(x, t)}{\partial x^2} = \frac{1}{K} \frac{\partial \theta(x, t)}{\partial t} \tag{2}$$

The boundary conditions that determine a solution of the differential equation (2) assume a variety of different forms of varying complexity. For the present analysis the ambient temperature change generating dynamic heat flow in the wall is represented in an idealized form, first as a sudden change or step function leading to a consideration of a regular series of rectangular pulses, and then as a sine wave. These are denoted in the analysis as Case 1 and Case 2. The boundary conditions are expressed in terms of air tempeature, but this may be replaced at the outside surface with equivalent outdoor temperature and the corresponding quantity at the inside surface if desired.

Case 1: Transient Heat Flow in a Single-layer Wall

Basically the analysis examines the linear flow of heat in a large homogeneous slab following a suddenly activated change in the ambient temperature at one face while at the opposite face the medium temperature is held constant.

The region considered is $0 \leqslant x \leqslant l$ within a large slab, thickness l, of an isotropic solid of thermal conductivity k and diffusivity K with heat transferred between an air medium and the solid at $x=0$, $x=l$ at the respective rates h_{so}, h_{si} times their temperature difference. For all values of time t the temperature of the medium in contact with the solid at $x=l$ is θ_i. Initially the medium temperature at $x=0$ is θ_o and the flow of heat is steady. Subsequently a step-function drop through θ_o is applied in the ambient temperature at the boundry $x=0$.

A solution is required to the heat equation (2) for the region satisfying the boundary conditions

$$k\frac{\partial\theta(x, t)}{\partial x} = h_{si}\{\theta_i - \theta(x, t)\}, \qquad x=l, \quad t>0 \tag{2.1.1}$$

$$k\frac{\partial\theta(x, t)}{\partial x} = h_{so}\theta(x, t), \qquad x=0, \quad t>0 \tag{2.1.2}$$

with

$$\theta(x, 0) = \theta_o + \frac{(1/h_{so} + x/k)(\theta_i - \theta_o)}{1/h_{so} + l/k + 1/h_{si}} \tag{2.1.3}$$

defining the initial condition.

By introducing the dimensionless quantities

$$\xi = x/l; \qquad \tau = Kt/l^2; \qquad B_{si} = h_{si}l/K; \qquad B_{so} = h_{so}l/k$$

the equations (2) through (2.1.3) are written more conveniently as

$$\frac{\partial\theta(\xi, \tau)}{\partial\tau} = \frac{\partial^2\theta}{\partial\xi^2}(\xi, \tau), \qquad 0<\xi<1, \quad \tau>0 \tag{2.1.4}$$

$$\frac{\partial\theta(\xi, \tau)}{\partial\xi} = B_{si}\{\theta_i - \theta(\xi, \tau)\}, \qquad \xi=1, \quad \tau>0 \tag{2.1.5}$$

$$\frac{\partial\theta(\xi, \tau)}{\partial\xi} = B_{so}\theta(\xi, \tau), \qquad \xi=0, \quad \tau>0 \tag{2.1.6}$$

$$\theta(\xi, 0) = \theta_o + \frac{(1/B_{so} + \xi)(\theta_i - \theta_o)}{1 + 1/B_{si} + 1/B_{so}} \tag{2.1.7}$$

The equation (2.1.4) is solved most readily by a routine application of the Laplace transform. The use of this method leads to the following subsidiary equations

$$\frac{d^2\bar{\theta}(\xi, p)}{d\xi^2} = p\bar{\theta}(\xi, p) - \theta(\xi, 0), \qquad 0<\xi<1 \tag{2.1.8}$$

$$\frac{d\bar{\theta}(\xi, p)}{d\xi} = B_{si}\{\theta_i/p - \bar{\theta}(\xi, p)\}, \qquad \xi=1 \tag{2.1.9}$$

$$\frac{d\bar{\theta}(\xi, p)}{d\xi} = B_{so}\bar{\theta}(\xi, p), \qquad \xi=0 \tag{2.1.10}$$

where $\bar{\theta}(\xi, p) = \mathscr{L}\{\theta(\xi, \tau)\}$, the symbol $\mathscr{L}$ implying the Laplace transform applied to temperature by the defining formula

$$\mathscr{L}\{\theta(\xi, \tau)\} = \int_0^\infty e^{-p\tau}\theta(\xi, \tau)\, d\tau, \qquad R(p)>0$$

From (2.1.8)

$$\bar{\theta}(\xi, p) = Ce^{\sqrt{p}\xi} + De^{-\sqrt{p}\xi} + \frac{\theta(\xi, 0)}{p}$$

The boundary conditions give

$$C = -\frac{B_{so}\theta_o}{2pf(p)}(\sqrt{p} - B_{si})e^{-\sqrt{p}}$$

$$D = -\frac{B_{so}\theta_o}{2pf(p)}(\sqrt{p} + B_{si})e^{\sqrt{p}}$$

so that by substitution

$$\bar{\theta}(\xi, p) = -B_{so}\frac{\theta_o}{p}\frac{g(p)}{f(p)} + \frac{\theta(\xi, 0)}{p} \tag{2.1.11}$$

where

$$g(p) = \sqrt{p}\cosh\sqrt{p}(1-\xi) + B_{si}\sinh\sqrt{p}(1-\xi)$$

$$f(p) = (p + B_{si}B_{so})\sinh\sqrt{p} + (B_{si} + B_{so})\sqrt{p}\cosh\sqrt{p}$$

The required solution $\theta(\xi, \tau)$ is obtained from $\bar{\theta}(\xi, p)$ with the inversion theorem for the Laplace transform which, in the customary notation, is written

$$\theta(\xi, \tau) = \frac{1}{2\pi i}\int_{\gamma-i\infty}^{\gamma+i\infty} e^{p\tau}\bar{\theta}(\xi, p)\, dp \tag{2.1.12}$$

The integrand is analytic in p with a simple pole at $p=0$, and others at $p=-\alpha_n^2$ $(n=1, 2, \ldots)$ representing the zeros of $f(p)$ which are given by the roots, all real, of

$$\alpha \tan\alpha = \frac{\alpha^2(B_{si} + B_{so})}{\alpha^2 - B_{si}B_{so}} \tag{2.1.13}$$

By Cauchy's theorem the line integral (2.1.12) is evaluated as $2\pi i$ times the sum of the residues at the poles of the integrand.

At $p=0$ the residue is

$$-B_{so}\theta_o\left\{\frac{g(p)}{f(p)}\right\}_{p=0}+\theta(\xi,0)$$

$$=-B_{so}\theta_o\left[\frac{\sqrt{p}\{1+p(1-\xi)^2/2+\cdots\}+B_{si}\{\sqrt{p}(1-\xi)+p^{3/2}(1-\xi)^3/3!+\cdots}{(p+B_{si}B_{so})(\sqrt{p}+p^{3/2}/3!+\cdots)+(B_{si}+B_{so})\sqrt{p}(1+p/2+\cdots)}\right]_{p=0}$$

$$+\frac{\theta_i/B_{so}+(1+1/B_{si})\theta_o+(\theta_i-\theta_o)\xi}{1+1/B_{si}+1/B_{so}}$$

$$=-B_{so}\theta_o\left\{\frac{1+B_{si}(1-\xi)}{B_{si}B_{so}+B_{si}+B_{so}}\right\}+\frac{\theta_i/B_{so}+(1+1/B_{si})\theta_o+(\theta_i-\theta_o)\xi}{1+1/B_{si}+1/B_{so}}$$

$$=\frac{\theta_i(\xi+1/B_{so})}{1+1/B_{si}+1/B_{so}} \tag{2.1.14}$$

representing the steady state temperature reached after a long time.

The other poles give residues

$$-B_{so}\theta_o\left\{\frac{g(p)e^{p\tau}}{p\dfrac{df(p)}{dp}}\right\}_{p=-\alpha_n^2}=2B_{so}\theta_o\sum_{n=1}^{\infty}\left(1-\frac{B_{si}B_{so}}{\alpha_n^2}\right)\sec\alpha_n\frac{g(\alpha_n)}{f(\alpha_n)}e^{-\alpha_n\tau}$$

where

$$g(\alpha_n)=\cos\alpha_n(1-\xi)+(B_{si}/\alpha_n)\sin\alpha_n(1-\xi) \tag{2.1.15}$$

$$f(\alpha_n)=\alpha_n^2+B_{si}(1+B_{si})+B_{so}(1+B_{so})$$
$$+(B_{si}B_{so}/\alpha_n^2)(B_{si}+B_{so}+B_{si}B_{so}) \tag{2.1.16}$$

Summing the residues leads to the final result

$$\theta(\xi,\tau)=\theta_i(\xi+1/B_{so})/(1+1/B_{si}+1/B_{so})$$
$$+2B_{so}\theta_o\sum_{n=1}^{\infty}\left(1-\frac{B_{si}B_{so}}{\alpha_n^2}\right)\sec\alpha_n\left\{\frac{q(\alpha_n)}{f(\alpha_n)}\right\}e^{-\alpha_n^2\tau} \tag{2.1.17}$$

Let $Q(t)$ represent the total quantity of heat passing through unit area of the slab at the surface $x=l$ over the time interval 0 to t. Then

$$Q(t)=k\int_0^t\left\{\frac{\partial\theta}{\partial x}(x,t)\right\}_{x=l}dt \tag{2.1.18}$$

or, in dimensionless notation

$$Q(\tau)=\int_0^\tau\left\{\frac{\partial\theta}{\partial\xi}(\xi,\tau)\right\}_{\xi=1}d\tau \tag{2.1.19}$$

where $Q(\tau)$ denotes $Q(t)K/lk$, a non-dimensional group. Obtaining the derivative $\{\partial\theta(\xi,\tau)/\partial\xi\}_{\xi=1}$ from equation (2.1.17), substituting the result in (2.1.19) and integrating as indicated, gives

$$Q(\tau)=\frac{\theta_{\rm i}\tau}{1+1/B_{\rm si}+1/B_{\rm so}}-\frac{\theta_{\rm o}}{1+1/B_{\rm si}+1/B_{\rm so}}\sum_{n=1}^{\infty}a_n(1-{\rm e}^{-\alpha_n^2\tau})$$

or

$$Q(\tau)=A\{\theta_{\rm i}\tau-\theta_{\rm o}\sum_{n=1}^{\infty}a_n(1-{\rm e}^{-\alpha_n^2\tau})\} \tag{2.1.20}$$

where

$$a_n=\frac{2(1-B_{\rm si}B_{\rm so}/\alpha_n^2)\sec\alpha_n}{\alpha_n^2 f(\alpha_n)/(B_{\rm si}+B_{\rm so}+B_{\rm si}B_{\rm so})}$$

and

$$A=(1+1/B_{\rm si}+1/B_{\rm so})^{-1} \tag{2.1.21}$$

representing thermal transmittance in non-dimensional form.

An expression for the integrated heat flow $Q(\tau)$ may also be obtained from the Laplace transform theorem which states that if $y(p)$ is the Laplace transform of $y(t)$, then the transform of $\int_0^t y(t')\,{\rm d}t'$ is $(1/p)y(p)$.

With this result it follows that the total amount of heat crossing unit area of the plane $\xi=1$ up to time τ is given by calculating the inversion

$$Q(\tau)=\frac{1}{2\pi i}\int_{\gamma-i\infty}^{\gamma+i\infty}\left\{\frac{1}{p}\frac{\partial\theta}{\partial\xi}(\xi,p)\right\}_{\xi=1}{\rm e}^{p\tau}\,{\rm d}p$$

Forming the integrand therefore from (2.1.11) and inverting gives

$$Q(\tau)=\frac{\theta_{\rm i}\tau}{1+1/B_{\rm si}+1/B_{\rm so}}-\frac{\theta_{\rm o}}{1+1/B_{\rm si}+1/B_{\rm so}}\left\{\frac{1+(B_{\rm si}+B_{\rm so})/2+B_{\rm si}B_{\rm so}/3!}{B_{\rm si}+B_{\rm so}+B_{\rm si}B_{\rm so}}\right\}$$
$$+\frac{\theta_{\rm o}}{1+1/B_{\rm si}+1/B_{\rm so}}\sum_{n=1}^{\infty}\frac{2(1-B_{\rm si}B_{\rm so}/\alpha_n^2)\sec\alpha_n{\rm e}^{-\alpha_n^2\tau}}{\alpha_n^2 g(\alpha_n)/(B_{\rm si}+B_{\rm so}+B_{\rm si}B_{\rm so})}$$

A comparison of this equation with (2.1.20) shows

$$\sum_{n=1}^{\infty}a_n=\frac{1+(B_{\rm si}+B_{\rm so})/2+B_{\rm si}B_{\rm so}/3!}{1+1/B_{\rm si}+1/B_{\rm so}}$$

By substitution therefore the expression for the total heat flow over the interval τ may be written alternatively as

$$Q(\tau)=\frac{\theta_{\rm i}\tau}{1+1/B_{\rm si}+1/B_{\rm so}}-\theta_{\rm o}\left\{\frac{1/3!+(1/B_{\rm si}+1/B_{\rm so})/2+1/B_{\rm si}B_{\rm so}}{(1+1/B_{\rm si}+1/B_{\rm so})^2}\right\}\sum_{n=1}^{\infty}(1-{\rm e}^{-\alpha_n^2\tau}) \tag{2.1.22}$$

The result (2.1.17) is fundamental in the calculation of temperature and heat flow in a large slab, enabling solutions to be written immediately when the condition applied at the boundary is expressible in step-function form.

Consider, for example, a change in the outdoor temperature represented by a rectangular pulse of height θ_o and duration t_1. This boundary condition may be written

$$\theta_o(t) = \theta_o\{H(t) - H(t-t_1)\} \tag{2.1.23}$$

where the Heaviside unit function $H(t)$ is introduced to define the temperature change θ_o 'on' for $0 < t \leq t_1$ and 'off' for $t > t_1$. Applying to (2.1.23) the well-known result

$$\mathscr{L}\{f(t-\alpha)H(t-\alpha)\} = e^{-\alpha p}\mathscr{L}\{f(t)\} \tag{2.1.24}$$

gives

$$\bar{\theta}_o(p) = \frac{\theta_o}{p}(1 - e^{-pt_1}) \tag{2.1.25}$$

In (2.1.11) replacing θ_o/p by $(\theta_o/p)(1-e^{-pt_1})$, the Laplace transform of the function at the outside surface now generating the temperature change in the slab gives immediately, via (2.1.17)

$$\theta(\xi, \tau) = \begin{cases} \theta_o + (\theta_i - \theta_o)(\xi + 1/B_{so})/(1 + 1/B_{si} + 1/B_{so}) \\ \quad -2B_{so}\theta_o \sum_{n=1}^{\infty} (1 - B_{si}B_{so}/\alpha_n^2) \sec \alpha_n \left\{\dfrac{g(\alpha_n)}{f(\alpha_n)}\right\} e^{-\alpha_n^2 \tau}, \qquad 0 < \tau \leq \tau_1 & (2.1.26) \\ \theta_i(\xi + 1/B_{so})/(1 + 1/B_{si} + 1/B_{so}) \\ \quad -2B_{so}\theta_o \sum_{n=1}^{\infty} (1 - B_{si}B_{so}/\alpha_n^2) \sec \alpha_n \left\{\dfrac{g(\alpha_n)}{f(\alpha_n)}\right\} \{e^{-\alpha_n^2 \tau} - e^{-\alpha_n^2(\tau - \tau_1)}\}, \\ \qquad \tau > \tau_1 & (2.1.27) \end{cases}$$

The results (2.1.26) and (2.1.27) for a single pulse may now be used to obtain the solution for a steady periodic condition at $x = 0$, represented as a regular series of rectangular pulses, periodicity T, and defined as follows

$$\begin{aligned} \theta_o(t) &= 0, & t &\leq 0, \\ \theta_o(t) &= \theta_o, & nT < t &\leq (nT + t_1), \quad n = 0, 1, 2, \ldots \\ \theta_o(t) &= 0, & (nT + t_1) < t &\leq (n+1)T \end{aligned}$$

Writing the sequence of pulses as

$$\theta_o(t) = \theta_o\{H(t) - H(t - t_1) + H(t - T) - H(t - \overline{T + t_1}) + H(t - 2T) - \cdots\}$$

and applying (2.1.24) to each term in the series gives

$$\begin{aligned} \bar{\theta}_o(p) &= \frac{\theta_o}{p}\{1 - e^{-pt_1} + e^{-pT} - e^{-p(T+t_1)} + e^{-2pT} - e^{-p(2T+t_1)} + \cdots\} \\ &= \frac{\theta_o}{p}\left\{\frac{1 - e^{-pt_1}}{1 - e^{-p(t_1+t_2)}}\right\} \end{aligned} \tag{2.1.28}$$

in which $t_1 + t_2 = T$.

As with the single pulse, it follows immediately from (2.1.11) by substituting the expression (2.1.28) for θ_o/p and paying due attention to the preceding sign that the solution for the temperature in the slab exposed at $x=0$ to a regular series of rectangular pulses is

$$\theta(\xi,\tau)=\frac{1}{2\pi \mathrm{i}}\int_{\gamma-\mathrm{i}\infty}^{\gamma+\mathrm{i}\infty} B_{\mathrm{so}}\frac{\theta_o}{p}\left\{\frac{1-\mathrm{e}^{-p\tau_1}}{1-\mathrm{e}^{-p(\tau_1+\tau_2)}}\right\}\frac{g(p)}{f(p)}\mathrm{e}^{p\tau}\,\mathrm{d}p+\theta(\xi,0) \quad (2.1.29)$$

Taking the inversion and following Carslaw and Jaeger (1959) the solution may be written

$$\theta(\xi,\tau)=\theta_s(\xi,\tau)-2B_{\mathrm{so}}\theta_o\sum_{n=1}^{\infty}\left(1-\frac{B_{\mathrm{si}}B_{\mathrm{so}}}{\alpha_n^2}\right)\sec\alpha_n\frac{g(\alpha_n)}{f(\alpha_n)}$$
$$\times\left\{\frac{\mathrm{e}^{-\alpha_n^2\tau}-\mathrm{e}^{-\alpha_n^2(\tau-\tau_1)}}{1-\mathrm{e}^{\alpha_n^2(\tau_1+\tau_2)}}\right\} \quad (2.1.30)$$

consisting of a steady periodic part $\theta_s(\xi,\tau)$ contributed by the simple poles at $p=2m\pi\mathrm{i}/(\tau_1+\tau_2)$, $m=\pm1, \pm2, \ldots,$ and a transient part. Present interest will be confined to $\theta_s(\xi,\tau)$; equating the expresions for $\theta(\xi,\tau)$ in equations (2.1.26) and (2.1.30) gives

$$\theta_s(\xi,\tau)=\theta_o+(\theta_i-\theta_o)\left(\xi+\frac{1}{B_{\mathrm{so}}}\right)\bigg/\left(1+\frac{1}{B_{\mathrm{si}}}+\frac{1}{B_{\mathrm{so}}}\right)$$
$$-2B_{\mathrm{so}}\theta_o\sum_{n=1}^{\infty}\left(1-\frac{B_{\mathrm{si}}B_{\mathrm{so}}}{\alpha_n^2}\right)\sec\alpha_n\left\{\frac{g(\alpha_n)}{f(\alpha_n)}\right\}\left\{\frac{\mathrm{e}^{\alpha_n^2(\tau_1-\tau)}-\mathrm{e}^{\alpha_n^2(\tau_1+\tau_2-\tau)}}{1-\mathrm{e}^{\alpha_n^2(\tau_1+\tau_2)}}\right\},$$
$$0<\tau\leqslant\tau_1 \quad (2.1.31)$$

Similarly, equating (2.1.27) with (2.1.30) gives

$$\theta_s(\xi,\tau)=\theta_i\left(\xi+\frac{1}{B_{\mathrm{so}}}\right)\bigg/\left(1+\frac{1}{B_{\mathrm{si}}}+\frac{1}{B_{\mathrm{so}}}\right)$$
$$-2B_{\mathrm{so}}\theta_o\sum_{n=1}^{\infty}\left(1-\frac{B_{\mathrm{si}}B_{\mathrm{so}}}{\alpha_n^2}\right)\sec\alpha_n\left\{\frac{g(\alpha_n)}{f(\alpha_n)}\right\}\left\{\frac{\mathrm{e}^{\alpha_n^2(2\tau_1+\tau_2-\tau)}-\mathrm{e}^{\alpha_n^2(\tau_1+\tau_2-\tau)}}{1-\mathrm{e}^{\alpha_n^2(\tau_1+\tau_2)}}\right\},$$
$$\tau_1<\tau\leqslant\tau_1+\tau_2 \quad (2.1.32)$$

The calculation of the rate of heat flow at any instant in the intervals indicated follows from these solutions.

The total heat flow $Q(\tau)$ in each interval through a unit area of the surface of the slab at $\xi=1$ is of greatest interest and is obtained for a period up to any time τ measured from the commencement of the interval by substituting the expressions for the instantaneous rates of heat flow in equation (2.1.18).

Putting τ equal to τ_1 and τ_2, respectively, gives the integrated heat flow over each interval as follows

$$Q(\tau_1) = A\left[(\theta_i - \theta_o)\tau_1 + \theta_o \sum_{n=1}^{\infty} 2a_n \frac{\{e^{\alpha_n^2\tau_1} - 1 + e^{\alpha_n^2\tau_2} - e^{\alpha_n^2(\tau_1+\tau_2)}\}}{1 - e^{\alpha_n^2(\tau_1+\tau_2)}}\right]$$

$$Q(\tau_2) = A\left[\theta_i\tau_2 - \theta_o \sum_{n=1}^{\infty} 2a_n \frac{\{e^{\alpha_n^2\tau_1} - 1 + e^{\alpha_n^2\tau_2} - e^{\alpha_n^2(\tau_1+\tau_2)}\}}{1 - e^{\alpha_n^2(\tau_1+\tau_2)}}\right] \quad (2.1.33)$$

where A and a_n are defined by (2.1.21).

By making the substitution

$$e^{\alpha_n^2\tau_1} - 1 + e^{\alpha_n^2\tau_2} - e^{\alpha_n^2(\tau_1+\tau_2)} = (1 - e^{-\alpha_n^2\tau_1})(e^{\alpha_n^2\tau_1} - e^{\alpha_n^2(\tau_1+\tau_2)}) \quad (2.1.34)$$

the expression in (2.1.33) may be written more conveniently as

$$Q(\tau_1) = A\left\{(\theta_i - \theta_o)\tau_1 + \theta_o \sum_{n=1}^{\infty} 2a_n H(\alpha_n)\right\}$$

$$Q(\tau_2) = A\left\{\theta_i\tau_2 - \theta_o \sum_{n=1}^{\infty} 2a_n H(\alpha_n)\right\} \quad (2.1.35)$$

where, from (2.1.34)

$$H(\alpha_n) = \frac{\sinh(\alpha_n^2\tau_1/2)\sinh(\alpha_n^2\tau_2/2)}{\sinh\{\alpha_n^2(\tau_1+\tau_2)/2\}} \quad (2.1.36)$$

From the expressions in (2.1.35) it follows that

$$A = \frac{Q(\tau_1) + Q(\tau_2)}{(\theta_i - \theta_o)\tau_1 + \theta_i\tau_2} \quad (2.1.37)$$

which identifies A as the mean rate of heat flow over a complete cycle through unit area of the slab exposed to a unit difference between the ambient temperatures on opposite sides.

The result, (2.1.37), although obtained from an analysis of heat transmission through a homogeneous wall, clearly applies also to multi-layer structures and is independent of the form, repeatedly applied, of the pattern of temperature change. Moreover, since the variation in the ambient temperature at the outside surface may be resolved into many cycles of different periodicity, it follows that the total amount of heat transferred over a long period through the external fabric of a building maintained at a constant indoor temperature is independent of the thermal capacity and is calculated simply as a steady state quantity.

The ambient temperature in the analysis of variable heat flow so far considered has been expressed as air temperature, although it was mentioned that this may be replaced by an equivalent temperature. Substituting therefore $\theta_{e.1}$ and $\theta_{e.2}$ to represent the equivalent outdoor temperature for the respective time intervals, the expressions (2.1.35) are written in a more general form as

$$Q(\tau_1)=A\left\{(\theta_i-\theta_{e.1})\tau_1+(\theta_{e.1}-\theta_{e.2})\sum_{n=1}^{\infty}2a_nH(\alpha_n)\right\}$$
$$Q(\tau_2)=A\left\{(\theta_i-\theta_{e.2})\tau_2-(\theta_{e.1}-\theta_{e.2})\sum_{n=1}^{\infty}2a_nH(\alpha_n)\right\} \qquad (2.1.38)$$

If τ_1 and τ_2 are assumed to represent successive day and night periods respectively, then by introducing equation (1.13)

$$Q(\tau_1)=A(\theta_i-\theta_o)_{\tau_1}\left\{\left[1-\frac{\{h_{ro}(\theta_{s.o}-\theta_o)+\eta I\}}{(\theta_i-\theta_o)_{\tau_1}}R_{so}\right]\tau_1\ \frac{(\theta_{e.1}-\theta_{e.2})}{(\theta_i-\theta_o)_{\tau_1}}\sum_{n=1}^{\infty}2a_nH(\alpha_n)\right\} \qquad (2.1.39)$$

$$Q(\tau_2)=A(\theta_i-\theta_o)_{\tau_2}\left\{\left[1-\frac{h_{ro}(\theta_{s.o}-\theta_o)}{(\theta_i-\theta_o)_{\tau_2}}R_{so}\right]\tau_2-\frac{(\theta_{e.1}-\theta_{e.2})}{(\theta_i-\theta_o)_{\tau_2}}\sum_{n=1}^{\infty}2a_nH(\alpha_n)\right\}$$

Now, defining thermal transmittance as the quantity of heat that passes through the inside surface of the structure per unit area and for unit air temperature difference in unit time, and applying this definition to (2.1.38) it follows that

$$A(\tau_1)=A\left[1-\frac{\{h_{ro}(\theta_{s.o}-\theta_o)+\eta I\}}{(\theta_i-\theta_o)_{\tau_1}}R_{so}+\frac{(\theta_{e.1}-\theta_{e.2})}{(\theta_i-\theta_o)_{\tau_1}\tau_1}\sum_{n=1}^{\infty}2a_nH(\alpha_n)\right] \qquad (2.1.40)$$

$$A(\tau_2)=A\left[1-\frac{h_{ro}(\theta_{s.o}-\theta_o)}{(\theta_i-\theta_o)_{\tau_2}}R_{so}-\frac{(\theta_{e.1}-\theta_{e.2})}{(\theta_i-\theta_o)_{\tau_2}\tau_2}\sum_{n=1}^{\infty}2a_nH(\alpha_n)\right] \qquad (2.1.41)$$

The results (2.1.40) and (2.1.41) demonstrate that the so defined thermal transmittance of a wall or roof facing a constant indoor temperature and naturally exposed on the other undergoes a diurnal swing about the shade transmittance, A, a coefficient of heat flow in the steady state, with the day time transmittance $A(\tau_1)$ higher than the night value $A(\tau_2)$. This pattern of diurnal variation in thermal transmittance has been observed in experiments on heat transmission through walls and gave measured U-values that accorded with calculated results (see Chapter 9).

A further result with a practical interest is the rate of change of temperature of the inside surface following a change in the ambient temperature at the outside surface. This is expressed most conveniently by calculating the ratio $\{\theta(l,0)-\theta(l,t)\}/\theta_o$ for the case of a single transient; putting $\xi=1$ in equation (2.1.17) the fractional change is

$$\frac{\theta(l,0)-\theta(l,t)}{\theta_o}=\frac{1/B_{si}}{1+1/B_{si}+1/B_{so}}\left(1-\sum_{n=1}^{\infty}a_n\alpha_n^2e^{-\alpha_n^2\tau}\right) \qquad (2.1.42)$$

Fig. 3 is a chart of equation (2.1.42) plotted for values of Kt/l^2 with l/k as

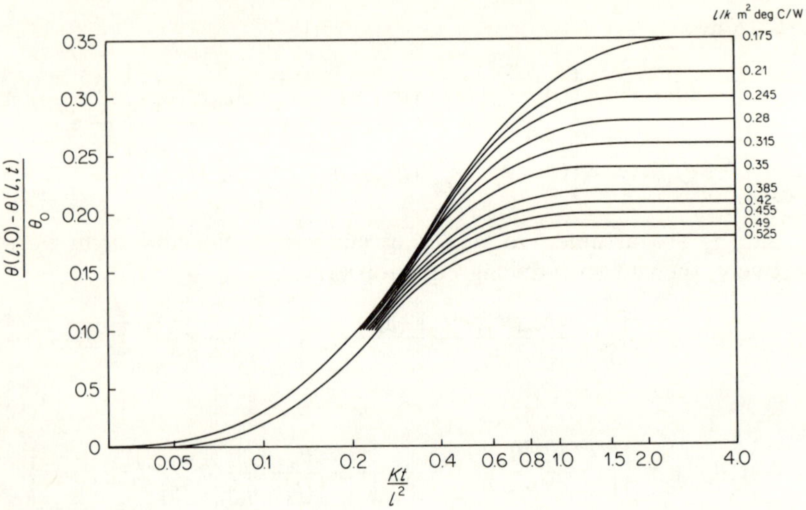

Fig. 3. Fractional temperature change $[\theta(l,0)-\theta(l,t)]/\theta_o$ plotted against Kt/l^2

the parameter, the range of the values of thermal conductance having been selected to include homogeneous walls of practical interest in building.

The curves shown in Fig. 4 for 225 mm brickwork and 25 mm thick fibreboard, selected to represent high and low thermal capacity media, respectively, have been constructed from Fig. 3 using the data given in Table 2.

It will be seen that with 25 mm fibreboard the steady state is restored slightly more than 1 h after applying the sudden change in the outdoor temperature.

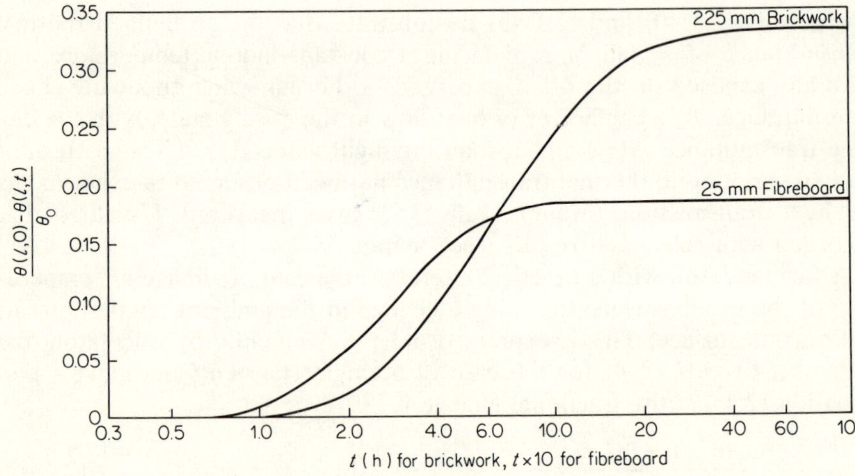

Fig. 4. $[\theta(l,0)-\theta(l,t)]/\theta_o$ plotted against t for brickwork, 225 mm thick, and fibreboard, 25 mm thick

Table 2 Values of l, k, and K for brickwork and fibreboard and surface coefficients, h_{si} and h_{so}

	Material	
	Brickwork	Fibreboard
Thickness l (mm)	225	25
Thermal conductivity k (W/m degC)	1.16	0.052
Thermal diffusivity K (m^2/s)	0.077×10^{-5}	0.016×10^{-5}
Outside surface transfer coefficient, h_{so} (W/m^2 degC)	18.7	
Inside surface transfer coefficient, h_{si} (W/m^2 degC)	7.95	

The transient change diffuses much more slowly through the heavier wall of brickwork, requiring about 40 h for the effect to become fully developed at the inside surface.

Data to facilitate numerical calculations are given in Table 3.

The solutions for variable heat flow define the temperature in a large finite slab exposed to step-function changes in the ambient temperature at the position $\xi = 0$, corresponding to the outside surface of a building, while at the opposite or inside surface, denoted by $\xi = 1$, the ambient temperature is constant. There are other situations of practical importance, for example intermittent space heating in buildings that require for their interpretation the corresponding solutions with the boundary conditions above reversed, i.e. with a step-function change applied at $\xi = 1$ and the ambient temperature constant at $\xi = 0$. The change may be in temperature or heat flux.

Table 3 Data for calculating variable heat flow

l/k (m^2 degC/W)	B_{si}	B_{so}	α_1	α_2
0.175	1.4	3.3	1.75	4.14
0.21	1.7	4.0	1.87	4.27
0.245	2.0	4.6	1.96	4.38
0.28	2.2	5.3	2.03	4.47
0.315	2.5	5.9	2.10	4.56
0.35	2.8	6.6	2.17	4.64
0.385	3.1	7.3	2.22	4.72
0.42	3.4	7.9	2.27	4.79
0.455	3.6	8.6	2.31	4.84
0.49	3.9	9.2	2.35	4.90
0.525	4.2	9.9	2.39	4.95

$B_{si} = h_{si} l/k = 7.95 l/k$; $B_{so} = h_{so} l/k = 18.7 l/k$

Consider the change at $\xi=1$ to be a rise in temperature and denote this by θ_i, then

$$\frac{\partial\theta(\xi,\tau)}{\partial\tau}=\frac{\partial^2\theta(\xi,\tau)}{\partial\xi^2} \qquad 0<\xi<1, \quad \tau>0$$

$$\frac{\partial\theta(\xi,\tau)}{\partial\xi}=B_{si}\{\theta_i-\theta(\xi,\tau)\}, \qquad \xi=1, \quad \tau>0$$

$$\frac{\partial\theta(\xi,\tau)}{\partial\xi}=B_{so}\theta(\xi,\tau), \qquad \xi=0, \quad \tau>0$$

$$\theta(\xi,\tau)=0, \qquad 0<\xi<1, \quad \tau\leq 0$$

The solution of these equations for the temperaure rise in the slab is

$$\theta(\xi,\tau)=\theta_i\left(\xi+\frac{1}{B_{so}}\right)\Big/\left(1+\frac{1}{B_{si}}+\frac{1}{B_{so}}\right)$$
$$-2B_{si}\theta_i\sum_{n=1}^{\infty}\left(1-\frac{B_{si}B_{so}}{\alpha_n^2}\right)\sec\alpha_n\left\{\frac{g(\alpha_n)}{f(\alpha_n)}\right\}e^{-\alpha_n^2\tau}, \qquad \tau>0 \qquad (2.1.43)$$

where

$$g(\alpha_n)=\cos\alpha_n\xi+(B_{so}/\alpha_n)\sin\alpha_n\xi \qquad (2.1.44)$$

and $f(\alpha_n)$ is defined by equation (2.1.16).

If after time t_1 the ambient temperature rise θ_i is suddenly removed, then

$$\theta(\xi,\tau)=2B_{si}\theta_i\sum_{n=1}^{\infty}\{e^{-\alpha_n^2\tau}-e^{-\alpha_n^2(\tau-\tau_1)}\}\left(1-\frac{B_{si}B_{so}}{\alpha_n^2}\right)\sec\alpha_n\left\{\frac{g(\alpha_n)}{f(\alpha_n)}\right\} \qquad (2.1.45)$$

By repeating this pattern of change in θ_i, a steady periodic condition is reached, and following (2.1.31) and (2.1.32) the temperature of the wall is then

$$\theta_s(\xi,\tau)=\theta_i\left(\xi+\frac{1}{B_{so}}\right)\Big/\left(1+\frac{1}{B_{si}}+\frac{1}{B_{so}}\right)$$
$$-2B_{si}\theta_i\sum_{n=1}^{\infty}\left(1-\frac{B_{si}B_{so}}{\alpha_n^2}\right)\sec\alpha_n\left\{\frac{g(\alpha_n)}{f(\alpha_n)}\right\}\left\{\frac{e^{\alpha_n^2(\tau_1-\tau)}-e^{\alpha_n^2(\tau_1+\tau_2-\tau)}}{1-e^{\alpha_n(\tau_1+\tau_2)}}\right\},$$
$$0<\tau\leq\tau_1 \qquad (2.1.46)$$

$$\theta_s(\xi,\tau)=2B_{si}\theta_i\sum_{n=1}^{\infty}\left(1-\frac{B_{si}B_{so}}{\alpha_n^2}\right)\sec\alpha_n\left\{\frac{g(\alpha_n)}{f(\alpha_n)}\right\}\left\{\frac{e^{\alpha_n^2(2\tau_1+\tau_2-\tau)}-e^{\alpha_n^2(\tau_1+\tau_2-\tau)}}{1-e^{\alpha_n(\tau_1+\tau_2)}}\right\},$$
$$\tau_1<\tau\leq\tau_1+\tau_2 \qquad (2.1.47)$$

Hence, the heat flow through the wall over any interval of time may be calculated; over a complete cycle it is simply $U\theta_i t_1$, a steady state quantity.

Intermittent space heating may include a period in which heat is applied at a constant rate to achieve a specified temperature rise in a given time.

This condition is analysed below by regarding the wall as enclosing a space to which a flux q per unit area of wall surface is added while at $x=0$ the ambient temperature is zero for all values of time. The space is unventilated, although a rate of air change may be included in the analysis without difficulty; also, the thermal capacity of air is assumed to be negligibly small. Then

$$\frac{\partial\theta(x,t)}{\partial t}=K\frac{\partial^2\theta}{\partial x^2}(x,t), \qquad 0<x<l, \quad t>0$$

$$k\frac{\partial\theta(x,t)}{\partial x}=q, \qquad x=l, \quad t>0$$

$$k\frac{\partial\theta(x,t)}{\partial x}=h_{so}\theta(x,t), \qquad x=0, \quad t>0$$

$$q=h_{si}\{\theta_i(t)-\theta(x,t)\}, \qquad x=l, \quad t>0$$

Putting $\xi=x/l$, $\tau=Kt/l^2$, $B_{so}=h_{so}l/k$, $B_{si}=h_{si}l/k$, $Q=ql/k$ and applying the Laplace transform gives

$$\frac{d^2\bar{\theta}(\xi,p)}{d\xi^2}-p\bar{\theta}(\xi,p)=0, \qquad 0<\xi<1, \quad \tau>0$$

$$\frac{d\bar{\theta}(\xi,p)}{d\xi}=\frac{Q}{p}, \qquad \xi=1, \quad \tau>0$$

$$\frac{d\bar{\theta}(\xi,p)}{d\xi}=B_{so}\bar{\theta}(\xi,p), \qquad \xi=0, \quad \tau>0$$

$$\frac{Q}{p}=B_{si}\{\bar{\theta}_i(p)-\bar{\theta}(\xi,p)\}, \qquad \xi=1, \quad \tau>0$$

The solution of the subsidiary equations is

$$\bar{\theta}(\xi,p)=\frac{Q}{p}\frac{g(p)}{f(p)}, \qquad \tau>0$$

where

$$g(p)=\sqrt{p}\cosh\sqrt{p}\,\xi+B_{so}\sinh\sqrt{p}\,\xi$$

$$f(p)=p\sinh\sqrt{p}+B_{so}\sqrt{p}\cosh\sqrt{p}$$

If, after time t_{q1} the flux is discontinued, then

$$\bar{\theta}(\xi,p)=\frac{Q}{p}(1-e^{-p\tau})\frac{g(p)}{f(p)}, \qquad \tau>\tau_{q1}$$

By the inversion theorem for the Laplace transform the expressions for

$\bar{\theta}(\xi, p)$ become

$$\theta(\xi,\tau)=\begin{cases} Q\left(\xi+\dfrac{1}{B_{so}}\right)-2Q\displaystyle\sum_{n=1}^{\infty}\frac{g(\alpha_n)}{f(\alpha_n)}\sec\alpha_n\cdot e^{-\alpha_n^2\tau}, & 0<\tau\leqslant\tau_{q1} \qquad (2.1.48)\\ 2Q\displaystyle\sum_{n=1}^{\infty}\frac{g(\alpha_n)}{f(\alpha_n)}\sec\alpha_n\{e^{-\alpha_n^2(\tau-\tau_{q1})}-e^{-\alpha_n^2\tau}\}, & \tau>\tau_{q1} \qquad (2.1.49)\end{cases}$$

where

$$g(\alpha_n) \text{ is given by (2.1.44)}$$
$$f(\alpha_n)=\alpha_n^2+B_{so}(1+B_{so})$$

and α_n are the roots ($n=1, 2, \ldots$), all real, of $\alpha_n \tan\alpha_n = B_{so}$. Roots of the transcendental equation are given in Appendix IV of Carslaw and Jaeger (1959).

From the boundary condition at $\xi=1$ the following expressions are obtained for the temperature of the air inside the enclosure

$$\theta_i(\tau)=\begin{cases} Q\left(1+\dfrac{1}{B_{si}}+\dfrac{1}{B_{so}}\right)-2Q\displaystyle\sum_{n=1}^{\infty}\frac{g(\alpha_n)_{\xi=1}}{f(\alpha_n)}\sec\alpha_n\cdot e^{-\alpha_n^2\tau}, & 0<\tau\leqslant\tau_{q1} \qquad (2.1.50)\\ 2Q\displaystyle\sum_{n=1}^{\infty}\frac{g(\alpha_n)_{\xi=1}}{f(\alpha_n)}\sec\alpha_n\{e^{-\alpha_n^2(\tau-\tau_{q1})}-e^{-\alpha_n^2\tau}\}, & \tau>\tau_{q1} \qquad (2.1.51)\end{cases}$$

The results obtained so far apply to a single step-function change in q followed by a reversal after time t_{q1} corresponding to a rectangular pulse in the applied flux. Assume the pulse is repeated with periodicity $T(=t_{q1}+t_{q2})$ represented by the function

$$\begin{aligned} q(t)&=0, && t\leqslant 0\\ &=q, && nT<t\leqslant(nT+t_{q1}),\quad n=0,1,2,\ldots\\ &=0, && (nT+t_{q1})<t\leqslant(n+1)T \end{aligned}$$

Then from (2.1.28)

$$\bar{\theta}(\xi,p)=\frac{Q}{p}\left\{\frac{1-e^{-p\tau_{q1}}}{1-e^{-p(\tau_{q1}+\tau_{q2})}}\right\}\frac{g(p)}{f(p)}$$

Inversion of this expression gives the temperature of the wall for all values of time including the transient due to the initial condition and may be written as

$$\theta(\xi,\tau)=\theta_s(\xi,\tau)-2Q\sum_{n=1}^{\infty}\left\{\frac{e^{-\alpha_n^2\tau}-e^{-\alpha_n^2(\tau-\tau_{q1})}}{1-e^{\alpha_n^2(\tau_{q1}+\tau_{q2})}}\right\}\frac{g(\alpha_n)}{f(\alpha_n)}\sec\alpha_n \qquad (2.1.52)$$

where $\theta_s(\xi, \tau)$ is the steady periodic temperature. By previous analysis it follows that

$$\theta_s(\xi,\tau)=\begin{cases} Q\left(\xi+\dfrac{1}{B_{so}}\right)-2Q\displaystyle\sum_{n=1}^{\infty}\left\{\dfrac{e^{\alpha_n^2(\tau_{q1}-\tau)}-e^{\alpha_n^2(\tau_{q1}+\tau_{q2}-\tau)}}{1-e^{\alpha_n^2(\tau_{q1}+\tau_{q2})}}\right\}\dfrac{g(\alpha_n)}{f(\alpha_n)}\sec\alpha_n, \\ \qquad\qquad 0<\tau\leqslant\tau_{q1} \quad (2.1.53) \\ 2Q\displaystyle\sum_{n=1}^{\infty}\left\{\dfrac{e^{\alpha_n^2(2\tau_{q1}+\tau_{q2}-\tau)}-e^{\alpha_n^2(\tau_{q1}+\tau_{q2}-\tau)}}{1-e^{\alpha_n^2(\tau_{q1}+\tau_{q2})}}\right\}\dfrac{g(\alpha_n)}{f(\alpha_n)}\sec\alpha_n, \\ \qquad\qquad \tau_{q1}<\tau\leqslant(\tau_{q1}+\tau_{q2}) \quad (2.1.54) \end{cases}$$

The form of the expressions (2.1.53) and (2.1.54) applies also to the air temperature with $\theta_{s.i}(\tau)$ the steady periodic air temperature replacing $\theta_s(\xi, \tau)$ to give

$$\theta_{s.i}(\tau)=\begin{cases} Q\left(1+\dfrac{1}{B_{si}}+\dfrac{1}{B_{so}}\right)-2Q\displaystyle\sum_{n=1}^{\infty}\left\{\dfrac{e^{\alpha_n^2(\tau_{q1}-\tau)}-e^{\alpha_n^2(\tau_{q1}+\tau_{q2}-\tau)}}{1-e^{\alpha_n^2(\tau_{q1}+\tau_{q2})}}\right\} \\ \qquad\times\dfrac{g(\alpha_n)_{\xi=1}}{f(\alpha_n)}\sec\alpha_n, \qquad 0<\tau\leqslant\tau_{q1} \quad (2.1.55) \\ 2Q\displaystyle\sum_{n=1}^{\infty}\left\{\dfrac{e^{\alpha_n^2(2\tau_{q1}+\tau_{q2}-\tau)}-e^{\alpha_n^2(\tau_{q1}+\tau_{q2}-\tau)}}{1-e^{\alpha_n^2(\tau_{q1}+\tau_{q2})}}\right\} \\ \qquad\times\dfrac{g(\alpha_n)_{\xi=1}}{f(\alpha_n)}\sec\alpha_n, \qquad \tau_{q1}<\tau\leqslant(\tau_{q1}+\tau_{q2}) \quad (2.1.56) \end{cases}$$

and a mean temperature of (qt_{q1}/UT) over a complete cycle.

Since the system is linear the solutions for both boundary conditions, namely constant temperature and constant flux, may be combined by superimposition to calculate the results of various patterns of intermittent space heating. Consider, for example, the pattern represented by the functions

$$\begin{aligned} q(t) &= 0, && t\leqslant 0 \\ &= q, && nT<t\leqslant(nT+t_{q1}) \\ &= 0, && (nT+t_{q1})<t\leqslant(n+1)T \\ \theta_i(t) &= 0, && t\leqslant t_{q1} \\ &= \theta_i, && (nT+t_{q1})<t\leqslant(nT+t_{q1}+t_1) \\ &= 0, && (nT+t_{q1}+t_1)<t\leqslant(n+1)T \\ T &= t_{q1}+t_{q2} \\ &= t_1+t_2 \end{aligned}$$

This defines a repeated cycle, periodicity T, composed of a steady application of a flux q that is 'on' for time t_{q1} and 'off' for time t_{q2} $(=T-t_{q1})$, and a step-rise θ_i in the indoor temperature that is 'on' for an interval t_1 immediately following t_{q1} and 'off' for t_2 $(=T-t_1)$. The outdoor temperature is invariable. Expressions for the temperature and heat flow over any selected interval of time may be obtained by superimposition of the solutions above. The total heat input, for example, to the system over a complete cycle is

$$qt_{q1} + U(\theta_i - \bar{\theta}_i)t_1 \tag{2.1.57}$$

where $\bar{\theta}_i$ is the contribution to the steady temperature θ_i over the time interval t_1 due to the constant rate of heat input to the enclosed space over the foregoing period t_{q1} and is given by

$$\bar{\theta}_i = \frac{1}{\tau_1} \int_{\tau_{q1}}^{\tau_{q1}+\tau_1} \theta_{s.i}(\tau)\, d\tau$$

in which $\theta_{s.i}(\tau)$ is obtained from (2.1.56).

In natural exposure the outdoor temperature is not constant, it varies in a characteristic diurnal fashion. This interacts with the thermal capacity of the wall with an effect on the thermal response of the system that may be included by substituting $U(t_1)$ for U in (2.1.57). The quantity $U(t_1)$ represents the mean transmittance of the wall over t_1 as defined by equations (2.1.40) and (2.1.41) which express the time-dependence of the air-to-air heat transmission coefficient over short intervals of time.

A matter of practical importance in the space heating of buildings is the amount of heat saved by intermittent heating expressed as a proportion of that required for continuous heating. Superimposition of the effects of simultaneous changes in the ambient temperature on both sides of the wall corresponding to the daily pattern of thermal conditions that normally occur with buildings gives the saving as

$$1 - \frac{qt_{q1} + U(t_1)(\theta_i - \bar{\theta}_i)t_1}{U\theta_i T}, \qquad 0 < t_1 < T$$

where U is the steady state transmittance. As the thermal capacity of the wall becomes smaller so do the values of qt_{q1} and $\bar{\theta}_i$, likewise the difference between $U(t_1)$ and U, with the result that the proportion of energy saved due to intermittent heating approximates to $1 - t_1/T$.

The above solutions for intermittent space heating assume that the space enclosed is sealed and has no internal structure. Clearly, ventilation and internal mass also affect the thermal behaviour of buildings. In Chapter 5 it is shown how these two additional items may be included in the calculation of $\theta(x, t)$, *et seq.*; although the boundary conditions are different, the method of analysis is similar and will not be repeated here. The solutions of the more comprehensive model representing a ventilated building with internal mass is more complex than those so far considered, but the added effect of ventilation alone may readily be determined.

For the problem considered in (2.1.43) through (2.1.47) the effect of ventilation is introduced simply by adding the product $v\theta_i$ to the energy balance of the enclosure. The symbol v denotes the thermal capacity of the enclosed air per unit area of the internal surface of the wall, and θ_i is the amount by which the temperature of the air is raised through a step-function change and maintained at this level.

When the enclosure is heated intermittently by the alternative method of injecting heat from a source of constant strength, the basic solution (2.1.48) for this condition is modified to include ventilation as follows

$$\theta(\xi,\tau)=Q\left(\xi+\frac{1}{B_{so}}\right)\Big/\left\{1+V\left(1+\frac{1}{B_{si}}+\frac{1}{B_{so}}\right)\right\}-\frac{2QB_{si}}{B_{si}+V}$$
$$\times\sum_{n=1}^{\infty}\frac{\left(\alpha_n^2-\dfrac{VB_{si}B_{so}}{B_{si}+V}\right)\left\{\cos\alpha_n\xi+\left(\dfrac{B_{so}}{\alpha_n}\right)\sin\alpha_n\xi\right\}\sec\alpha_n\cdot e^{-\alpha_n^2\tau}}{\alpha_n^2\left[\alpha_n^2+B_{so}(1+B_{so})+\dfrac{VB_{si}}{B_{si}+V}\right.}\,,$$
$$\left.\times\left\{2+B_{so}(1+B_{so})+\frac{VB_{si}}{B_{si}+V}(1+B_{so}+B_{so}^2)\right\}\right]$$
$$0<\xi\leq 1,\quad \tau>0\qquad(2.1.58)$$

where α_n $(n=1,2,\ldots)$ are the roots of

$$\{\alpha_n^2(B_{si}+V)-VB_{si}B_{so}\}\tan\alpha_n=\alpha_n\{B_{si}B_{so}+V(B_{si}+B_{so})\}$$

and $V=vl/k$.

Further solutions relating to this principle of space heating follow from (2.1.58), as shown in (2.1.49) through (2.1.56).

Case 2: Sinusoidal Heat Flow in a Single-layer Wall

An alternative periodic change that approximates to the diurnal variation in the outdoor temperature is the simple sine wave. Writing this as $\theta_o e^{i\omega t}$ in which ω is the angular velocity, and introducing the non-dimensional quantity $\Omega=l^2\omega/K$, then from equation (2.1.11) it follows that

$$\theta(\xi,\tau)=\frac{1}{2\pi i}\int_{\gamma-i\infty}^{\gamma+i\infty}\frac{B_{so}\theta_o}{(p-i\Omega)}\left\{\frac{\sqrt{p}\cosh\sqrt{p}(1-\xi)+B_{si}\sinh\sqrt{p}(1-\xi)}{(p+B_{si}B_{so})\sinh\sqrt{p}+(B_{si}+B_{so})\sqrt{p}\cosh\sqrt{p}}\right\}e^{p\tau}\,dp$$
$$+\theta(\xi,0)\qquad(2.2.1)$$

The residue at $p=i\Omega$ in the inversion integral (2.2.1) gives the required solution for the steady periodic component of temperature variation in the uniform solid which, at the inside surface, is calculated to be

$$\theta(1,\tau)=\frac{2\sqrt{\Omega}e^{-\sqrt{\Omega/2}}}{\sqrt{(a^2+b^2)}}\,\theta_o\exp i(\Omega\tau-\sqrt{\Omega/2}+\psi)\qquad(2.2.2)$$

where

$$a = B_{si} + \sqrt{\Omega/2}(1 + B_{si}/B_{so}) - [\{B_{si} - \sqrt{\Omega/2}(1 + B_{si}/B_{so})\} \cos \sqrt{2\Omega}$$

$$- \sqrt{\Omega/2}\{1 + B_{si}/B_{so} - (2/B_{so})\sqrt{\Omega/2}\} \sin \sqrt{2\Omega}]e^{-\sqrt{2\Omega}} \quad (2.2.3)$$

$$b = \sqrt{\Omega/2}\{1 + B_{si}/B_{so} + (2/B_{so})\sqrt{\Omega/2}\} + [\{B_{si} - \sqrt{\Omega/2}(1 + B_{si}/B_{so})\} \sin \sqrt{2\Omega}$$

$$+ \sqrt{\Omega/2}\{1 + B_{si}/B_{so} - (2/B_{so})\sqrt{\Omega/2}\} \cos \sqrt{2\Omega}]e^{-\sqrt{2\Omega}}$$

and

$$\tan \psi = (a - b)/(a + b)$$

For most purposes the solution may be simplified without significant loss of accuracy by omitting the exponential term in the expressions for a and b. This is certainly the case with solid masonry walls conducting a diurnal (24 h) temperature variation: the exact solution may be preferred for cyclic temperature variations of longer period. The solution defines a temperature oscillation that diminishes in amplitude as the thickness of the wall increases, and lags in phase by an increasing amount behind the outdoor temperature variation. The phase displacement is equivalent to a time lag measured as the interval between the outdoor variation and the resulting temperature change at the inside surface. Denoting the time lag as δ, and T the period, then

$$\delta = T\lambda/2\pi \quad (2.2.4)$$

in which, from equation (2.2.2), the phase angle λ takes the expression

$$\lambda = \sqrt{\frac{\Omega}{2}} - \tan^{-1}\left(\frac{a-b}{a+b}\right) \quad (2.2.5)$$

The amplitude attenuation ϕ defined as the ratio of the amplitudes is

$$\phi = \frac{2\sqrt{\Omega}e^{-\sqrt{\Omega/2}}}{\sqrt{(a^2 + b^2)}} \quad (2.2.6)$$

The quantities λ and ϕ are important properties in practical considerations of periodic heat flow in buildings, and may be read off directly from suitably prepared charts. Writing $C = k/l$ and $Q/C = l^2/2K$, where C is the thermal conductance and Q/C is the ratio of (heat stored/heat transmitted) in the steady state, then by substitution the above expressions for λ and ϕ may be written

$$\delta = \sqrt{\frac{T}{2\pi}}\sqrt{\frac{Q}{C}} - \frac{T}{2\pi}\tan^{-1}\left[\frac{h_{si}h_{so}/C^2 - (Q/C)4\pi/T}{\frac{h_{si}h_{so}}{C^2} + 2\frac{(h_{si} + h_{so})}{C}\sqrt{\frac{Q}{C}}\sqrt{\frac{2\pi}{T}} + \frac{Q}{C}\frac{4\pi}{T}}\right] \quad (2.2.7)$$

$$\phi=\frac{2\sqrt{2}(h_{so}/C)\sqrt{2\pi/T}\sqrt{Q/C}\exp\left(-\sqrt{2\pi/T}\sqrt{Q/C}\right)}{\sqrt{\left[\frac{2\pi}{T}\frac{Q}{C}\left(2\sqrt{\frac{2\pi}{T}}\sqrt{\frac{Q}{C}}+\frac{(h_{si}+h_{so})}{C}\right)^2+\left\{\sqrt{\frac{2\pi}{T}}\sqrt{\frac{Q}{C}}\frac{(h_{si}+h_{so})}{C}+\frac{h_{si}h_{so}}{C^2}\right\}^2\right]}} \tag{2.2.8}$$

in which the expressions for a and b in (2.2.3) have been introduced with the term involving the quantity $e^{-\sqrt{2\Omega}}$ omitted as being negligibly small.

Fig. 5 and Fig. 6 have been prepared from equations (2.2.7) and (2.2.8) giving values of the time-lag and amplitude attenuation at the inside surface in the passage of a sinusoidal variation, period 24 h, in the outdoor temperature through a solid wall, or roof, represented as a large slab of isotropic material of thickness l (mm), conductivity k (W/m degC) and diffusivity K (m^2/s), with h_{si} and h_{so} equal to 7.95 and 18.7 (W/m^2 degC), respectively. These values for the surface transfer coefficients correspond to shade conditions and at normal rates of air movement. The parameter to the curves is the shade thermal transmittance, the coefficient most widely used for defining the heat insulation property of a structure and related to C as

$$\frac{1}{U}=\frac{1}{h_{si}}+\frac{1}{C}+\frac{1}{h_{so}}$$

The quantity Q/C is easily calculated as $l^2/2K$ and the charts enable the characteristic properties λ and ϕ to be determined most readily.

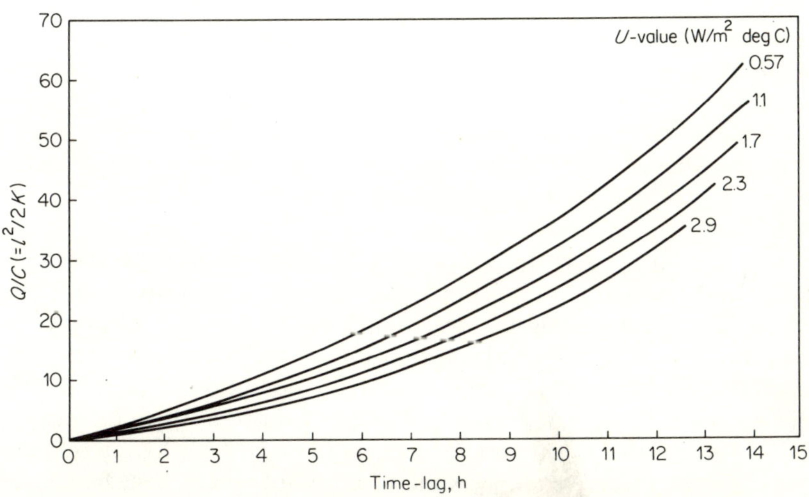

Fig. 5. Periodic heat flow through a unform slab; Q/C versus time-lag (outside air—inside surface)

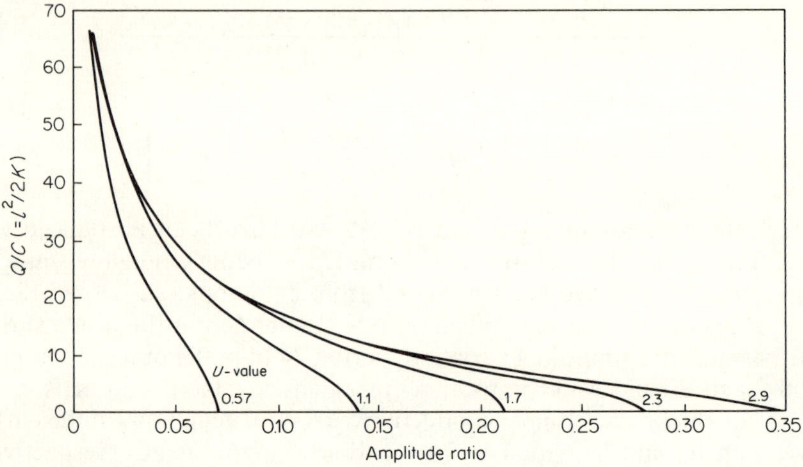

Fig. 6. Periodic heat flow through a uniform slab; Q/C versus amplitude ratio (outside air—inside surface)

3. SUMMARY

A determination of the space heating load in a building due to conduction depends essentially on calculating the heat flux at the interior surfaces of the external sections. In natural exposure the exact expression for the instantaneous heat flux consists of a steady state term and a complex quantity, the value of which varies with time. The steady state term is the product of the thermal transmittance (U-value), defined as the reciprocal of the air-to-air resistance of the structure, and the air-to-air temperature difference. The form of the time-dependent quantity is determined by the boundary conditions; its magnitude diminishes as the thermal capacity of the conducting medium is reduced and with time increasing. It follows that the variable state may be neglected for computing the instantaneous flow through thin walls of material of high thermal diffusivity, and the mean rate of heat flow over long periods in all forms of construction computed as a steady state quantity.

Steady heat flow may be calculated to take account of the weather elements and ventilation in walls of cavity construction in which the airspace is not sealed. The calculations are lengthy but for practical purposes computed results may be tabulated for a variety of different structures exposed to a comprehensive range of ambient climatic conditions. For solid walls of single- and multi-layer composition, also cavity walls in which the airspace is sealed, a straightforward method of calculation may be adopted by defining a 'shade' thermal transmittance as the unit rate of steady heat flow from air to air through a structure shaded from solar radiation and exposed to an environment at a mean radiant temperature equal to the air temperature.

The 'shade' value of thermal transmittance is readily adjusted for determining heat flow under other environmental conditions.

The majority of external walls in buildings are more complex than the plane homogeneous slab considered in this chapter, and in the main consist of two slabs, often of dissimilar thermal properties and separated by an air space. The corresponding rsults for transient and periodic heat flow through a cavity structure are obtained in Chapter 6 with further applications of the Laplace transform method of solution.

SYMBOLS

a_n	Quantity involving α_n
C	Thermal conductance
c	Volumetric specific heat of air
h_{c1}, h_{c2}	Cavity surfaces inner and outer leaf convection coefficients, respectively
h_{ci}	Inside surface convection transfer coefficient
h_{co}	Outside surface convection transfer coefficient
h'_{co}	Outside surface convection transfer coefficient with ambient medium at equivalent outdoor temperature
h_{ro}	Outside surface radiation transfer coefficient
h'_{ro}	Outside surface radiation transfer coefficient with ambient medium at equivalent outdoor temperature
h_{si}	Inside surface transfer coefficient
h_{so}	Outdoor surface transfer coefficient
I	Incident short wave radiation flux at outside surface
K	Thermal diffusivity
k	Thermal conductivity
l	Thickness
n	Ventilation rate of cavity
$Q(t)$	Total heat flow over interval duration t
q	Heat flux
R	Surface-to-surface resistance of single leaf structure
R_1	Inner leaf thermal resistance
R_2	Outer leaf thermal resistance
$R_{a.a}$	Indoor air to outdoor air thermal resistance
R_c	Thermal resistance of unventilated cavity
R'_c	Thermal reistance of ventilated cavity
R_{si}	Indoor surface thermal resistance
R_{so}	Outdoor surface thermal resistance
R'_{so}	Outdoor surface thermal resistance with ambient medium at equivalent outdoor temperature
T	Period of outdoor periodic temperature function
t	Time
U	Thermal transmittance

U'	Thermal transmittance, shade value
v	Wind velocity
x	Position
α_n, $n = 1, 2, 3, \ldots$	Successive roots of relevant function
β	Short wave radiation flux absorbed by glass in single glazing
β_1, β_2	Short wave radiation flux absorbed by inner and outer panes of glass, respectively, in double glazing
δ	Time-lag
η	Absorptivity of outside surface for short wave radiation
$\theta(x, t)$	Temperature at position x and time t
θ_1	Cavity surface inner leaf temperature
θ_2	Cavity surface outer leaf temperature
θ_e	Equivalent outdoor temperature
θ_i	Indoor air temperature
θ_o	Outdoor air temperature
θ_{si}	Indoor surface temperature
θ_{so}	Outdoor surface temperature
$\theta_{s.o}$	Outdoor temperature in long-wave radiation heat transfer
$\bar{\theta}$	Cavity air temperature
λ	Angular phase displacement
σ	Stefan–Boltzmann constant for radiation heat exchange
ϕ	Amplitude attenuation
ω	Angular frequency of sinusoidal temperature variation at outdoor surface

Dimensionless quantities

A	$(1 + 1/B_{si} + 1/B_{so})^{-1}$
B_{si}	$h_{si}l/k$
B_{so}	$h_{so}l/k$
$Q(\tau)$	$Q(t)K/lk$
τ	Kt/l^2
ξ	x/l
Ω	$l^2\omega/K$

REFERENCES

Carslaw, H. E. and Jaeger, J. C. (1959) *Conduction of Heat in Solids*, Second Edition, Oxford University Press.

Hutchinson, F. W. (1962) *Heating and Humidifying Load Analysis*, The Ronald Press Co., New York.

Jurges, W. (1924) *Beihefte zum Gesundheits-Ingenieur*, **1**, no. 19.

CHAPTER 2

Thermal Conductivity

The choice of material for heat insulation purposes is determined essentially, though not exclusively, by its ability to restrict the flow of heat. This is measured by the k-value or thermal conductivity of the material, a specific property defined as follows:

> The thermal conductivity, k-value, of a uniform substance is the quantity of heat which passes in unit time through unit area of a slab of indefinite extent and of unit thickness when unit difference of temperature is established between its faces.

The units of measurement most commonly used are W/m degC; other combinations of units are tabulated with conversion factors in BS 874 : 1973.

With the exception of bright aluminium foil, materials used in buildings as heat insulators may be regarded as mixtures of gases and solid bodies, or as hollow solid bodies filled with gases. They have low k-values, say below 0.15 W/m degC. They owe their insulating property to the low thermal conductivity of air which, when dry and free of convective movement, transmits heat at the rate of 0.0245 W/m degC at 273 K. In general the higher the fractional void volume, or porosity, the lower will be the conductivity of the material. This effect is illustrated in Fig. 1 which shows the increase in k of a heat insulating concrete with density at atmospheric temperature. Other factors, such as the arrangement of the solid bodies, their size, shape, and surface emissivities, also have an influence on the thermal conductivity. These are considered later in the chapter.

Heat insulators may be classified as fibrous, granular, and cellular according to their physical structure. Familiar examples of fibrous insulation are asbestos wool, glass wool, slag wool, and others made from plants as raw products such as fibreboard and compressed strawboard. Normally the fibres have a random orientation, although specialized products of this kind, with the fibres arranged uniformly, may be manufactured having a greater insulating value.

Granulated materials include pumice, cork, diatomaceous powder, and

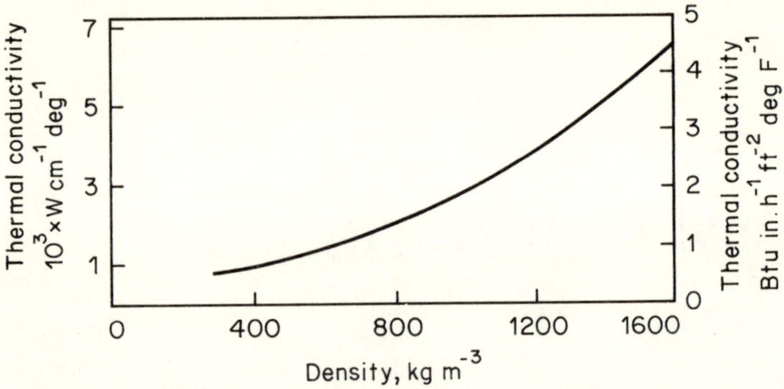

Fig. 1. Thermal conductivity of cellular concrete as a function of density

lightweight concretes. The latter are manufactured with various aggregates and are used very largely for the insulation of buildings as the inner leaf in the traditional form of cavity wall construction. Accordingly they are required to have a minimum compressive strength and their bulk densities therefore are relatively high (around 1000 kg/m^3) resulting in conductivities of the order of 0.215 W/m degC, or more, as compared, say, with 0.039 W/m degC for granulated cork.

Cellular materials have a structure similar to that of a foam. The continuous solid contains cells, large or small, which may be closed or open. In open-cell materials the voids are mostly interconnected, permitting free movement of air through the volume of the material. The voids in closed-cell materials are substantially airtight. Cork is a naturally occurring cellular material. Others of this type are products manufactured by means of foaming agents, or by the evolution of gas in a liquid which is then set hard as a porous mass.

The most effective as thermal insulators, and possessing other useful properties, are a group of synthetic materials known as cellular plastics. The term is used for both foamed and expanded products. The former may have an open- or closed-cell structure while that of the latter is substantially of closed cells. Foamed rigid polyurethane has by far the lowest thermal conductivity of all the heat insulating materials suitable for use in buildings with values ranging from 0.020 to 0.025 W/m degC. Table 1 gives some properties of cellular plastics.

Many attempts have been made to obtain formulae that express the conductivity of a material in terms of porosity and the conductivities of the component phases of solid and gas. The simplest models for theoretical prediction represent the two phases as separate resistances in the form of slabs arranged in series or in parallel to the direction of heat flow. In its general form the structure will consist of a network of both arrangements.

Other models consider porous bodies as structures with a continuous solid

Table 1 Properties of cellular plastics

Material	Approximate density ($kg\,m^{-3}$)	Thermal conductivity at 283 K ($10^4\,W\,cm^{-1}\,deg^{-1}$)	Max. temperature recommended for continuous operation (283 K)	7 days' water absorption (vol. %)	Behaviour in fire
Expanded polystyrene	16	3.46	352.5	3.0	Softens and collapses. Flame retardent grade available
Expanded polystyrene	24	3.31	352.5	2.5	
Expanded polyvinyl chloride	40	3.46	337.5	3.0	Collapses and generally burns with difficulty
Expanded polyvinyl chloride	72	3.46	337.5	3.8	
Foamed urea-formaldehyde (U.F.)	8	3.46	373	Fairly high	Resistant to ignition
Foamed phenol-formaldehyde (P.F.)	32	4.05	403	High	Highly resistant to ignition
Foamed phenol-formaldehyde (P.F.)	56	3.89	403	60	
Foamed polyurethane (carbon dioxide blown)	40	3.89	380	4.3	Generally inflammable but relatively flame retardent forms are available
Foamed polyurethane (fluorinated hydrocarbon blown)	24	2.45	372	5.5	

phase and structures with a continuous air phase. Fibrous and loose-fill materials approximate to the latter, while foamed materials may be regarded as examples of a continuous solid phase structure.

1. THERMAL CONDUCTIVITY OF FIBROUS MATERIAL

From his work on the conductivity of textile fabrics, Schuhmeister (1877) gave the following expression for the conductivity k_m of the mixture of air and fibre

$$k_m = x(k_1 v_1 + k_2 v_2) + y \frac{k_1 k_2}{k_1 v_1 + k_2 v_2} \tag{1.1}$$

where k_1, k_2, and v_1, v_2 denote the conductivity and fractional volume of the air and fibre, respectively, and $x+y=1$. He assumed $x=\frac{1}{3}$ and $y=\frac{2}{3}$ by regarding the mass of fibres as plane slabs, one-third parallel to the direction of heat flow and the remainder at right angles to the flow. These proportions represent a weighted average of the series and parallel arrangements in a random mixture of fibres.

Baxter (1946) obtained empirical values of $x=0.21$ and $y=0.79$ as giving the closest agreement between (1.1) and measured values of the conductivities of thin samples of cotton, wool, and rayon fabrics. These data were found to fit equally well the expression

$$k_m = k_1^{v_1} k_2^{v_2} \tag{1.2}$$

derived by Lees (1900) by considering the fabric constituents as long prisms with their axes perpendicular to the direction of heat flow and arranged in a draught-board pattern with their edges touching along their length. The close agreement between these formulae and the experimental values of k_m was found to apply to the fabrics indicated within the density range 283–531 kg/m³. At densities below 100 kg/m³ there was an apparent increase in conductivity, due probably to the enhanced effect of convection in the larger air spaces at the lower bulk densities, and neither formula fitted the experimental data.

Baxter concluded that the formula (1.1) with $x=0.21$ and $y=0.79$ may be used to estimate the conductivity of the fibre from measurements of the conductivity of thin fabrics of cotton, wool, and rayon.

A later investigation by Strong, Bundy, and Bovernkerk (1960) gave a theoretical estimate of the heat conduction along the solid paths in a random arrangement of fibres. The model for this calculation is shown in Fig. 2(a) as a symmetrical lattice of fibres all with the same diameter. At the points of intersection of the fibres there is a tiny circular contact area through which heat is transmitted in equal amounts to the right and to the left (Fig. 2(b)). Considering the flow to the right only, the contact unit indicated in Fig. 2(b) is redrawn in Fig. 2(c) to represent a thermal resistance due to one-half of the actual contact area and a length of fibre which is one-half of the distance between successive fibres.

The total rate of heat flow Q through a 1 cm² area of the fibre mat for a temperature difference ΔT across each contact unit is gien by

$$Q = \frac{4\pi r^2 k}{r \ln(4r/\Delta r) + 2l} n\Delta T \tag{1.3}$$

where n is the number of fibre junctions. The quantity Δr represents the compression of the radius corresponding to the pressure flattening at the area of contact and is equal to (area of contact/$2\pi r$). For a typical glass fibre mat having fibres with an average diameter of 36×10^{-4} mm and density 250 kg/m³, the formula (1.3) gives for Q the value 42 μW. It will be apparent that this may be regarded as the effective value for the coefficient

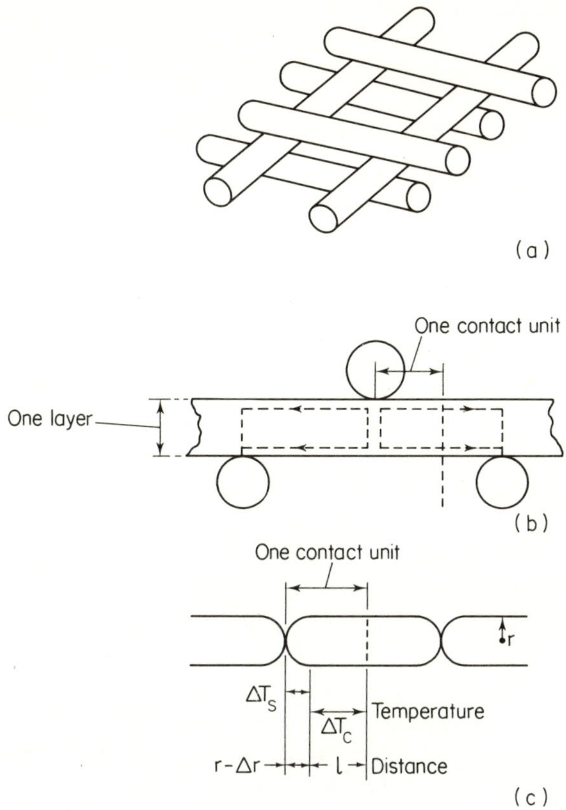

Fig. 2. Ideal glass fibre lattice. (a) A symmetrical lattice of glass fibres; (b) detail of a series of two fibre junctions showing the heat flow path; (c) an idealized fibre junction which is approximately thermally equivalent to (b)

of heat conduction in the fibres. For the fibre mat specified, the calculated figure compares with 19–22 μW obtained experimentally. Tests on mats with thicker fibres gave measured values that diverged increasingly from those calculated. Strong and his co-workers explain the difference between measured and calculated values as being due to fewer fibre contacts in an actual mat compared with the model, variation in the diameter of the fibres, and the greater lateral stiffness of those of larger size. It may be inferred that fibre breakage will cause the effective conductivity of the fibres to increase.

The k-value of a heterogeneous material is sometimes expressed as the 'apparent' conductivity in order to indicate that convection and radiation as well as true conduction contribute to the overall transfer process. Results in convection transfer are reported by Allcut (1951) from a comprehensive

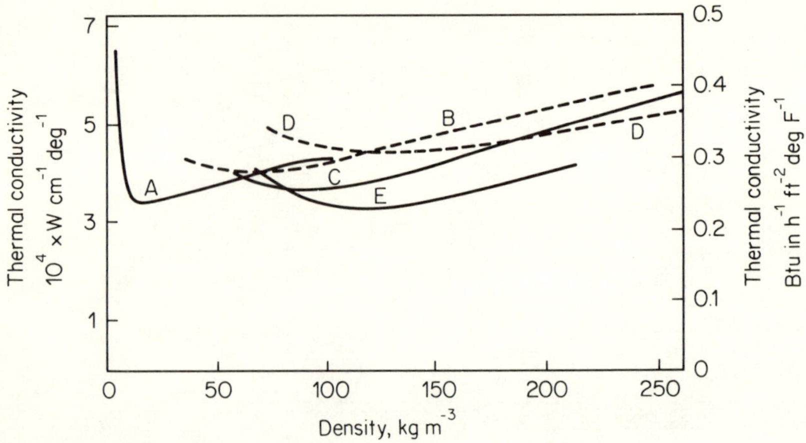

Fig. 3. 'Apparent' thermal conductivity of various materials at different densities. A, kapok; B, bagasse; C, cork board; D, slag wool; E, mineral wool

series of conductivity measurements on some common heat insulating materials using a 300 mm guarded hot plate apparatus. The materials studied were of powdered, granular, and fibrous composition and included commercial products such as expanded polystyrene, silica aerogel, and mineral wool. The effect on the thermal conductivity of varying the density of fibrous material is shown by the curves for kapok, slag wool, and mineral wool marked A, D, and E in Fig. 3. The curves have a characteristic hook shape indicating that below certain optimum densities an appreciable amount of heat is transmitted by convection, as observed earlier by Baxter.

Fishenden and Saunders (1961) give a theoretical interpretation of this phenomenon by relating the Nusselt number to the product of the Grashof and Prandtl numbers for natural convection across a layer of fluid. The theory predicts that by dividing a single air layer into a series of thinner layers the convection currents may be suppressed completely and at the same time greatly decrease the exchange of heat by radiation. The transfer is then mainly by conduction in air. This explains the low thermal conductivity of many fibrous and porous materials and forms the basis for the manufacture of many common heat insulating products.

Allcut estimated the contribution by convection to the overall transfer of heat in glass wool with the aid of a result obtained by Kannuluik and Martin (1933) from experiments on the conduction of heat in powders. These experiments showed that k may be expressed as a function of the air pressure p with the formula

$$p/k = ap + b \tag{1.4}$$

Deviations from the straight line relationship develop as convection in the

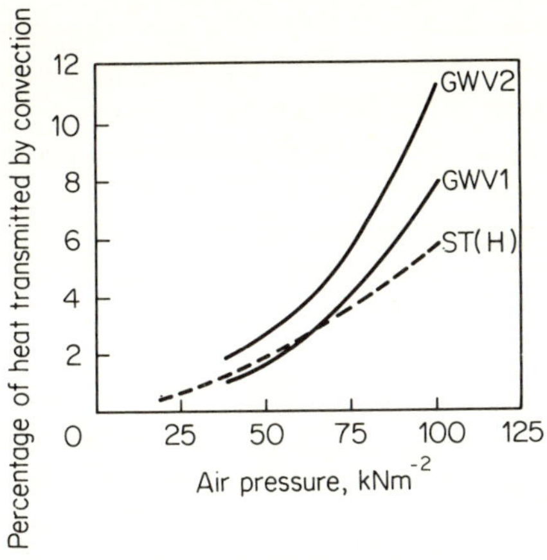

Fig. 4. Percentages of total heat transferred by convection (after Allcut, 1951)

air spaces or pores increases. This behaviour is observed as the air pressure is increased and enables the contribution due to this mode of heat transfer to be deduced. Results thus obtained by Allcut are shown in Fig. 4.

Allcut also demonstrated an increase in heat transfer due to convection in glass wool insulation as the density is reduced but with the pressure remaining constant (see Fig. 5).

An insight into the mechanism of heat transmission in porous material first became possible with the development of the kinetic theory of gases.

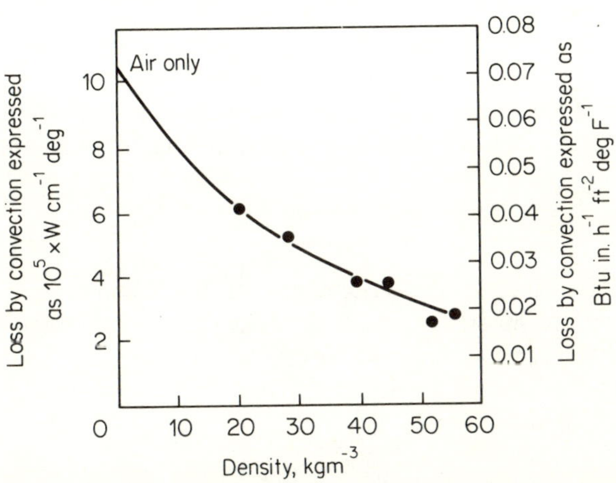

Fig. 5. Convection through glass wool at different densities with standard atmospheric pressure (after Allcut, 1951)

The concept of the mean free path of a gas molecule led to results of fundamental importance in the theory of gas conduction in fibrous insulation. In this connection the work of Verschoor and Greebler (1952) is particularly noteworthy and will be described to illustrate how the kinetic theory may be applied to determine theoretical design requirements for improving the heat insulating properties of fibrous material.

Consider a mat of a fibrous material, glass wool for example, in which the fibres, all of the same thickness, are randomly oriented in planes that lie parallel to the mat, and assume that heat flows through the mat in a direction perpendicular to these planes. Let f be the volume fraction of fibre. Then the volume of fibre in an elementary thickness Δx of unit cross-section area is $f\Delta x$ and the total projected area of fibres in this volume element is $4f\Delta x/\pi D$, where D is the fibre diameter. This projected area is the probability that a molecule of gas will collide with a fibre in the interval Δx. Now, by the kinetic theory of gases, the probability ψ_g of the molecules in a gas colliding with one another before moving a distance x is

$$\psi_g = 1 - e^{-x/l_g} \tag{1.5}$$

where l_g is the mean free path, dependent on temperature and pressure. Similarly, the probability of a gas molecule colliding with a fibre before moving a distance x may be written

$$\psi_f = 1 - e^{-x/l_f} \tag{1.6}$$

where l_f is the mean free path for molecule-fibre collisions. Verschoor and Greebler regard l_f as an equivalent pore size for fibrous insulation and an important quantity therefore in the theory of heat transmission in such material. It may be calculated by setting x equal to Δx so that ψ_f then equates to the fibre collision cross-section to give

$$l_f = 0.785(D/f) \tag{1.7}$$

Consideration of the probability of an intermolecular collision for all values of x gives as the mean free path l for such collisions in an assembly of fibres

$$l = \frac{l_g l_f}{l_g + l_f} \tag{1.8}$$

Now, by kinetic theory, the coefficient of heat conduction of a gas is proportional to the mean free path of its constituent molecules. Assuming that a relationship of the same form may be used to calculate the thermal conductivity of the gas within the insulation, then representing this by k_{cd} it follows from (1.8) that

$$k_{cd} = k_g \frac{l_f}{l_f + l_g} \tag{1.9}$$

The magnitude of k_{cd} may be estimated by considering fibrous insulation at, say, 330 K; the thermal conductivity of air at this temperature is 0.029 W/m degC and $l_g = 81\ \mu$m, where p is pressure in mm Hg. Then by substitution, the formula (1.9) becomes

$$k_{cd} = \frac{0.157(Dp/f)}{81 + 0.785(Dp/f)} \tag{1.10}$$

It will be noted from (1.10) that gas conduction in fibrous insulation is determined by factors for which some measure of control is possible during manufacture. The significance of this is apparent from Table 2 which shows gas conduction at atmospheric pressure as the most important of the heat transfer processes.

The solid curve in Fig. 6 shows the variation of the apparent or measured conductivity of fibrous insulation with pressure. Clearly, little further reduction in the conductivity can be gained by reducing the air pressure below 13.3 N/m^2. It is inferred that the asymptotic k-value of 0.016 represents the conductivity due to radiation and to conduction in the solid part of the material. By adding to this amount the theoretical value of k_{cd} calculated from (1.10) and dividing the sum by $(1-f)$, representing the volume fraction of air, the dashed curve in Fig. 6 is obtained showing the variation in the

Table 2 Thermal conductivity of type A fibre insulation with different gases at a mean temperature of 338 K and for a temperature gradient of 26 deg cm^{-1}

Gas	Insulation density kg m^{-3}	Pressure kN m^{-2}	Thermal conductivity[a] $10^4 \times$ W cm^{-1} deg^{-1} λ	λ_{cd}	λ_{cv}
Helium	8.73	101	21.9	16.2	2.28
Helium	24.0	101	19.2	16.2	2.08
Helium	24.0	10.1	18.9	15.9[b]	2.00
Helium	74.0	101	18.1	16.1	1.17
Air	8.73	101	5.69	2.92	0.91
Air	24.0	101	4.25	2.92	0.50
Air	24.0	10.1	4.22	2.91	0.49
Air	74.0	101	3.56	2.92	0.30
Carbon dioxide	8.73	101	4.70	1.96	0.88
Carbon dioxide	24.0	101	3.25	1.96	0.49
Carbon dioxide	24.0	10.1	3.25	1.96	0.49
Carbon dioxide	74.0	101	2.55	1.96	0.27
Freon-12	8.73	101	3.76	1.20	0.71
Freon-12	24.0	101	2.33	1.20	0.35
Freon-12	24.0	10.1	2.33	1.20	0.35
Freon-12	74.0	101	1.67	1.20	0.20

[a] The difference between the values of λ listed and the sum of λ_{cd} and λ_{cv} is the contribution by radiation and solid conduction.

[b] For helium there is a noticeable difference between λ_{cd} and λ_g at 10.1 kN m^{-2} and even some difference at atmospheric pressure (101 kN m^{-2}), since the mean free path and the thermal conductivity of this gas are large.

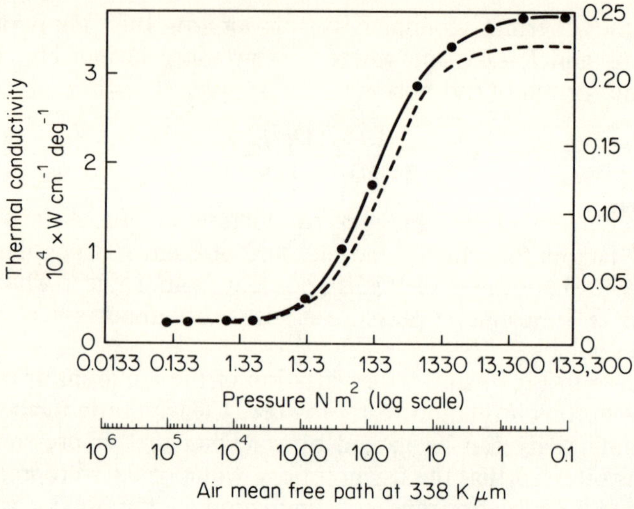

Fig. 6. Thermal conductivity versus air pressure for 74 kg m type A insulation at 65 degC mean temperature; solid line, experimental; dashed line, theoretical (theoretical curve neglects contribution by air convection) (after Verschoor and Greebler, 1952)

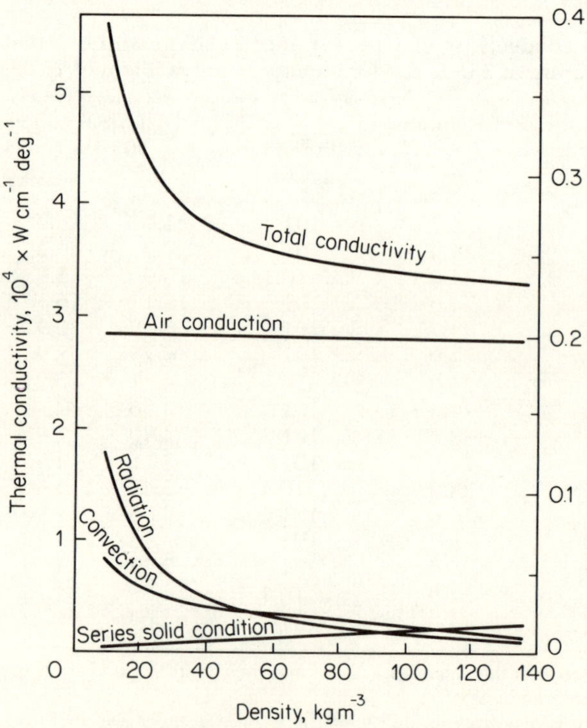

Fig. 7. Contribution by each of mechanisms of heat transfer in type A insulation at atmospheric pressure versus insulation density at 338 K mean temperature

theoretical conductivity with pressure. At intermediate pressures $l_f \simeq l_g$ and k_{cd} increases with p until l_g is relatively small; gas conduction then becomes independent of pressure. At normal atmospheric pressure the enclosed air is essentially free air, i.e. the mean free path of the gas molecules is very small compared with the average spacing of the fibres. Fibre insulating materials in common use have $l_f \simeq l_g$. By using a finer fibre, say 0.01 μm diameter in a bulk density of 160 kg/m^3, the conductivity of such material may be reduced substantially by 50 percent or more.

The difference between the experimental and theoretical curves in Fig. 6 is attributed to convection and has been used with several gases to measure convection in fibrous insulation. This and other processes in the mechanism of heat transmission are compared in Fig. 7 in their relation to density, the property that may readily be controlled in the manufacture of fibre insulating material. From this diagram emerge the following criteria for achieving a low conductivity fibre insulation. The bulk density should be reduced to minimize solid conduction but not beyond the stage when the improvement in insulating value is offset by the resulting increase in radiation. A considerable reduction in gas conduction follows by evacuating the system to a

Table 3 Formulae to determine typical values for the thermal conductivity of wood perpendicular to grain (units Btu in. h^{-1} ft^{-2} deg F^{-1} where not stated)

	Reference
A. For oven-dry wood:	
$$\lambda = 1.39d_1 + 0.165$$	Rowley (1933)
where d_1 is the dry bulk density in g cm^{-3} based on the oven-dry weight and the volume when dry.	
B. For wood containing up to 40% moisture by weight:	
$$\lambda = d_2(1.39 + 0.028M) + 0.165$$	Rowley (1933)
where d_2 is the dry bulk density in g cm^{-3} based on the oven-dry weight and the volume at current moisture content and M is moisture content.	
C. For wood at any given moisture content:	
$$\lambda = \frac{d_2[1.39 + 0.028M]}{1 - d_2[0.009(30 - M)]} + 0.165$$	Rowley (1933)
where d_2 is the dry bulk density in g cm^{-3} based on the oven-dry weight and the volume when green.	
D. For pine and spruce only, knowing the conductivity at 'room moisture content'	
$$\lambda_u = \lambda_t + 0.02u$$	
where λ_u is the conductivity in k cal m^{-1} h^{-1} deg^{-1} at a percentage moisture content by volume u above the moisture content when conditioned to the room; and λ_t is the conductivity under room conditions in the same units.	*A Handbook of Hardwoods* (1956)

pressure of about 10^{-1} Torr which can be achieved in practice. A further reduction in pressure gives no significant improvement. Alternatively, the insulating value may be improved at atmospheric pressure by using the smallest diameter fibres, or increasing the density to the point where this begins to yield a diminishing return due to solid conduction. At higher temperature (above 420 K) the conductivity increases rapidly; it is reduced by increasing the density using fine fibres so that the solid phase is in a finely divided state.

A material that also has a fibre composition and one that is used extensively in building, is timber. Although the main application is for structural purposes, the medium values of thermal conductivity give it, to a moderate degree, the added property of thermal insulation. Wood is a strongly anisotropic material with respect to heat conduction, the conductivity parallel with the grain (longitudinal) being about twice that in the radial and tangential directions. It is usual to express conductivity values for heat flow perpendicular to the direction of the grain. Charts have been prepared whereby k-values may be determined when the dry density and moisture content are known (Rowley, 1933; Wangaard, 1940; Davies, 1943). Table 3

Table 4 Comparison of typical values for the thermal conductivity of wood perpendicular to the grain from four sources

Dry density ($kg\,m^{-3}$) (for volume at current moisture content)	Moisture content (% dry weight)	Thermal conductivity ($10^4 \times W\,cm^{-1}\,deg^{-1}$) Wangaard (1933) Rowley (1940) from nomograph	MacLean, from 1941 curves	NPL from past results	Mean
160	0	5.76	5.47	(5.47)*	5.61
	10	6.48	6.20	6.20	6.34
	20	7.2	6.91	(6.91)*	7.05
320	0	8.21	8.80	(7.64)*	8.20
	10	9.36	10.1	8.80	9.35
	20	10.65	11.5	(9.94)*	10.65
480	0	10.8	12.1	(9.94)*	10.95
	10	12.5	14.0	11.7	12.7
	20	14.4	15.8	(13.4)*	14.5
640	0	13.8	15.1	(12.1)*	13.7
	10	16.3	17.7	14.5	16.1
	20	18.7	20.3	(17.0)*	18.7
800	0	17.3	18.4	(15.0)*	16.8
	10	19.6	21.6	17.3	19.4
	20	(21.7)*	24.8	(19.4)*	22.0

* Assessed value to avoid unnecessary irregularity in last column.

gives formulae expressing these relationships and in Table 4 typical values reported from various sources for the conductivity of wood perpendicular to the grain are compared. The formulae are sufficiently accurate for practical purposes.

2. THERMAL CONDUCTIVITY OF GRANULAR MATERIAL

There are many formulae relating the thermal conductivity of a granular material to the conductivities and fractional volumes of its components. The simplest of these consider a mixture of two phases, solid and fluid, arranged in series or in parallel with respect to the direction of heat flow. The parallel arrangement gives the maximum conductivity

$$k = fk_s + (1-f)k_m \tag{2.1}$$

where k_s is the conductivity of the solid component occupying a volume fraction f of the material, and k_m is the conductivity of the fluid occupying the pore space.

The series arrangement has a minimum conductivity

$$k = \frac{k_s k_m}{fk_m + (1-f)k_s} \tag{2.2}$$

The general structure of actual materials corresponds to combinations of both arrangements which, for a random distribution, give the simplest formula

$$k = k_m^{(1-f)} k_s^f \tag{2.3}$$

A model combining the series and parallel arrangements was proposed by Wyllie and Southwick (1954) for calculating the electrical conductivity of a medium of conductive particles saturated with a conducting elecrolyte. For the analogous thermal problem the model gives

$$k = \frac{ak_s k_m}{k_s(1-d) + dk_m} + bk_s + ck_m \tag{2.4}$$

in which $a+b+c=1$ and $ad+b=1-f$.

Woodside and Messmer (1961) compared (2.4) with experimental results and found fair agreement when used with $b=0$, $c=\phi-0.03$, $a=1-c$, and $d=(1-\phi)/a$, where ϕ is the fractional void volume.

The theory of the conductivity of heterogeneous media is complex and has been attempted in a few simple cases only. The problem essentially is to calculate the disturbance to the steady flow of heat in a uniform medium due to the presence of an object of different thermal conductivity. Mathematically it is analogous to calculating the induced magnetism in a body placed in a uniform external field. Solutions for certain arrangements in induced magnetism were first obtained by Maxwell (1904) and provide the results for the corresponding thermal cases. The formula thus obtained by Brailsford and Major (1964) from Maxwell's result for the average conductivity k of a random distribution of solid spheres conductivity k_1 in a continuous medium of conductivity k_2 is

$$\frac{k}{k_2}=\left\{1-2f\left(\frac{1-k_1/k_2}{2+k_1k_2}\right)\right\}\Big/\left\{1+f\left(\frac{1-k_1/k_2}{2+k_1k_2}\right)\right\} \tag{2.5}$$

where f is the volume fraction of phase 1. The spheres are assumed to be far enough apart to avoid mutual interaction. Interchanging the suffixes 1 and 2 reverses the distribution of the phases and gives for phase 1 continuous

$$\frac{k}{k_2}=\frac{k_1}{k_2}\left\{3\frac{k_2}{k_1}+2f\left(1-\frac{k_2}{k_1}\right)\right\}\Big/\left\{3-f\left(1-\frac{k_2}{k_1}\right)\right\} \tag{2.6}$$

Further examination of Maxwell's formulae in their application to heat transmission has been made by De Vries (1952).

The results (2.5) and (2.6) are illustrated by curves 3 and 4, respectively, in Fig. 8 together with the curves for k_{max} and k_{min} representing (2.2) and (2.3). The curves are calculated in all cases with $k_1/k_2=0.1$.

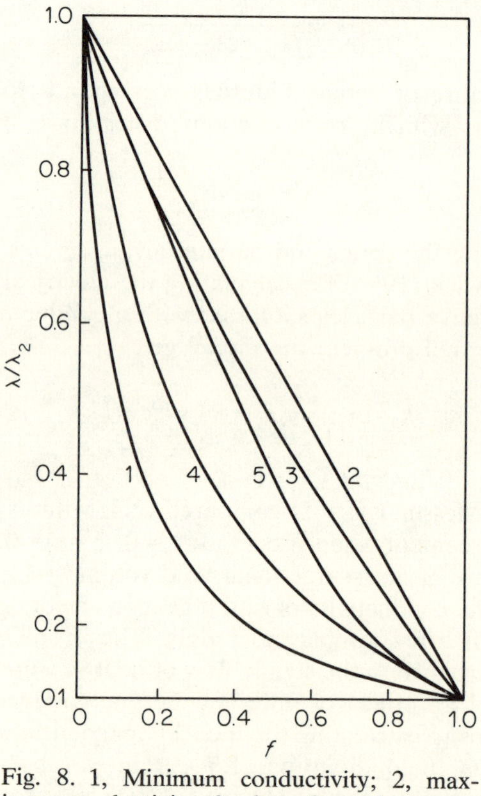

Fig. 8. 1, Minimum conductivity; 2, maximum conductivity; 3, phase 2 continuous; 4, phase 1 continuous; 5, random mixture

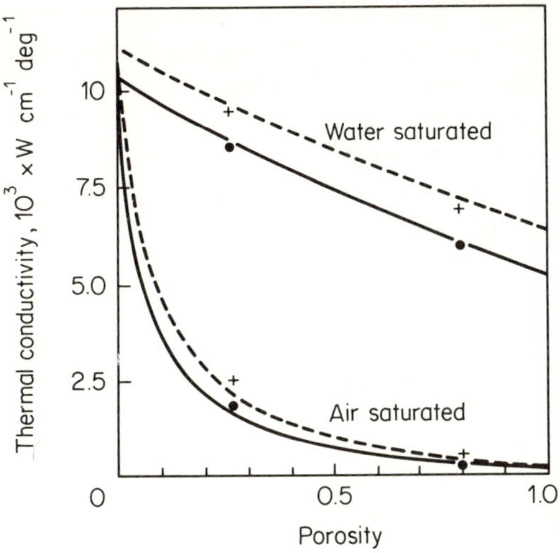

Fig. 9. Comparison of theory with experiment (Sugawara and Yoshiza, 1961) for glass balls surrounded by air and water at 373 K (dashed line) and 273 K (solid line)

Sugawara and Yoshizawa (1961) found good agreement (Fig. 9) for equations (2.5) and (2.6) with measured values of conductivity. The measurements were made on specimens specially constructed as two-phase assemblies to conform with the distribution of phases corresponding to these equations. A similar investigation by Pratt and Ball (1956) with a steel shot aggregate concrete gave measured and calculated conductivity values of 1.99 and 1.70 W/mK respectively. In view of the experimental difficulties involved the agreement was considered not unsatisfactory. For natural porous material comparison is more difficult and insufficient is known about their structure to permit general predictions of thermal conductivity.

3. THERMAL CONDUCTIVITY OF DAMP MATERIAL

The thermal conductivity of dry air at 10 degC is 0.025 W/m degC; at the same temperature water conducts heat at about 25 times this rate. The influence of moisture on heat transmission in porous material may be considered by examining some of the difficulties involved in measuring thermal conductivity by the steady state method. These are demonstrated by the experiments of Woodside and Cliffe (1959) to determine the conductivity of samples of dry, moist, and wet sand. The apparatus used was a 300×300 mm guarded hot plate with a 150 mm square central test area and mounted vertically so that heat flowed horizontally through the 50 mm thick specimens. The dry sand reached equilibrium after 60 h giving a k-value of

2.64×10^{-3} W/cm degC. Saturated sand, moisture content 24 percent by weight, took 4–8 h to reach equilibrium and gave a conductivity value approximately ten times higher. Moist sand, water content 2 percent, took the longest time, namely 300 h, to reach equilibrium, giving a conductivity of 5.6×10^{-3}. At 4 percent water content the equilibrium time was 170 h and the conductivity increased to 10.4×10^{-3} W/cm degC. An empirical relationship due to Kersten (1949) predicted a value of 11.8 for this particular specimen but estimated a figure about 55 percent too high at the 2 percent moisture content. An examination of the moist specimens at the completion of the tests showed a high concentration of water in the lower guard ring area where, for the 4 percent sample, near saturation conditions were found. The results of these tests, plotted in Fig. 10, indicate a movement of moisture from the hot to the cold region where, above a water content of 2.95 percent, gravity drainage began to develop causing water to accumulate outside the test area. The tests were repeated at 4 percent

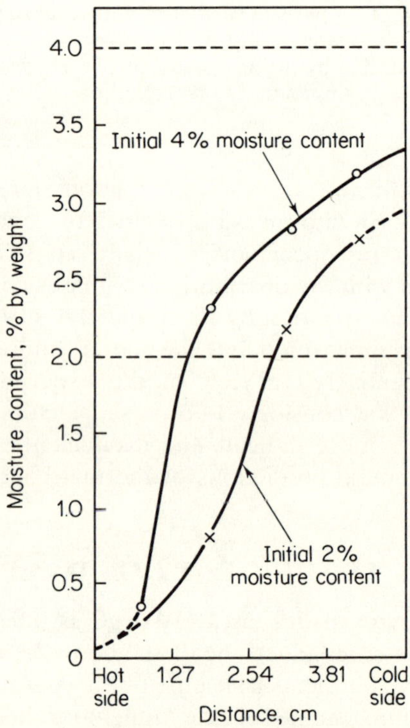

Fig. 10. Equilibrium moisture distribution: initial 2% and 4% moist Ottawa sand; $160\,kg\,m^{-3}$ dry density; mean temperature, 297 K; average temperature gradient $3.3\,deg\,cm^{-1}$, determined by hot plate apparatus (from Woodside and Cliffe, 1959)

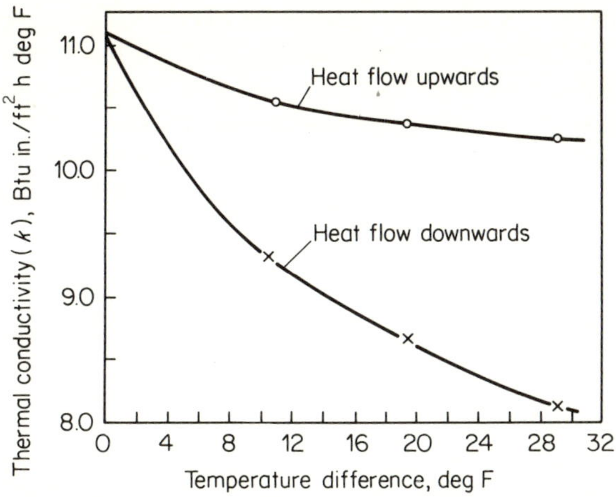

Fig. 11. Conductivity of 4% moisture content Ottawa sand at various temperature gradients and heat flow directions determined by 8 in. uniflow apparatus with 1 in. thick specimen; density 102 lb/ft^3; mean temperature 74.5 degF

moisture content with upward and downward heat flow, i.e. with the test assembly arranged horizontally. In this position the moisture content in the central test portion remained constant and a conductivity value higher than with the previous arrangement was obtained. It is interesting to note also that the top specimen gave a higher k-value than the bottom specimen, as shown in Fig. 11.

Moisture in porous solids increases the thermal conductivity with sensible heat, transmitted largely by conduction, and latent heat transferred by water vapour movement. Considering first the sensible heat component, its effect will depend on the manner in which the water is distributed. Joy (1957) proposed the following arrangements in an attempt to interpret practical results.

1. Series arrangement: water located in several layers perpendicular to the direction of heat flow. This would be expected to give the minimum conductivity.

2. Parallel arrangement: water located in continuous paths that by-pass the less conducting air cells, thus giving the maximum rate of heat trnsmission.

3. Bead arrangement: water existing in small beads located uniformly through the insulation and considered to be the dispersed phase.

4. Foam arrangement: water films surrounding the insulating particles, the films connecting in a foam or honeycomb structure so that the water forms the continuous phase.

Assuming immobility of the water, Joy represented each arrangement by a formula and compared calculated values of conductivity for each of the four separate arrangements with test figures. The results generally indicated a combination of different arrangements in a given insulation with a given water content, causing the conductivity to vary over a range of about ±6 percent. A similar examination by Krischer and Esdorn (1956) for predicting the conductivity of damp material is described below, but latent heat transfer in damp pores, investigated by Krischer (1941) in damp pores, will first be considered.

Latent heat of evaporation of water is conveyed by water vapour diffusion in air from a region of high to low partial vapour pressure. Stefan's law defines this movement of moisture with the formula

$$\omega_v = -\frac{D_{air}}{R_v T}\frac{p}{p-p_v}\frac{dp_v}{dx} \tag{3.1}$$

where

ω_v = amount of water vapour diffusing,
D_{air} = coefficient for the diffusion of water vapour in air,
p = air pressure,
p_v = partial pressure of the water vapour,
R_v = gas constant for water vapour,
T = absolute temperature, and
x = length co-ordinate

The perm in. is the most commonly used measure for D_{air} with units grains in/ft^{2}h in Hg. At room temperature the value of D_{air} is 120 perm in. Other units with conversion factors are given in BS 2972:1975.

If the walls of the pores are wet, saturation pressure exists, which in turn depends upon temperature, and (3.1) may be written as

$$\omega_v = -\frac{D_{air}}{R_v T}\frac{p}{p-p_v}\frac{dp_v}{dT}\frac{dT}{dx} \tag{3.2}$$

Let q_v denote the amount of heat transferred by water vapour diffusion in air, then

$$q_v = \omega_v L, \quad \text{where } L \text{ is the latent heat of evaporation}$$

$$= -\frac{D_{air}}{R_v T}\frac{p}{p-p_v}\frac{L\,dp_v}{dT}\frac{dT}{dx}$$

or, more conveniently

$$q_v = -k_v\, dT \tag{3.3}$$

where

$$k_v = \frac{D_{air}}{R_v T}\frac{p}{p-p_v}\frac{L\,dp}{dx} \tag{3.4}$$

It will be apparent from (3.4) that k_v represents the equivalent conductivity

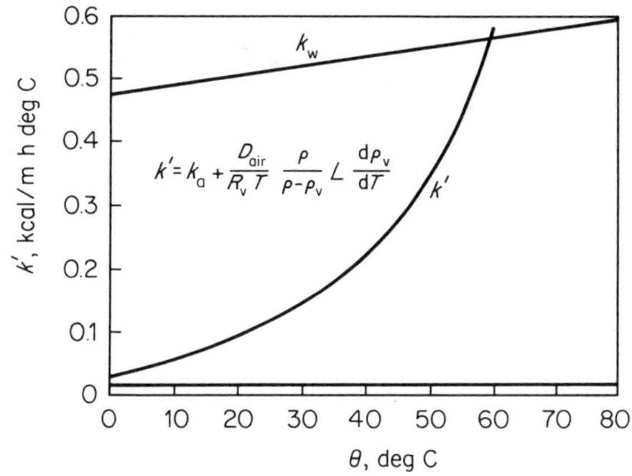

Fig. 12. Equivalent conductivity k', of porous material (conduction and diffusion)

of a damp pore due to latent heat transfer. The overall conductivity k^1 of the pore may be represented therefore as

$$k^1 = k_a + k_v$$

where k_a is the thermal conductivity of air. Fig. 12, due to Krischer (1941), shows a rapid increase in k^1 due to water vapour diffusion in air with rise in temperature. The corresponding curves for water and air are included. When the rising temperature reaches 333 K the graph shows that the rate of heat transfer through the wet pores is as great as if these pores were entirely filled with water. It follows that at 333 K the conductivity of damp porous material becomes independent of the extent to which the pores are filled with water. Moreover, above this temperature heat is transmitted at a higher rate in the pores of a moist material filled with air than if completely filled with water. Further consequences of (3.4) are:

(1) an increase in the water content above 333 K will cause the conductivity to fall by diminishing the volume fraction of air space, and

(2) at low moisture the overall conductivity will rise more rapidly with temperature than at high moisture content. Compared, however, with its strong influence on the conductivity of a damp pore, latent heat transfer by diffusion of water vapour in air has, depending on the temperature, a considerably lower effect on the overall conductivity of the porous material. For example, for a moist sand at 23.8 degC equation (3.4) gives $k_v = 0.014$ W/m degC, a relatively small amount.

As already indicated, the value of k_v will depend upon the porosity of the mateial. From observations of the diffusion of gases through dry and moist soils, Penman (1940) obtained the relationship

$$D = 0.66 P D_{air} \tag{3.5}$$

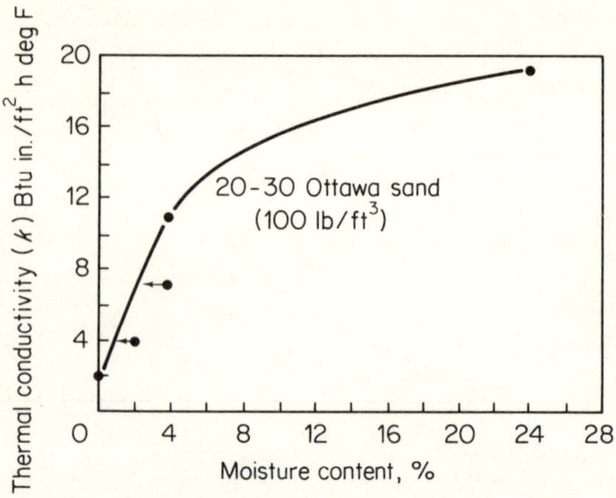

Fig. 13. Conductivities of Ottawa sand determined by guarded hot plate apparatus; mean temperature, 75 degF; temperature gradient 15 degF/in

for the vapour diffusion of the material. The contribution by water vapour diffusion in air to the total transfer of heat in a damp material of porosity *P* may be calculated therefore by substituting that result in (3.4).

A further process accompanying the transfer of heat in a damp material is the circulation of water in the pores leading to the formation of zones of different water content. During the period of establishing a steady temperature gradient in test specimens when measuring thermal conductivity by the conventional guarded hot plate method, water vapour diffusion causes differences in wetness. If the transfer by diffusion exceeds the rate at which capillary forces return the displaced water then a part of the test specimen may become completely dry. Fig. 13 shows the variation of conductivity with moisture content obtained by Woodside and Cliffe with a wet sand.

It will be seen that the conductivity decreases more rapidly in the warmer region than it increases in the colder region, these regions being indicated by the moisture distribution as described below. The net result of this rearrangement in the initially uniform moisture distribution is that the steady state method when used with a damp material measures an apparent conductivity smaller than the 'true' value defining an isotropic specimen. Referring to Fig. 11, this non-uniform distribution of the water content also explains the drop in the measured conductivity with increasing temperature and the higher conductivity of the upper specimen where the moisture redistributes itself more uniformly.

The formation of zones is a feature of capillary water movement. Its significance in the measurement of conductivity may be examined by considering the volume flow of liquid through a tube, the pores in the specimen material being likened to an assembly of tubes of varying diameter. The capillary forces in the tubes increase as these become narrower so that the finest capillaries will draw water from the others. According to Poiseuille's equation for the flow of liquid through a tube, the resistance to flow through a tube of diameter d varies inversely as d^4, whereas capillary forces varies inversely as d. It follows that the force opposing capillary movement is greater the lower the water content, for when the water content is lowest the moisture can only be located in the narrowest pores. Applied to the measurement of thermal conductivity this means that the capillary return suction diminishes as the water content of the specimen becomes lower and zoning then develops more readily.

The foregoing discussion identifies four processes of heat transfer in a damp porous solid; these are summarized as follows:

(1) heat transfer in the solid, k_s,

(2) heat transfer in the liquid, k_w

(3) heat transfer by combined molecular conduction and latent heat transfer in air contained in pores with damp walls, $k_a + k_v$, and

(4) molecular heat conduction in pores with dry walls, k_a.

Fig. 14, due to Krischer and Esdorn (1956), is a schematic representation of a damp porous solid, overall conductivity k, in which the components of heat transfer listed above are shown as plates arranged in series and parallel with equivalent thermal conductivities k_2 and k_1, respectively.

It will be seen from the diagram that the following relationships apply

$$k_1 = (1-\varepsilon)k_s + \psi_\omega k_\omega + b(\varepsilon - \psi_\omega)(k_a + k_v) + (1-b)(\varepsilon - \psi_\omega)k_a \tag{3.6}$$

$$k_2 = \left\{\frac{1-\varepsilon}{k_s} + \frac{\psi_\omega}{k_\omega} + b\frac{(\varepsilon - \psi_\omega)}{k_a + k_v} + \frac{(1-b)(\varepsilon - \psi_\omega)}{k_a}\right\}^{-1} \tag{3.7}$$

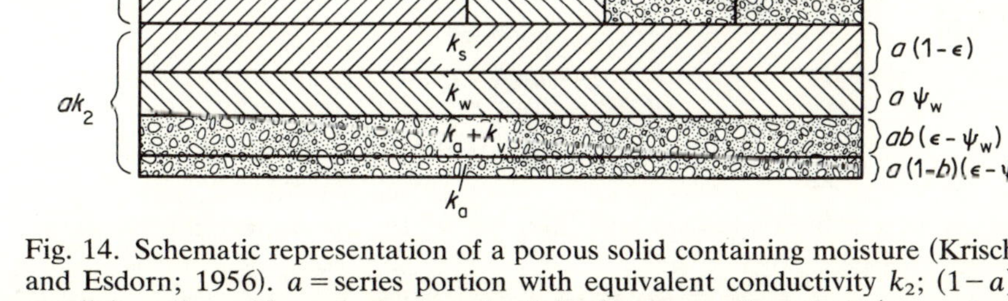

Fig. 14. Schematic representation of a porous solid containing moisture (Krischer and Esdorn; 1956). a = series portion with equivalent conductivity k_2; $(1-a)$ = parallel portion with equivalent conductivity k_1; b = portion of air (as a volume) contained in pores with damp walls; ε = porosity; Ψ = moisture content of the material (vol. %)

and for the overall conductivity

$$k=\left(\frac{1-a}{k_1}+\frac{a}{k_2}\right)^{-1} \tag{3.8}$$

For a specific material the quantities k_s and a are obtained by writing

$$k_{\text{dry}}=\left(\frac{1-a}{k_{1\,\text{dry}}}+\frac{a}{k_{2\,\text{dry}}}\right)^{-1}$$

where

$$k_{1\,\text{dry}}=(1-\varepsilon)k_s+\varepsilon k_a$$

$$k_{2\,\text{dry}}=\left(\frac{1-\varepsilon}{k_s}+\frac{\varepsilon}{k_a}\right)^{-1}$$

with similar expressions for the material saturated. The two equations for k_{dry} and k_{sat} determine k_s and a. The quantity b is obtained as the value that gives the best fit of theoretical to experimentally determined values of k. Krischer and Esdorn applied the method to certain lightweight concretes and obtained satisfactory agreement between measured and calculated values for different values of porosity and moisture content.

4. MEASUREMENT OF THE THERMAL CONDUCTIVITY OF BUILDING MATERIAL

Values of thermal conductivity are measured with experimental arrangements that simulate solutions of the differential equation defining the conduction of heat in a homogeneous isotropic solid

$$\nabla^2\theta=\frac{1}{K}\frac{\partial\theta}{\partial t} \tag{4.1}$$

The general solution of (4.1) consists of a steady state term defining temperature as a function of position only, and a dynamic term relating temperature additionally with time according to the boundary conditions. For undirectional flow in steady state through a plane slab of area A and thickness l

$$q=-k\frac{\{\theta(0)-\theta(l)\}}{\displaystyle\int_0^2 \mathrm{d}l/A} \tag{4.2}$$

For flow in a cylinder between radii r_1 and r_2 the corresponding result is

$$q=-k\frac{\{\theta(r_1)-\theta(r_2)\}}{\displaystyle\int_{v_2}^{v_1} \mathrm{d}r/2\pi r} \tag{4.3}$$

The usual procedure for measuring the thermal conductivity of solids for use in buildings is based on equation (4.2). This relationship may be simulated experimentally with little difficulty within the range of temperatures that obtain in buildings, and has been adopted in many countries as the principle on which standard methods of measurement are designed, although details of the experimental arrangement differ from one specification to another. For example, differences occur in the size and shape of the sample, also in the orientation of the equipment. In all cases the test specimen is sandwiched in slab form between a hot plate and a cold plate and conditions arranged so that the flow of heat through the central test area of the specimen is in one direction only. When a state of thermal equilibrium is reached the conductivity is calculated from measurements of the heat flux at the hot face and the mean temperature gradient through the specimen. The experimental arrangement is shown in Fig. 15.

A typical hot-plate/guard-plate assembly consists of a hot-plate 200 mm square co-planar with a guard area formed by a plate 300 mm square with a central hole 206 mm square. The hot plate and guard plate are of paxolin 0.8 mm thick wound uniformly with a heater tape 1.5 mm by 0.15 mm thick, electrical resistance 1.08 Ω/300 mm, and clamped between two sheets of aluminium 3 mm thick. The hot plate and guard plate are wired so that their temperatures may be varied independently to prevent lateral heat flow. This is monitored to within 0.01 degC with two pairs of differentially connected thermocouples, one on either side of the plate assembly. The thermocouples (nichrome-eureka) are laid in channels grooved in the surface of the plates parallel to, and at a distance of 15 mm from, the edge.

The mean temperature gradient through the specimen is obtained from the readings of three pairs of thermojunctions wired differentially to give the temperature difference between opposite faces. Each pair may be selected and the potential difference measured separately, thereby affording a check on the uniformity of the gradient. The range of the readings should not exceed 5 percent of the average.

Two cold plates complete the test assembly; a typical design is shown in Fig. 15. The plate is a hollow aluminium casting 300 × 300 × 46 mm containing a labyrinthine channel through which the coolant liquid is circulated.

One of the principal sources of error in conductivity determinations is the measurement of the surface temperature of the specimens. To obtain a reliable measure of the temperature of a non-yielding surface it may be necessary to embed the thermocouples with leads in shallow grooves in the surface of the material. Contact is improved by sandwiching a 3 mm thick sheet of a foamed silicone rubber between the specimens and the plates and loading the test assembly with a pressure of not less than 3 kN/m^2, equivalent to a mass of 30 kg acting uniformly over the plates.

For each determination of thermal conductivity two identical specimens are required, measuring 300 mm square, and thickness not exceeding 50 mm under test. The maximum deviation from flatness and uniform thickness of

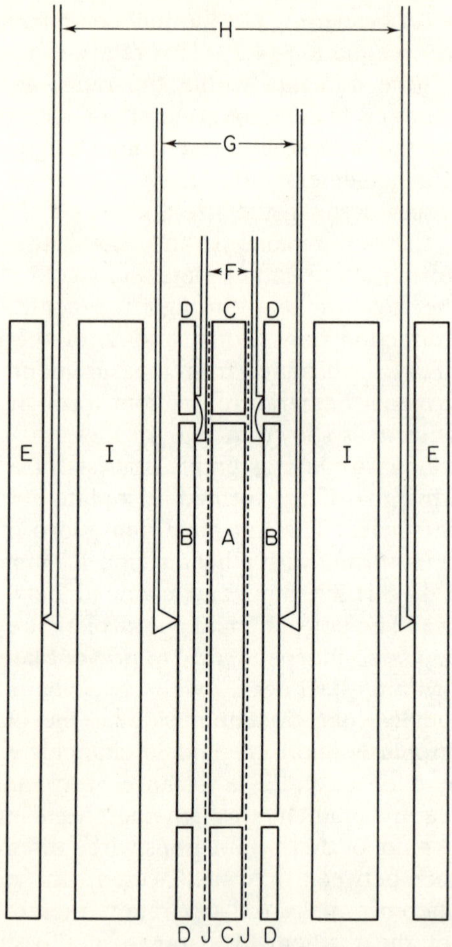

Fig. 15. Low temperture guarded hot plate: schematic assembly. *Heating unit.* Central section of heating unit: A, central heater core; B, central surface plates. Guard section of heating unit: C, guard heater; D, guard surface plates. E, cooling units with surface plates; F, differential thermocouples; G, heating unit surface thermocouples; H, cooling unit surface thermocouples; I, test specimens; J, insulation sheet. (Courtesy ASTM)

the specimens when in position in the hot-plate assembly should not exceed 0.2 mm over their full width. When the steady state in heat flow is reached readings of temperature and power consumption in the heater are made at regular intervals over a period of 8 h, the maximum permissible range of surface temperature readings during this time being 5 percent of the average.

The guarded hot-plate method is normally used with specimens of the test material in slab form, 300 mm square. The thickness may vary to conform with the requirements for controlling lateral heat flow, as discussed in detail in Chapter 4.

Criteria for the measurement of the thermal conductivity of thermal insulating materials, and guidance notes, have been published by the British Calibration Service as follows

1. 'Supplementary criteria for laboratory approval—the measurement of thermal conductivity by the guarded and plain hot-plate methods' (thermal conductivity in the range 0.01 to 2 W/mK, temperature 77 K to 373 K)—BCS 0911, October 80.

2. 'Preparation of thermal conductivity test specimens from masonry materials'—BCS 5561, October 80.

3. 'Uncertainties associated with imperfect surfaces in thermal conductivity measurements using standard hot-plate equipment'—BCS 5562, October 80.

For routine testing by the steady state method using the guarded hot-plate apparatus, specimens should be dry or near-dry. If the material is damp during the period of a test (up to 30 h or more) moisture will transfer from the warm towards the cold region as a result of the temperature gradient, thereby leading to the formation of wet and dry zones within the specimen. The net result of this rearrangement of moisture is that, for a damp material, the steady state method measures an apparent conductivity less than the 'true' conductivity, i.e. with the moisture uniformly distributed through the specimen. It is difficult to interpret the meaning of such a value for making practical calculations of heat transmission through naturally exposed structures. This uncertainty over the use of the guarded hot-plate method with damp material may be reduced by working with a lower temperature gradient than is normally the case, but this in turn demands refinements in experimental technique that are unlikely to be acceptable for routine measurements. The considerable care necessary in such cases for measuring accurately the small heat flux and temperature gradient is illustrated by the experimental studies described by Jesperson (1949) on the moisture dependence of the thermal conductivity of building materials.

There is no universally accepted specification for the pre-conditioning of conductivity specimens. Some codes specify that specimens shall be tested in an over-dry condition by heating to a constant weight at 100 degC. The values of conductivity thus obtained are more easily understood and provide reliable data for comparing the heat insulating value of different materials.

Where the variation of conductivity with moisture content is known an adjustment to the dry value may then be made; a relationship useful for this purpose between the thermal conductivity of porous inorganic materials and moisture content is given by Jakob (1949):

Water content (volume %)	1	2.5	5	10	15	20	25
Factor, *f*, to correct *k*-value of dry material	1.30	1.55	1.75	2.10	2.35	2.55	2.75

This is an average result obtained with a variety of different materials and gives generally satisfactory agreement with published data, as shown in Chapter 9. From the discussion earlier of moisture movement in porous material it follows that care should be taken in adjusting conductivity figures measured with damp specimens to correspond with other values of water content. In many cases specimens for routine measurement are prepared simply by leaving them to settle to a constant weight under normal atmospheric conditions in the laboratory; this appears to be satisfactory at 18 degC and 40 percent relative humidity.

In recent years non-steady state methods have been developed for making rapid measurements of the thermal conductivity of building material. In certain cases, depending upon the boundary conditions, the thermal diffusivity may be determined also. The heat flux entering the specimen integrated over the few minutes duration of these experiments is about a one-thousandth part of that required in a steady state determination. Of the two, the dynamic method is expected to cause less disturbance to the initial moisture distribution and to be the more promising therefore for dealing with damp specimens. Even so, it is doubtful whether the *k*-value may be accepted as a property of an isotropic material. Briefly, the principle of the dynamic method of measurement is to heat the specimen in a prescribed manner for a short period and record the temperature rise with time. From these readings the conductivity is obtained using the theoretical temperature versus time relationship calculated for the particular experimental arrangement. Various of these have been devised and fall broadly into two groups depending upon the form of the heat source; this may be a straight wire embedded in the specimen, or a plane strip of thin metal or other suitable material sandwiched between the test slabs. The experiments are completed in a few minutes compared with several hours required for a determination by the steady state, although in both methods the strength of the heat source is similar.

The line heat source (and probe) method was first proposed by Stalhane and Pyk (1931). Ideally, when an infinitely long thin wire is embedded in an infinite homogeneous medium and heated then after an initial transient change the rise in temperature $\theta_2 - \theta_1$, at a point in the medium for a time interval $t_2 - t_1$, will be

$$\theta_2 - \theta_1 = \frac{Q}{4\pi k} \ln \left(\frac{t_2}{t_1}\right) \tag{4.4}$$

Thus, a plot of temperature rise versus the logarithm of time should give a straight line with slope $Q/4\pi k$. From this and the applied power Q per unit length of the heat source, the conductivity is easily calculated.

Several probes representing practical forms of the line heat source have been developed and used for measurements on many types of low conductivity material (Hooper and Lepper, 1950; D'Eustachio and Schreiner, 1952; Woodside and Messmer, 1961; Wechsler and Glaser, 1965). Vos and Erkelens (1958) used the method in an interesting field application to determine the content and distribution of moisture in walls. The design and use of thermal probes require consideration (Blackwell, 1953, 1956; Jaeger, 1958; de Vries and Peck, 1958) of the following interrelated factors:

1. Finite probe length and diameter.
2. Thermal properties of the probe material.
3. Contact resistance between probe and sample.
4. Location of temperature measurement device.
5. Effect of sample boundaries.

The difficulty of ensuring good thermal contact with the sample uniformly along the length of the probe is probably the largest source of error in the use of these devices, especially with hard material.

The second of the two dynamic methods mentioned, using a plane heat source, was first proposed by Vernotte (1937) and developed by Clarke and Kingston (1950). The boundry conditions define a constant heat flux at one face of the test slab and at the other a non-conducting plane where the rise in temperature is measured. The conductivity and diffusivity of the material of the slab are calculated from the solution for the temperature rise derived below.

Consider the test specimen a plane slab, uniform thickness l, of a material with conductivity k, density ρ, specific heat s, and diffusivity K. These properties are assumed to be independent of the rise in temperature, $\theta(x, t)$, for all values of position x and time t. Initially the temperature is uniform throughout. For $t > 0$ the face at $x = 0$ is assumed to be in perfect thermal contact with a plane heat source of constant strength ϕ and thermal capacity w, while the face of the specimen at $x = l$ is arranged to be non-conducting. Neglecting edge effects it is required to calculate the temperature rise $\theta(x, t)$ that satisfies

$$\frac{\partial \theta(x, t)}{\partial t} = K \frac{\partial^2 \theta(x, t)}{\partial x^2}, \qquad 0 < x < l, \quad t > 0 \tag{4.5}$$

with

$$-k \frac{\partial \theta(x, t)}{\partial x} = \phi - w \frac{\partial \theta(x, t)}{\partial t}, \qquad x = 0, \quad t > 0 \tag{4.6}$$

$$-k \frac{\partial \theta(x, t)}{\partial x} = 0, \qquad x = l, \quad t > 0 \tag{4.7}$$

and

$$\theta(x, 0) = 0, \qquad \text{all } x, \quad t = 0$$

For convenience introduce dimensionless groups by writing

$$\xi = x/l; \qquad \tau = Kt/l^2; \qquad \psi = \phi l/k; \qquad W = w/\rho s l$$

and apply the Laplace transform to the resulting equations to give

$$\frac{d^2\bar{\theta}(\xi, p)}{d\xi^2} - p\bar{\theta}(\xi, p) = 0, \qquad 0 < \xi < 1, \quad \tau > 0 \tag{4.8}$$

$$-\frac{d\bar{\theta}(\xi, p)}{d\xi} = \frac{\psi}{p} - Wp\bar{\theta}(\xi, p), \qquad \xi = 0, \quad \tau > 0 \tag{4.9}$$

$$-\frac{d\bar{\theta}(\xi, p)}{d\xi} = 0, \qquad \xi = 1, \quad \tau > 0 \tag{4.10}$$

where

$$\bar{\theta}(\xi, p) = \int_0^\infty \theta(\xi, \tau) e^{-p\tau} \, d\tau, \qquad R(p) > 0$$

These equations are solved to give

$$\bar{\theta}(\xi, p) = \frac{\psi}{p} \frac{\cosh \sqrt{p}(1-\xi)}{(\sqrt{p} \sinh \sqrt{p} + Wp \cosh \sqrt{p})}$$

Then, by the inversion theorem for the Laplace transform the solution for the temperature rise is

$$\theta(\xi, \tau) = \frac{1}{2\pi i} \int_{\gamma - i\infty}^{\gamma + i\infty} \frac{\psi}{p} \frac{\cosh \sqrt{p}(1-\xi)}{(\sqrt{p} \sinh \sqrt{p} + Wp \cosh \sqrt{p})} e^{p\tau} \, dp \tag{4.11}$$

If the thermal capacity of the heater is small compared with that of the specimen, the solution for $\theta(\xi, \tau)$ may be written

$$\theta(\xi, \tau) \simeq \frac{1}{2\pi i} \int_{\gamma - i\infty}^{\gamma + i\infty} \frac{\psi}{p^{3/2}} \frac{\cosh \sqrt{p}(1-\xi)}{\sinh \sqrt{p}} e^{p\tau} \, dp$$

$$- \frac{1}{2\pi i} \int_{\gamma - i\infty}^{\gamma + i\infty} W \frac{\psi}{p} \frac{\cosh \sqrt{p}(1-\xi) \cosh \sqrt{p}}{\sinh^2 \sqrt{p}} e^{p\tau} \, dp \quad \text{to } \mathcal{O}(W^2) \tag{4.12}$$

Both integrands in (4.12) are analytic functions of p. Consider the first of these; it has a double-order pole at the origin of the p-plane, and simple poles at points $p = -n^2\pi^2$, $n = 1, 2, \ldots$, located on the negative real axis.

The residues at the poles are easily calculated to give

$$L^{-1}\left\{\frac{\psi}{p^{3/2}}\frac{\cosh\sqrt{p}(1-\xi)}{\sinh\sqrt{p}}\right\}$$

$$=\psi\left\{\frac{(1-\xi)}{2}-\frac{1}{6}+\tau-\frac{2}{\pi^2}\sum_{n=1}^{\infty}\frac{(-)^n}{n^2}\cos n\pi(1-\xi)\mathrm{e}^{-n^2\pi^2\tau}\right\} \quad (4.13)$$

where L^{-1} denotes the inverse Laplace operation. The second integral in (4.12) is seen to be the correction to the temperature rise for the capacity of the heater. To obtain this write

$$\cosh\sqrt{p}(1-\xi)\cosh\sqrt{p}=\cosh^2\sqrt{p}\cosh\sqrt{p}\xi-\sinh\sqrt{p}\cosh\sqrt{p}\sinh\sqrt{p}\xi \quad (4.14)$$

and consider the quantity

$$\frac{1}{p}\frac{\cosh^2\sqrt{p}\cosh\sqrt{p}\xi\mathrm{e}^{p\tau}}{\sinh^2\sqrt{p}} \quad (4.15)$$

Put $q=p+n^2\pi^2$ in (4.15) to give

$$-\frac{\mathrm{e}^{-n^2\pi^2\tau}}{(q-n^2\pi^2)}\frac{\cos^2 n\pi\sqrt{1-q/n^2\pi^2}\cos n\pi\xi}{\sin^2 n\pi\sqrt{1-q/n^2\pi^2}}\cdot\mathrm{e}^{q\tau} \quad (4.16)$$

and rewrite in the form

$$4\mathrm{e}^{-n^2\pi^2\tau}\cos n\pi\xi\left\{1+\frac{1}{2n^2\pi^2}\left(\tau-\frac{5}{8n^2\pi^2}\right)+\left(\tau+\frac{1}{2n^2\pi^2}\right)\frac{1}{q}+\cdots\right\} \quad (4.17)$$

The expression (4.17) is the Laurent expansion of the function (4.15) near the simple pole at $p=-n^2\pi^2$, n integral, at which the sum of residues is calculated to be

$$\frac{2}{\pi^2}\sum_{n=1}^{\infty}(1+2n^2\pi^2\tau)\cos n\pi\xi\frac{\mathrm{e}^{-n^2\pi^2\tau}}{n^2} \quad (4.18)$$

Another component of the integral (4.12) associated with W is the quantity

$$-W\frac{\psi}{p}\frac{\sinh\sqrt{p}\cosh\sqrt{p}\sinh\sqrt{p}\xi}{\sinh^2\sqrt{p}}\mathrm{e}^{p\tau} \quad (4.19)$$

This has simple poles at $p=-n^2\pi^2$, n integral, at which the sum of residues is

$$-\frac{2}{\pi}\sum_{n=1}^{\infty}\frac{\mathrm{e}^{-n^2\pi^2\xi}}{n}\sin n\pi\xi \quad (4.20)$$

Finally, the simple poles at the origin in (4.15) and (4.19) contribute the residue

$$-W\psi\left\{\frac{(1-\xi)^2}{2}+\frac{1}{6}+\tau\right\} \tag{4.21}$$

Summing the residues gives the solution for the temperature rise in the test specimen as follows

$$\begin{aligned}\theta(\xi,\tau)\simeq\psi\Bigg[&\frac{(1-\xi)^2}{2}-\frac{1}{6}+\tau-\frac{2}{\pi^2}\sum_{n=1}^{\infty}\frac{(-)^n}{n^2}\cos n\pi(1-\xi)\mathrm{e}^{-n^2\pi^2\tau}\\&-W\left\{\frac{(1-\xi)^2}{2}+\frac{1}{6}+\tau-\frac{2}{\pi^2}\sum_{n=1}^{\infty}[(1+2n^2\pi^2\tau)\cos n\pi\xi\right.\\&\left.-n\pi\sin n\pi\xi]\frac{\mathrm{e}^{-n^2\pi^2\tau}}{n^2}\right\}\Bigg]\quad\text{to } \mathcal{O}(W^2)\end{aligned} \tag{4.22}$$

The contribution due to the capacity of the heater will be apparent by comparing (4.22) with (4.13).

With time increasing the temperature difference between the surfaces at $x=0$ and $x=l$ approaches an asymptotic value which provides a simple expression for calculating k. Following the assumption in (4.12) concerning relative thermal capacities, this constant temperatue difference is given approximately by the residue at the origin of the p-plane in the inversion integral

$$\frac{1}{2\pi\mathrm{i}}\int_{\gamma-\mathrm{i}\infty}^{\gamma+\mathrm{i}\infty}\frac{\psi}{p^{3/2}}\frac{(\cosh\sqrt{p}-1)}{\sinh\sqrt{p}}\left(1-W\sqrt{p}\frac{\cosh\sqrt{p}}{\sinh\sqrt{p}}\right)\mathrm{e}^{p\tau}\,\mathrm{d}p\quad\text{to } \mathcal{O}(W^2)$$

The calculation yields

$$\underset{t\to\infty}{Lt}\ \{\theta(0,\tau)-\theta(1,\tau)\}\simeq\frac{\psi(1-W)}{2} \tag{4.23}$$

The time taken to reach this quasi-steady state condition may be calculated from the sum of the transient terms in the inversion integral. With 12 mm thick fibreboard, for example, it is about 30 min compared with 3 min required with the short-time solution.

Calculation of Thermal Conductivity and Diffusivity

Using the solution (4.22), values of k and K are calculated from readings of the temperature rise with time taken at the non-conducting surface. Thus, pulling $\xi=1$, and writing (4.22) as

$$\theta(1,\tau)\simeq\psi[F(\tau)-Wf(\tau)] \tag{4.24}$$

then

$$\frac{\theta(1, 2\tau)}{\theta(1, \tau)} = \frac{F(2\tau) - Wf(2\tau)}{F(\tau) - Wf(\tau)}$$

which may be rewritten

$$\frac{\theta(1, 2\tau)}{\theta(1, \tau)} \simeq \frac{F(2\tau)}{F(\tau)}\left[1 + W\left\{\frac{f(\tau)}{F(\tau)} - \frac{f(2\tau)}{F(2\tau)}\right\}\right] \quad \text{to } \mathcal{O}(W^2)$$

or

$$\log\left\{\frac{\theta(1, 2\tau)}{\theta(1, \tau)}\right\} \simeq \log\left\{\frac{F(2\tau)}{F(\tau)}\right\} + W\left\{\frac{f(\tau)}{F(\tau)} - \frac{f(2\tau)}{F(2\tau)}\right\} \tag{4.25}$$

where

$$F(\tau) = \tau - \frac{1}{6} - \frac{2}{\pi^2} \sum_{n=1}^{\infty} \frac{(-)^n}{\pi^2} e^{-n^2\pi^2\tau} \tag{4.26}$$

and

$$f(\tau) = \tau + \frac{1}{6} - \frac{2}{\pi^2} \sum_{n=1}^{\infty} (1 + 2n^2\pi^2\tau) \frac{e^{-n^2\pi^2\tau}}{n^2} \tag{4.27}$$

The thermal conductivity and diffusivity are then calculated by proceeding as follows.

1. In (4.25) neglect the W term and determine the value of τ that satisfies the logarithm of the temperature-rise ratio; then from $\tau = Kt/l^2$ a first estimate of K is obtained.
2. In (4.24) neglect the W term and, using the value of τ from (1) obtain a first estimate for ψ; k then follows from $\psi = \phi l/k$.
3. From the definition $K = k/ps$ an estimate for s, the specific heat of the test material, is obtained which in turn provides an estimate for W.
4. Using these estimates for τ and W, calculate

$$W\{f(\tau)/F(\tau) - f(2\tau)/F(2\tau)\}$$

and subtract from $\log\{\theta(1, 2\tau)/\theta(1, \tau)\}$ to give a revised estimate for τ. These values then lead to revised estimates for ψ and k, as indicated above.

Numerical result

The method of calculation is illustrated by an example worked out in detail in Table 5 from the point where a value for $W(=0.01045)$ is obtained from steps 1, 2, and 3 in the procedure for computation described above. The calculation uses the following additional notation

$$m = \pi^2 k/\rho s l^2; \qquad G(m) = W\{f(\tau)/F(\tau) - f(2\tau)/F(2\tau)\}$$

Table 5 Calculation of k and K from test data for an insulation board

1	2	3	4	5	6	7	8	9	10	11	12	13
Time t (secs)	Temperature rise (degC)	mt	$W \times G(mt)$	log $\left\{\frac{\theta(1, 2\tau)}{\theta(1, \tau)}\right\}$	(5)–(4)	Antilog (6)	Corrected mt	$F(2mt)$	(9)–$Wf(2mt)$	$\frac{\pi^2 k}{2\phi l}$	$k \times 10^5$ (cal/cm s degC)	$m \times 10^5$
155	0.2499	1.598	0.0185	1.5206	1.5021	4.491	1.633	0.8483	0.8256	3.304	19.0	1054
157.5	0.2568	1.648	0.0180	1.4940	1.4760	4.375	1.683	0.8954	0.8720	3.396	19.5	1069
160	0.2637	1.696	0.0175	1.4697	1.4522	4.273	1.731	0.9408	0.9167	3.476	20.0	1082
162.5	0.2715	1.722	0.0173	1.4570	1.4397	4.220	1.758	0.9652	0.9407	3.465	19.9	1082
165	0.2801	1.727	0.0172	1.4375	1.4375	4.210	1.763	0.9700	0.9455	3.376	19.4	1068
167.5	0.2887	1.731	0.0172	1.4525	1.4353	4.201	1.768	0.9745	0.9499	3.290	18.9	1056
170	0.2986	1.739	0.0171	1.4488	1.4317	4.186	1.775	0.9818	0.9571	3.205	18.4	1044
172.5	0.3052	1.781	0.0168	1.4293	1.4125	4.106	1.819	1.0235	0.9982	3.269	18.8	1054
175	0.3123	1.823	0.0164	1.4112	1.3948	4.034	1.861	1.0629	1.0370	3.321	19.1	1063
177.5	0.3196	1.846	0.0162	1.4012	1.3850	3.995	1.885	1.0856	1.0594	3.315	19.1	1062
180	0.3274	1.879	0.0160	1.3875	1.3715	3.941	1.919	1.1188	1.0921	3.336	19.2	1066
182.5	0.3343	1.904	0.0158	1.3772	1.3614	3.901	1.946	1.1439	1.1168	3.341	19.2	1066
185	0.3411	1.930	0.0156	1.3676	1.3520	3.865	1.971	1.1675	1.1401	3.342	19.2	1065
187.5	0.3489	1.960	0.0154	1.3558	1.3404	3.820	2.002	1.1980	1.1702	3.354	19.3	1068

Test Data

Heater current	11.86 amps
Heater resistance	0.007643 ohms/cm^2
Specimen thickness	$l = 0.91$ cm

The result is as follows:

$k = 19.2 \times 10^{-5}$ cal/cm s degC

$s = 0.352$ cal/g degC

$K = 0.000897$ cm^2/s

Conductivity values obtained by this method agree satisfactorily with those determined with the guarded hot-plate apparatus. A disadvantage, however, is the large number of specimens that are needed to ensure a non-heat conducting plane. By arranging instead for the temperature at this position to be held constant throughout the experiment fewer specimens are required and the test procedure is simplified. This alternative arrangement is represented mathematically as follows

$$\frac{\partial \theta(x, t)}{\partial t} = K \frac{\partial^2 \theta(x, t)}{\partial x^2}, \qquad 0 < x < l, \quad t > 0 \tag{4.28}$$

$$-k \frac{\partial \theta(x, t)}{\partial x} = \phi - w \frac{\partial \theta(x, t)}{\partial x}, \qquad x = 0, \quad t > 0 \tag{4.29}$$

$$\theta(x, t) = 0, \qquad x = l, \quad \text{all } t$$

$$\theta(x, t) = 0, \qquad 0 < x < l, \quad t > 0$$

where $\theta(x, t)$ is the rise in temperature of the test slab due to a constant flux ϕ of heat applied at $x = 0$. For convenience substitute

$$\xi = x/l; \qquad \pi = Kt/l^2; \qquad \psi = \phi l/k; \qquad W = w/\rho sl$$

and apply the Laplace transform to the resulting equations to give, in the customary notation

$$\frac{d^2 \bar{\theta}(\xi, \rho)}{d\xi^2} - p\bar{\theta}(\xi, p) = 0, \qquad 0 < \xi < 1, \quad \tau > 0 \tag{4.30}$$

$$\left.\begin{aligned} -\frac{d\bar{\theta}(\xi, p)}{d\xi} &= \frac{\psi}{p} - Wp\bar{\theta}(\xi, p), \qquad \xi = 0, \quad \tau > 0 \\ \bar{\theta}(\xi, p) &= 0, \qquad \xi = 1, \quad \tau \geqslant 0 \end{aligned}\right\} \tag{4.31}$$

Solving these equations and applying the inversion theorem for the Laplace transform gives

$$\theta(\xi, \tau) = \frac{1}{2\pi i} \int_{\gamma - i\infty}^{\gamma + i\infty} \frac{\psi}{pg(p)} \sinh \sqrt{p}(1 - \xi) e^{p\tau} \, dp \tag{4.32}$$

where

$$g(p) = Wp \sinh \sqrt{p} + \sqrt{p} \cosh \sqrt{p} \tag{4.33}$$

Taking the inversion gives the solution for the temperature rise of the slab as

$$\theta(\xi, \tau) = \psi\left\{1 - 2 \sum_{n=1}^{\infty} \frac{\sin \alpha_n (1-\xi) \operatorname{cosec} \alpha_n}{\alpha_n^2 (1 + W - W^2)} e^{-\alpha_n^2 \tau}\right\}, \qquad 0 < \xi < 1 \tag{4.34}$$

where

$$\alpha_n \tan \alpha_n = 1/W \tag{4.35}$$

The roots α_n are tabulated (Carslaw and Jaeger, 1959). If W is negligibly small, then

$$\theta(\xi, \tau) = \psi\left\{1 - \frac{8}{\pi^2} \sum_{n=1}^{\infty} \frac{\sin (2n+1)(\pi/2)(1-\xi)}{(2n+1)^2} e^{-\alpha_n^2 \tau}\right\} \tag{4.36}$$

where

$$\alpha_n = (2n+1)(\pi/2), \qquad n = 0, 1, \ldots \tag{4.37}$$

Ball (1967) describes an experimental application of equation (4.36) in which the temperature rise is measured at the heated face of the test specimen ($\xi = 0$). The method uses electrically conducting graphite-treated paper to obtain a uniform and constant plane heat source of negligible thermal capacity. This is sandwiched between two identical test specimens each with its opposite face maintained at a constant temperature by contact with a thick steel plate and a slab of dense concrete to form a heat sink. It will be noted that the steady state guarded hot-plate apparatus also provides these boundary conditions, thus extending its application to dynamic measurements of thermal conductivity.

As a brief conclusion to this chapter it may be stated that high accuracy and ease of measurement have established the steady state method as standard for the determination of the conductivity of dry or near-dry materials up to 2 W/m degC. It is particularly suitable for routine measurement and the results obtained offer a reliable basis for the selection of material for heat insulating purposes. Dynamic methods, although not restricted for use with dry material, nevertheless require additional refinements in experimental technique that may cause them to be unsuitable at the present state of development for the routine application which industrial testing services prefer.

REFERENCES

Allcut, E. A. (1951) *Proceedings, General Discussion on Heat Transfer,* Institute of Mechanical Engineers.

Ball, E. F. (1967) *Proc. Inst. Refrig.,* **Feb. 1967**.

Baxter, S. (1946) *Proc. Phys. Soc.*, **48,** 105.
Blackwell, J. H. (1953) *Can. J. Phys.*, **31,** 472.
Blackwell, J. H. (1954) *J. Appld. Phys.*, **25,** 137.
Blackwell, J. H. (1956) *Can. J. Phys.*, **34,** 412.
Brailsford, A. D. and Major, K. G. (1964) *Br. J. Appld. Phys.*, **15,** 313.
BS 874:1973, British Standards Institution, London.
Carslaw, H. E. and Jaeger, J. C. (1959) *Conduction of Heat in Solids*, Clarendon Press, Oxford.
Clarke, L. N. and Kingston, R. S. T. (1950) *Aust. J. Appld. Sci.*, **1** (2), 172.
Davies, D. S. (1943) *Ind. Eng. Chem.*, **35,** 673.
D'Eustachio, D. and Schreiner, R. E. (1952) *Trans. Am. Soc. Heat. Vent. Engrs.*, **58,** 331.
de Vries, D. A. (1952) *Meded. Landb. Hogesch. Wageningen*, **52,** 1.
de Vries, D. A. and Peck, A. J. (1958) *Aust. J. Phys.*, **11,** 255.
Fishenden, M. and Saunders, O. A. (1961) *An Introduction to Heat Transfer*, Oxford University Press, London.
Hooper, F. C. and Lepper, F. R. (1950) *Trans. Am. Soc. Heat. Vent. Engrs.*, **56,** 309.
Jaeger, J. C. (1958) *Aust. J. Phys.*, **11,** 167.
Jakob, M. (1949) *Heat Transfer*, Vol. 1, Chapman and Hall, London.
Jesperson, H. B. (1949) *Varme*, **14** (4), 77.
Joy, F. A. (1957) *ASTM Special Technical Publication*, no. 217.
Kannuluik, W. G. and Martin, L. H. (1933) *Proc. R. Soc. A*, **141,** 144.
Kersten, M. S. (1949) *Univ. Minnesota Eng. Exp. Sta. Bull.*, **28**.
Krischer, O. (1941) *Wärme and Kaltetechnik*, **43** (1).
Krischer, O. and Esdorn, H. (1956) *Forsch. Ing. Wes.*, **22** (1).
Lees, C. H. (1900) *Phil. Mag.*, **49,** 221.
Maxwell, J. C. (1904) *A Treatise on Electricity and Magnetism*, Clarendon Press, Oxford.
Penman, H. L. (1940) *J. Agric. Sci.*, **30,** 438, 570.
Pratt, A. W. and Ball, J. M. E. (1956) *JIHVE*, **24,** August, 201.
Rowley, F. B. (1933) *Heat. Pip. Air Condit.*, **12,** 459.
Schuhmeister (1877) *Ber. K. Akad. Wein.*, **76,** 283.
Stalhane, B. and Pyk, S. (1931) *Tekn. Tidskr. Stockh.*, **61,** (28), 389.
Strong, H. M., Bundy, E. P., and Bovernkerk, H. P. (1960) *J. Appld. Phys.*, **31,** 39.
Sugawara, A. and Yoshizawa, Y. (1961) *Aust. J. Phys.*, **14,** 469.
Vernotte, P. (1937) *C.R. Acad. Sci., Paris*, **204.**
Verschoor, J. D. and Greebler, P. (1952) *Trans. Am. Soc. Mech. Engrs.*, **74,** 961.
Vos, B. H. and Erkelens, H. J. (1958) *Allg. Wärmetech.*, **8,** 211.
Wangaard, E. F. (1940) *Heat. Pip. Air Condit.*, **12,** 459.
Wechsler, A. E. and Glaser, P. E. (1965) *Icarus*, **4,** 335.
Woodside, W. and Cliffe, J. B. (1959) *Soil Sci.*, **87** (2), 75.
Woodside, W. and Messmer, J. H. (1961) *J. Appld. Phys.*, **32,** 1688.
Wyllie, M. R. J. and Southwick, J. (1954) *Petrol Technol., London*, **6,** 44.

CHAPTER 3

The Thermal Resistance of Airspaces in Cavity Building Structures

The transfer of heat between surfaces separated by an airspace takes place by radiation, convection, and conduction. If the airspace is closed at the edges to form a sealed cavity all the heat leaving the hot face is received by the cold face. A cavity that is not sealed loses a proportion of this heat by ventilation. The thickness of the airspace in cavity construction in buildings is small compared with the edge dimensions of the boundary surfaces which are normally plane. For practical calculations of heat flow through structures a cavity is assumed to be sealed and is represented as a resistance in a simple series network of thermal resistances. The resistance of a sealed airspace is defined as the reciprocal of the quantity of heat transferred in the steady state in unit time between unit areas of the boundary surfaces when their temperatures differ by one degree. When the cavity is ventilated surface temperatures and heat flow are no longer uniquely related and the foregoing definition does not then conform with the conventional method for calculating the thermal transmittance of a structure.

The IHVE (Institution of Heating and Ventilating Engineers) *Guide Book A* (1970) recommends that the thermal resistance of a 5 mm thick airspace, unventilated and with high emissivity surfaces, shall be taken as 0.11 m^2K/W irrespective of the direction of heat flow. A figure of 0.18 is recommended when one or both surfaces have a low emissivity. For a thicker airspace, 20 mm or more, the corresponding values are 0.18 and 0.35 except when the heat flow is downwards for which the recommended values are 0.21 and 1.06, respectively. The IHVE *Guide* includes ventilated airspaces with reistance values that are specific to certain wall and roof construction in which the minimum thickness of the airspace is 20 mm.

Tabulation of design data in the *Guide* published by the American Society of Heating and Air-conditioning Engineers, New York, distinguishes between the position of the airspace (horizontal, vertical, or 45° slope), between summer and winter, and whether the heat flow is up or down. Different thicknesses of airspace are also considered. The results are based on National Bureau of Standards data presented in *Housing Research Paper*, no. 32 (1954), and are calculated from the set of temperatures to represent average winter conditions and a different set for summer conditions. Altogether thirteen different cases are considered giving cavity resistance values ranging from 0.14 (horizontal, up, summer) to 0.22 (horizontal, down, winter). In particular, for heat flowing upwards in winter through an inclined roof the resistance is given as $0.16\,\mathrm{m^2K/W}$.

There appears to be no systematic tabulation of resistance values for airspaces that are ventilated. During experiments on heat transfer through sheeted roofs naturally exposed in the Roof Laboratory at the Building Research Station, ventilation was observed in cavities which, in industry, are normally considered to be sealed. This chapter includes an examination in detail of the effect of ventilation on the insulating value of airspaces in naturally exposed structures.

1. MEASUREMENTS OF THE INSULATING PROPERTIES OF AN AIRSPACE

The thermal conductance C of a sealed airspace is the sum of two mutually independent coefficients: h_r, the radiation coefficient, and h_c, the convection–conduction coefficient. The dimensions of the cavity in a building structure are such that the radiation heat transfer may be assumed to take place between parallel planes of infinite extent, and the radiation coefficient has the expression

$$h_r = \frac{\sigma(T_1^4 - T_2^4)}{(1/e_1 + 1/e_2 - 1)(\theta_1 - \theta_2)} \tag{1.1}$$

where σ is the Stefan–Boltzmann constant, and e_1 and e_2 are the emissivities of the boundary surfaces at temperatures θ_1 and θ_2 with T_1 and T_2 the corresponding absolute temperatures. The quantity $(1/e_1 + 1/e_2 - 1)$ is the effective emissivity for radiation transfer between the two surfaces and is normally abbreviated as E. The radiation coefficient is independent of the slope of the cavity and may be calculated without difficulty. The convection–conduction coefficient, on the other hand, displays an important variation with the direction of heat flow.

General mathematical solutions are not available for calculating the convection heat transfer across a ventilated cavity bounded by parallel surfaces at different temperatures. Semi-empirical methods based on dimensional analysis are commonly used to evaluate heat transfer under these conditions.

Foremost among the earliest investigations of heat transfer in cavities is the work of Griffiths and Davis (1922). This consisted of a comprehensive series of experiments on the transfer of heat by radiation and convection primarily to obtain data on problems of insulation connected with the construction of cold-stores. The experimental arrangement is shown diagrammatically in Fig. 1. The airspace was enclosed between two vertical

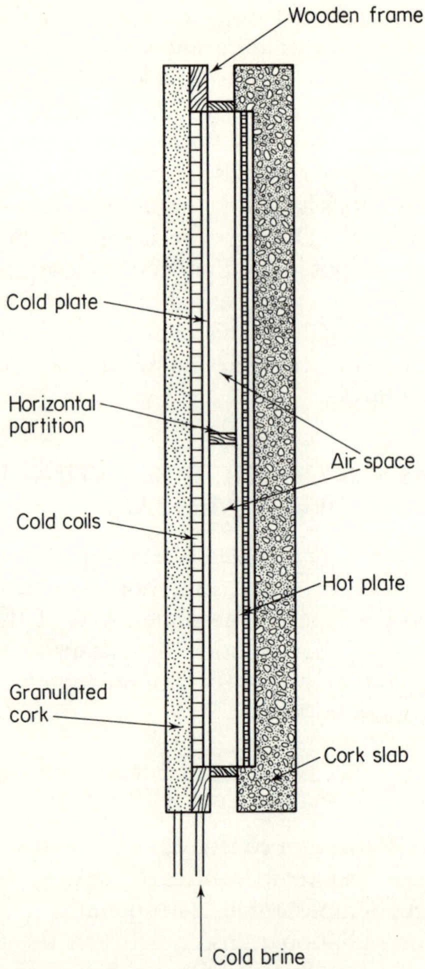

Fig. 1. Diagram of experimental arrangement for studying heat transfer by convection in air between vertical plates forming the sides of a closed chamber (Griffiths and Davis, 1922)

metal plates suspended parallel to each other. One of these was heated electrically and formed the hot plate, the other was maintained at a lower temperature by circulating cooled brine through a lead coil placed in contact with its back surface. Since convection heat transfer formed the major study in this investigation it was necessary to reduce the radiation transfer as much as possible. This was done by making the two metal plates of polished aluminium. Readings of temperatures were taken, and heat flow across the cavity calculated, when the apparatus had attained a steady state. The data

Table 1 Heat transfer across a vertical cavity: data due to Griffiths and Davis (1922), cavity resistance values added to original summary

Distance between plates (in.)	Height of each space (ft in.)	Mean temperature difference between plates θ (degF)	Heat Transfer (Btu/ft^{2}h) Radiation, H_r	Convection, H_c	$H_c/\theta^{5/4}$	Cavity resistance $R_c=\theta/[H_c+H_r]$
1	4 1.6	57.8	3.3	30.5	0.193	1.71
	4 1.6	99.7	6.8	60.5	0.193	1.48
	2 0.8	80.9	5.3	45.8	0.190	1.58
	2 0.8	48.8	2.9	22.9	0.179	1.89
$\frac{1}{2}$	4 1.6	101.2	6.9	58.8	0.184	1.54
	4 1.6	58.8	3.6	29.2	0.180	1.79
	4 1.6	14.8	0.8	6.1	0.191	2.14
	2 0.8	29.5	1.6	13.0	0.183	2.02
	2 0.8	102.3	7.2	58.4	0.180	1.56
2	4 1.6	102.8	7.2	60.7	0.185	1.51
	4 1.6	59.9	3.6	30.0	0.179	1.78
	4 1.6	46.1	2.8	22.2	0.185	1.84
	4 1.6	28.6	1.6	11.3	0.166	2.22
	4 1.6	28.1	1.6	11.0	0.164	2.23
	*	63.4	4.5	56.3	0.316	1.04
	*	65.4	4.4	55.2	0.297	1.10

* No boundaries to space between plates. In these cases θ = temperature excess of hot plate over air of room, and convecting area = 17.3 ft^2 instead of 17.2 ft^2.

obtained are shown reproduced from the original paper in Table 1 with an extra column added on the extreme right giving the cavity resistance, R_c, calculated from the expression

$$R_c=\theta/(H_r+H_c) \tag{1.2}$$

where θ is the mean temperature difference in degrees Fahrenheit between the plates, and H_r and H_c represent heat transfer measured in Btu/ft^{2}h due to radiation and convection, respectively.

With two exceptions, the results were obtained with a sealed airspace. Within the limits of experimental error it was concluded that the heat transfer by convection across a closed vertical airspace, of the dimensions specified, is independent of the height and thickness of the space. Theory indicates that for streamline flow in natural convection between a plane surface and air the quantity $H_c/\theta^{5/4}$, in which θ denotes the temperature difference between the surface and the air, is almost constant except for a slight variation with the air temperature. In the case of heat transfer across a sealed cavity the same expression but with θ representing the temperature difference between the bounding surfaces would also be expected to remain almost constant. The data in Table 1 demonstrate this.

Some of the variation in the experimental results may be due to variation in the temperature of the air, but as the original paper gives no information about airspace temperatures this cannot be checked. The cavity resistance is seen to vary inversely as the mean temperature difference between the plates in a manner predicted by theory, as noted above. The corresponding quantity for radiation transfer, H_r/θ, increases with mean temperature, but, in these experiments, the effect of this on the value of the cavity resistance is reduced to an almost negligible proportion by the low effective emissivity ($E = 0.05$). When the boundaries at the top and bottom of the cavity were removed to permit natural air exchange between the airspace and the room, measurements indicated that the hot plate lost heat by convection in a manner similar to that of a plate suspended vertically in free air. Clearly some of this heat is removed from the system by the stream of air moving through the open space, resulting in lower values of cavity resistance as defined by (1.2). This particular aspect of heat transfer in cavities is considered in detail later.

Metal plates enclosed the airspaces in the experiments of Griffiths and Davis. A later investigation by Rowley and Algren (1929) considered airspaces between building boards commonly used in built-up wall sections; they included fibreboard, flaxboard, plasterboard, and others of similar surface characteristics. The measurements were made with the hot plate and hot box, apparatus which is generally accepted for standards purposes as methods for measuring the flow of heat through building materials and structures. Airspace coefficients were determined for sealed cavities varying in thickness from 0.051 to 0.713 in. by the hot plate, and from 0.5 to 1.5 in. by the hot box. The results for vertical airspaces at different mean temperatures are shown in Fig. 2. It is reasonable to assume that the curves may be extrapolated below the lower limit of the observations to pass as shown through the zero point. For thicker airspaces they approach horizontal lines and indicate that the maximum resistance is reached as from 0.7 to 0.8 in.

Wilkes and Peterson (1937) extended the investigations of Griffiths and Davis to measurements on full-sized airspaces in various positions bounded by surfaces of high and low emissivity. The airspace measured 8 ft high by 2 ft 8 in. wide and $3\frac{5}{8}$ in. across and the whole arrangement was pivoted at the centre so that it could be tipped at any angle to permit measurements with heat flowing upwards or downwards. Fig. 3 shows how the quantity $H_c/\theta^{5/4}$, described by Wilkes and Peterson as the coefficient of convection in which θ denotes the temperature difference between the surfaces, varied according to position. The value 0.18 obtained for horizontal flow (vertical position) compares favourably with the group of values reported by Griffiths and Davis (see Table 1). Fig. 4 is a plot of conductance versus temperature difference between the cavity surfaces. The values are slightly below those of Griffiths and Davis (the reciprocal of R_c in Table 1) for a vertical cavity which may be due partly to unknown differences in temperature level.

Variation in the experimental values of the coefficient of convection h_c with height is indicated in Fig. 5. According to Wilkes and Peterson the

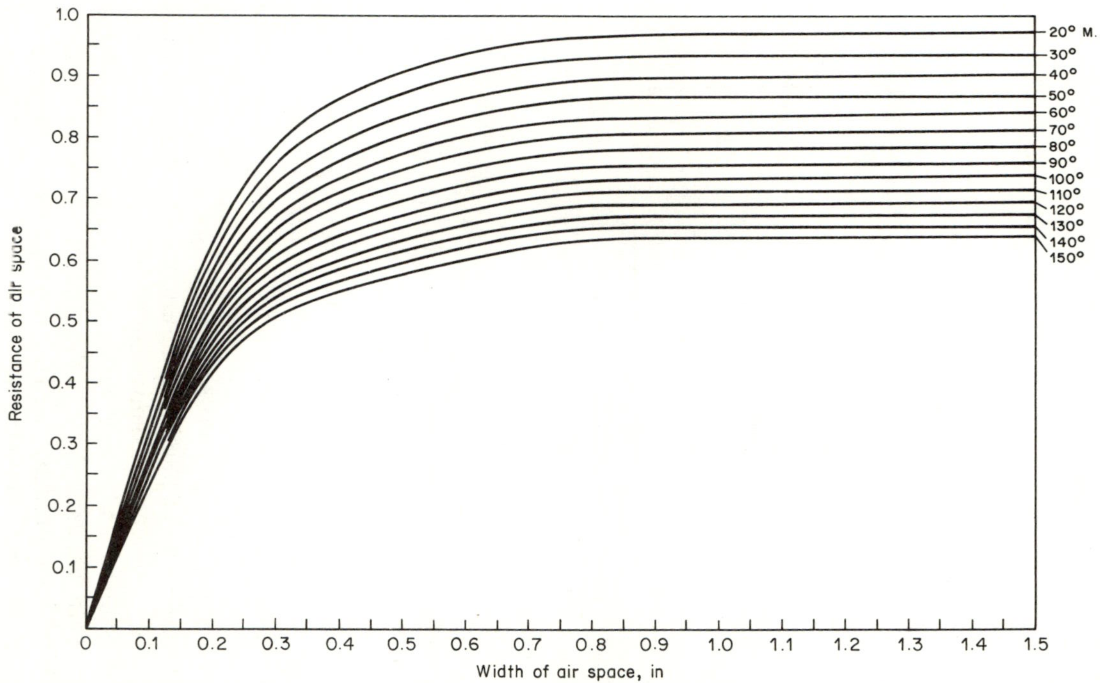

Fig. 2. Resistance of vertical airspaces for different mean temperatures, degF (Rowley and Algren, 1929)

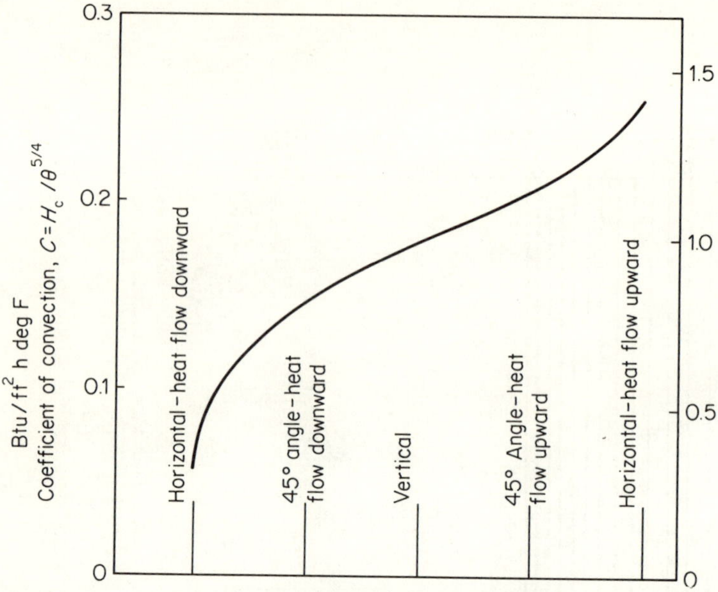

Fig. 3. Curve indicating coefficient of convection for air spaces versus position (Wilkes and Peterson, 1937)

pattern of variation suggests streamline motion up to about 15 in. above the bottom edge of the airspace. At this height h_c reaches a minimum value; thereafter the motion becomes turbulent and the coefficient assumes a steady value, irrespective of height, above 24 in. An interpretation likewise had been proposed earlier by Griffiths and Davis (1922) to explain a pattern of similar form obtained by them with a heated plate suspended vertically in free air.

Robinson, Powlitch, and Dill (1954) carried out experiments designed to resolve inconsistencies in published information on the conductance of airspaces and to provide data encompassing the range of airspace thickness, temperature difference, and mean temperature encountered in building insulation. The work was carried out at the National Bureau of Standards as a research project sponsored by the Housing and Home Finance Agency and is described by that Authority in *Housing Research Paper.* no. 32 (1954). The experiments were made on airspaces in test panels using a guarded hot box apparatus that conformed substantially to the requirements of ASTM 236–53, *Method of Test for Thermal Conductance and Transmittance of Built-up Sections by means of the Guarded Hot Box*. Provision was made for the whole arrangement to be rotated on horizontal trunnions as in the experiments of Wilkes and Peterson (1937). The test panels measured

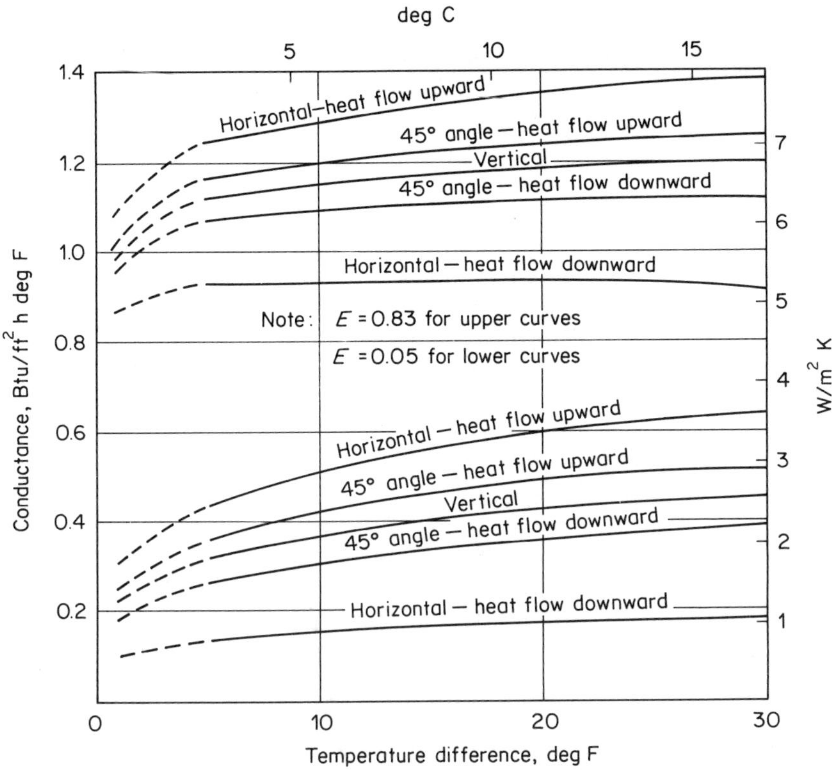

Fig. 4. Curves of conductance versus temperature difference between cavity surfaces (Wilkes and Peterson, 1937)

$5\frac{1}{2}$ ft wide × 8 ft high and enclosed sealed airspaces up to 2 ft across. Each set of temperature and heat flow measurements was made in the steady state on an enclosed airspace measuring 2 ft 8 in. by 5 ft at the centre of the panel. Airspaces of various thicknesses were examined with boundary surfaces of different emissivity.

It may be shown by the methods of dimensional analysis that the heat transferred by convection and conduction combined across a sealed air layer satisfies an expression of the form

$$H = (k\theta/l)F(c^2 l^3 ag\theta/k^2) \tag{1.3}$$

where θ is the temperature difference between the surfaces and the characteristic dimension l denotes the thickness of the airspace; the other quantities are as defined in the list of symbols. If heat flow measurements are made on a system in which the temperature and size only are allowed to

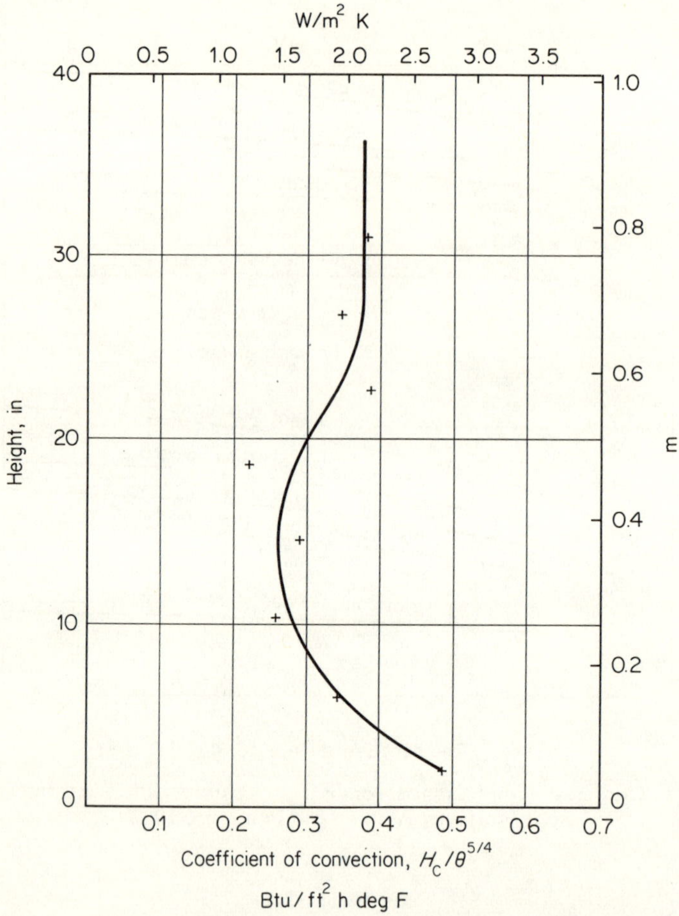

Fig. 5. Curves indicating variation in coefficient of convection versus height (Wilkes and Peterson, 1937)

vary, all other quantities remaining constant, then (1.3) relating the data may be written

$$H = (\theta/l)F(\theta l^3) \tag{1.4}$$

or

$$h_c l = F(\theta l^3)$$

where h_c is defined as the ratio of the heat transfer (omitting radiation) to the temperature difference between the surfaces. This form of expression neglects the slight variation in a, the coefficient of thermal expansion in air, with temperature. Robinson applied a correction for this effect and expressed the results for a mean cavity air temperature of 50 degF using the formula

$$(h_c)_\theta = (h_c)_{50}\{1 - 0.001(\theta - 50)\} \tag{1.5}$$

The experimental results, adjusted as described to an air temperature of 50 degF, are shown in Fig. 6 plotted according to (1.4). Each curve may be expanded into a family of curves relating h_c to l, and with θ as the parameter representing heat flow for a particular orientation. One such group of curves is shown in Fig. 7 for a horizontal airspace and in Fig. 8 for a vertical airspace. This method of correlating the data follows that used by Griffiths and Davis (see above) in a study of convection heat loss from a vertical plane surface except that in their analysis the characteristic dimension l represented the height of the vertical surface in the direction of the convection current and θ was the temperature difference between the plate and the air. Corresponding curves for other relationships in both series of experiments are similar; the results for the cavity only are considered here.

The thermal conductance, C, of an airspace was calculated by Robinson from the expression

$$C = h_c + Eh_r$$

with values of h_c and h_r taken from the graphs in Fig. 7 and Fig. 2, Chapter 1, respectively. Numerical examples of the calculation are worked later in this Chapter.

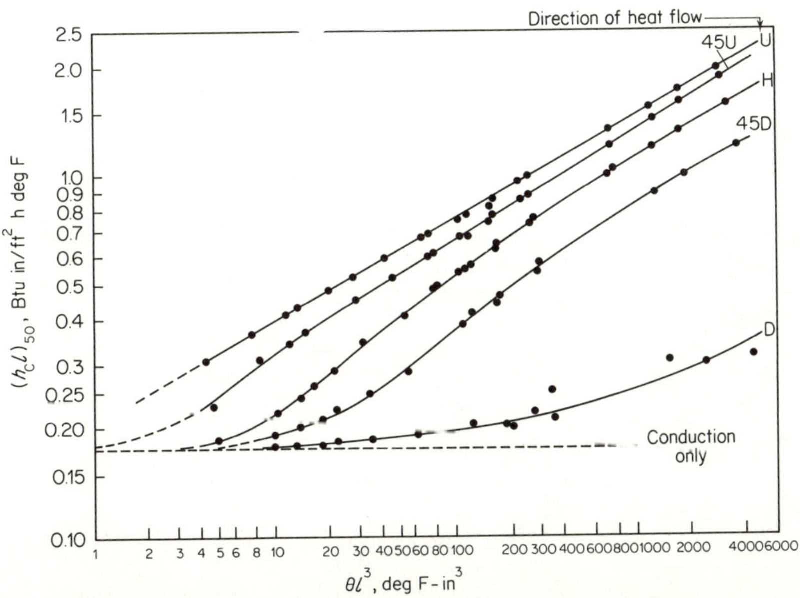

Fig. 6. Natural convection across an air layer: variation of $(h_c l)_{50}$ with θl^3 for different orientations (Robinson, Powlitch, and Dill, 1954)

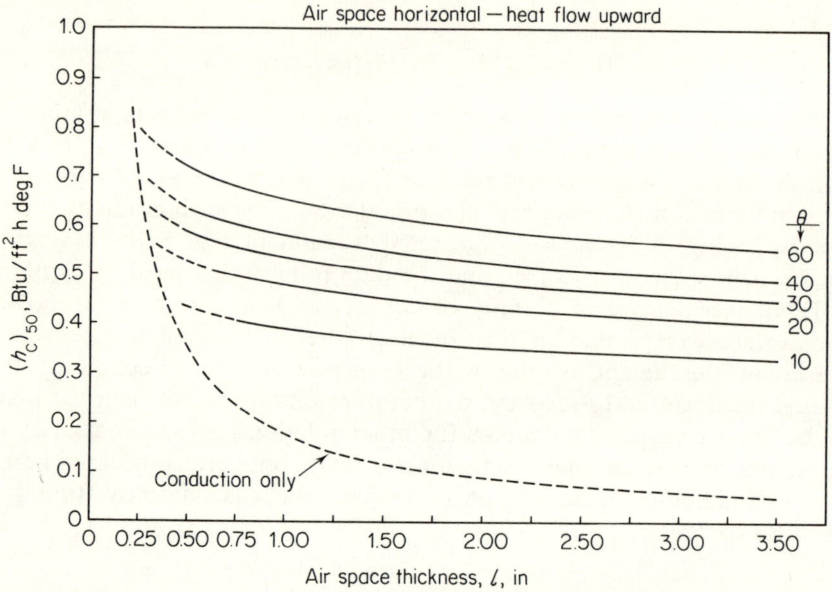

Fig. 7. Natural convection across an air layer: variation of convection coefficient with airspace thickness (Robinson, Powlitch, and Dill, 1954)

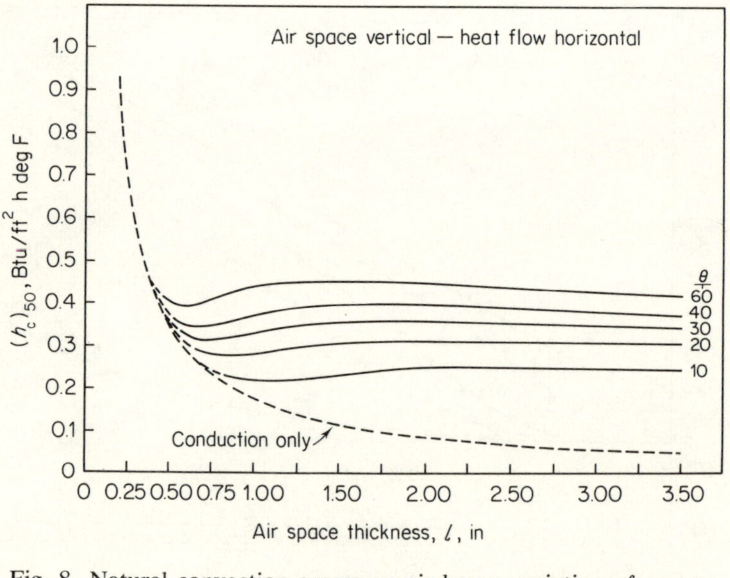

Fig. 8. Natural convection across an air layer: variation of convection coefficient with airspace thickness (Robinson, Powlitch, and Dill)

It is seen in Fig. 6 that for all the directions of heat flow except upwards, the curves are asymptotic to a value of $(h_c l)_{50}$ equal to 0.175. Writing

$$h_c l = \frac{Hl}{\theta k} k = \frac{H}{H_k} k$$

where H_k denotes heat transfer by conduction only, and noting that the conductivity of dry air at 50 degF is 0.175 Btu in/ft^{2}h degF, it follows that under conditions limited by this value the transfer of heat across the layer takes place solely by conduction. The suppression of convection currents in this region is illustrated in Fig. 9 (Fishenden and Saunders, 1959) in which the Nusselt number $\mathrm{Nu} = H/H_k$ is plotted against log (Gr Pr) for natural convection across a closed air layer between two parallel vertical plane surfaces. The product of the Grashof number, Gr, and the Prandtl number, Pr, is called the Rayleigh number, Ra. Until Ra exceeds about 10^3, Nu is unity denoting heat transfer by conduction only. The change in the slope of

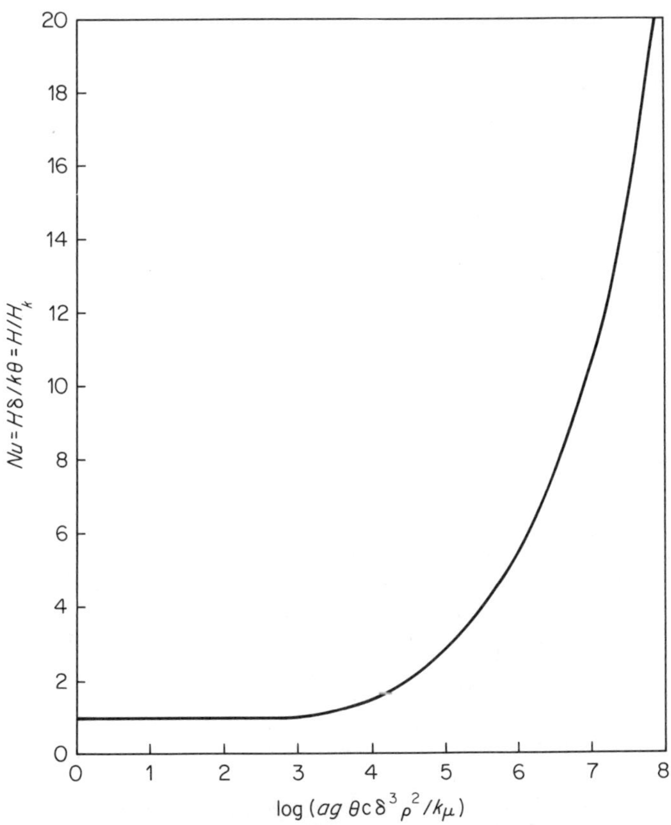

Fig. 9. Natural convection across a fluid layer between two parallel vertical plane surfaces, closed at the edges (Fishenden and Saunders, 1950)

the curve as the value of Ra increases is used to explain heat transfer phenomena in fibrous and porous insulating materials.

A summary of formulae defining convective heat flow between parallel plates arranged horizontally and vertically that is useful to the present considerations is given by Bosworth (1952) and reproduced below.

For horizontal parallel plates, distance apart l.

(i) No convection, $\mathrm{Ra} < 1620$

$$\mathrm{Nu} = \frac{hl}{k} = 1$$

(ii) $8000 < \mathrm{Ra} < 300\,000$

$$\mathrm{Nu} = 0.208\,(\mathrm{Ra})^{1/4}$$

(iii) $\mathrm{Ra} > 300\,000$

$$\mathrm{Nu} = 0.074\,(\mathrm{Ra})^{1/3}$$

For vertical plates, height L, distance apart l.

(i) As above.

(ii) $15\,000 < \mathrm{Ra} < 150\,000$

$$\mathrm{Nu} = 0.19\,(\mathrm{Ra})^{1/4}(L/l)^{-1/9}$$

$\mathrm{Ra} > 150\,000$

$$\mathrm{Nu} = 0.071\,(\mathrm{Ra})^{1/3}(L/l)^{-1/9}$$

This summary is based on the results of experiments reported by Mull and Reiher (1930).

2. CALCULATION OF THE THERMAL RESISTANCE OF A CAVITY IN A ROOF NATURALLY EXPOSED

The thermal resistance of a ventilated cavity may be calculated from a consideration of the heat balance of an element of space occupied by an air stream bounded between a pair of parallel planes and open at the edges. This elementary volume is represented in Fig. 10 by the dimensions x and y, and unit length in the z-direction. Heat transfer by conduction is assumed to take place in the y-direction only. Equating the net rate of gain of heat by conduction into the volume to that removed by convection due to the flow of air gives the heat balance equation

$$\frac{\mathrm{d}}{\mathrm{d}y}\left\{k(y)\frac{\mathrm{d}\theta}{\mathrm{d}y}\right\}\delta y\delta x = \rho v(y)\frac{\mathrm{d}\theta}{\mathrm{d}x}\,\delta x\delta y \tag{2.1}$$

This expression may be integrated over the width of the airspace to introduce the surface heat exchange at the lower and upper planes, denoted by subscripts 1 and 2, respectively. It is often extremely difficult to define practical surface boundary conditions exactly and to manipulate them

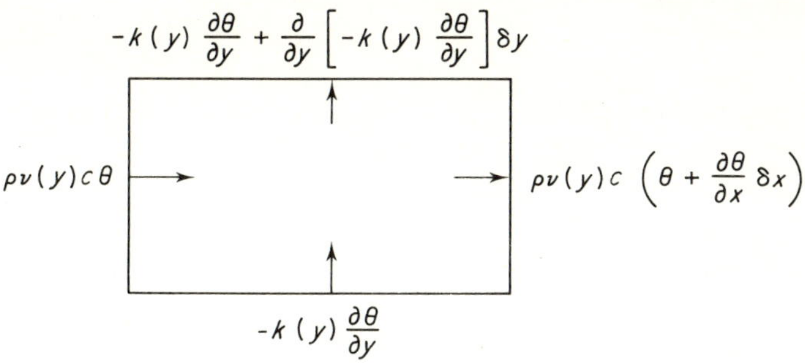

Fig. 10. Heat transfer in an air layer bounded by parallel planes

mathematically; this is particularly true of a roof exposed to the weather. In the present analysis actual conditions at the boundaries are simplified by assuming that the surfaces are isothermal, that k is independent of position, and that the velocity profile is flat. The boundary conditions may then be written

$$\left.\begin{aligned} &\text{surface 1} \quad k\frac{\mathrm{d}\theta}{\mathrm{d}y}+h_{c1}\{\theta_1-\theta(x)\}=0 \\ &\text{surface 2} \quad k\frac{\mathrm{d}\theta}{\mathrm{d}y}-h_{c2}\{\theta_2-\theta(x)\}=0 \end{aligned}\right\} \tag{2.2}$$

Integrating (2.1) and introducing the boundary conditions (2.2) gives

$$h_{c1}\{\theta_1-\theta(x)\}+h_{c2}\{\theta_2-\theta(x)\}=\rho vcl\frac{\mathrm{d}\theta}{\mathrm{d}x} \tag{2.3}$$

and finally

$$(h_{c1}\theta_1+h_{c2}\theta_2)L-(h_{c1}+h_{c2})L\bar{\theta}=\rho vcl\{\theta(L)-\theta(0)\} \tag{2.4}$$

where L represents the length of the cavity in the direction of the flow of air, $\theta(0)$ and $\theta(L)$ are the temperatures of the air as it enters and leaves the cavity, and $\bar{\theta}$ is the mean temperature of the air stream defined as

$$\bar{\theta}=\frac{1}{L}\int_0^L\theta(x)\,\mathrm{d}x \tag{2.5}$$

where from (2.3)

$$\left.\begin{aligned} &\theta(x)=\psi+\{\theta(0)-\psi\}\exp\{-x(h_{c1}+h_{c2})/\rho vcl\} \\ &\text{and} \\ &\psi=(h_{c1}\theta_1+h_{c2}\theta_2)/(h_{c1}+h_{c2}) \end{aligned}\right. \tag{2.6}$$

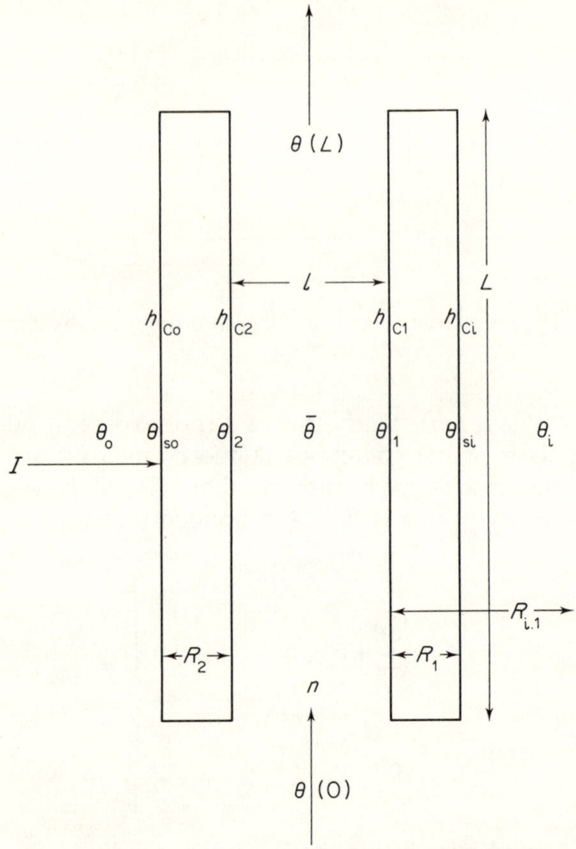

Fig. 11. Representation of cavity structure

For the purpose of the analysis the steady transfer of heat is considered in a cavity roof represented as a pair of plane parallel surfaces bounding an open airspace whose thickness is small compared with the edge dimensions (Fig. 11).

The transfer of heat through this system, neglecting edge effects, under steady state conditions is represented by the following heat balance equations

$$\begin{aligned} \frac{1}{R_{i.1}}(\theta_i - \theta_1) &= E\sigma(T_1^4 - T_2^4) + h_{c1}(\theta_1 - \bar{\theta}) \\ h_{c1}(\theta_1 - \bar{\theta}) &= nlc\{\theta(L) - \theta(0)\} + h_{c2}(\bar{\theta} - \theta_2) \\ h_{c2}(\bar{\theta} - \theta_2) + E\sigma(T_1^4 - T_2^4) &= \frac{1}{R_2}(\theta_2 - \theta_{so}) \\ &= h_{co}(\theta_{so} - \theta_0) + \sigma e T_{so}^4 - I \end{aligned} \tag{2.7}$$

where

$R_{i.1}$ = resistance from inside air to inner cavity surface;
R_2 = resistance of outer leaf;
h_{co} = convection coefficient at outside surface;
n = ventilation rate of cavity expressed as number of air changes per hour;
θ_i = inside air temperature;
I = total intensity of radiation absorbed at the outside surface;

and the other symbols are as shown in Fig. 11.

The resistance of the cavity is calculated from the values of θ_1, θ_2, θ_{so} and $\bar{\theta}$ obtained by solving equations (2.7) for specified conditions.

In a closed cavity with isothermal boundary surfaces the heat flux is uniform throughout. With such a system the airspace resistance is uniquely defined as the ratio of the temperature difference between the surfaces to the heat flux. If the cavity is not closed then some heat is removed by convection due to the flow of ventilating air, and values of the heat flux at the two boundary surfaces differ. When this occurs an alternative expression for the resistance of the cavity may be obtained from writing

$$\theta_1 - \theta_2 = \theta_i - \theta_0 - (\theta_i - \theta_1) - (\theta_2 - \theta_{so}) - (\theta_{so} - \theta_0) \tag{2.8}$$

Introducing resistances gives the equivalent expression

$$\theta_1 - \theta_2 = QR_{a.a} - QR_{i.1} - (Q - \delta Q)R_2 - (Q - \delta Q)R_{so} \tag{2.9}$$

or, with slight rearrangement

$$\frac{\theta_1 - \theta_2}{Q} = R_{a.a} - (R_{i.1} + R_2 + R_{so}) + \frac{\delta Q}{Q}(R_2 + R_{so}) \tag{2.10}$$

where

$R_{a.a}$ = air-to-air resistance;
R_{so} = outside surface resistance;
Q = heat flux at surface 1;
δQ = rate of heat removal from cavity by ventilating air per unit area of hot surface.

The value of the cavity resistance, R_c, used in practical calculations of heat transfer in buildings is determined by the expression

$$Q = (\theta_i - \theta_0)/(R_{i.1} + R_c + R_2 + R_{so}) = (\theta_i - \theta_0)/R_{a.a} \tag{2.11}$$

By definition $I/R_{a.a}$ is the thermal transmittance or U-value of the structure.

The quantity R_c, as defined by (2.11), complies with conventional heat transfer calculations for sealed and ventilated cavity structures and may be called the effective cavity resistance.

It was stated above that the resistance of a sealed airspace is uniquely defined as $(\theta_1 - \theta_2)/Q$. If this expression is used also to calculate the resistance, denoted R'_c, of a ventilated cavity then from (2.10) it follows that

$$R'_c = R_c + \frac{\delta Q}{Q}(R_2 + R_{so}) \tag{2.12}$$

The effective airspace resistance is calculated from the distribution of temperature in the structure and therefore depends on the resistance of the structural layers and the ambient climatic conditions. In the numerical analysis that follows the data governing heat exchange at the exposed outer surface have been selected to represent a wide range of influence due to climate in mid-winter. The equations represent a general formulation and may be evaluated for any set of conditions in the steady state.

Discussion of Data for Numerical Calculations

Structural Resistances $R_{i.1}$ and R_2

For illustrative purposes the thermal resistance of a sheeted roof is represented by putting $R_{i.1} = 0.35$ and $R_2 = 0\ \mathrm{m^2K/W}$. These resistance values apply to a common form of sheeted roof structure consisting of a metal cladding lined with 12 mm insulation board to enclose a cavity normally 75–125 mm thick. An alternative cladding material is sheet asbestos in corrugated form for which R_2 may be taken to be $0.03\ \mathrm{m^2K/W}$; this resistance however is comparatively small and may be neglected for the present analysis.

The calculations are not restricted to sheeted roofs clad with thin, highly conducting sheets of impermeable material the sole purpose of which is to give protection against the weather. In this, the more usual form of building construction, the cavity lies on the 'cold' side of the insulation and therefore the mean temperature of the air in the cavity may be expected to differ by no more than a slight amount from that of the outside air ventilating the cavity. This will result in ventilation having little effect on the rate at which heat is conducted through the insulating lining and hence on the cavity resistance. From this reasoning it follows that other proportional arrangements of structural resistance about the cavity may lead to different results. With a higher proportion of the air-to-air resistance of the structure on the 'cold' side of the cavity, which is equivalent to placing the cavity more towards the higher indoor temperature, ventilation may be expected to have a greater influence on the mean temperature of the cavity and produce a correspondingly greater change in the cavity resistance. This feature of the problem is examined in the analysis below by repeating the calculations with $R_{i.1} = 1.0$ and $R_2 = 2.8$ representing a roof construction with the airspace on the 'warm' side of the insulation.

Effective Emissivity, E

The general expression for the net exchange of heat by radiation, H_r, between two plane surfaces of areas A_1 and A_2, at temperatures θ_1 and θ_2, with emissivities e_1 and e_2, and separated by a non-absorbing medium is

$$H_r = \sigma A_1 \frac{(e_1 T_1^4 f_2 - e_2 T_2^4 f_1 A_2/A_1 + r_1 e_2 T_2^4 f_1 f_2 A_2/A_1 - r_2 e_1 T_1^4 f_1 f_2)}{1 - r_1 r_2 f_1 f_2} \tag{2.13}$$

where r_1 and r_2 are the reflectivities of the respective surfaces and f_1 is the fraction of the energy emitted by the element dA_2 that falls on dA_1, and conversely for f_2. For convenience (2.13) is generally written in the abbreviated form

$$H_r = \sigma F_a F_e A_1 (T_1^4 - T_2^4) \tag{2.14}$$

where F_a and F_e are factors that take account of the configuration and emissivities of the surfaces.

The dimensions of a cavity in a roof are such that the exchange of heat by radiation may be assumed to take place between parallel planes of infinite extent in which case $A_1 = A_2$ and $f_1 = f_2 = 1$ to give $F_a = 1$. Making these substitutions and assuming that absorptivity is equal numerically to emissivity so that $r_1 = 1 - e_1$ and $r_2 = 1 - e_2$ then it follows from (2.13) and (2.14) that

$$F_e = \frac{e_1 e_2}{e_1 + e_2 - e_1 e_2} = E \tag{2.15}$$

The following pairs of emissivity values have been selected for the present calculations from data given by Fishenden and Saunders (1932)

$$e_1 = 0.9 = e_2; \qquad E = 0.84$$

$$e_1 = 0.05, \quad e_2 = 0.9; \qquad E = 0.05$$

A value of 0.05 for the effective emissivity, E, corresponds to the use of low emissivity surfaces for improving the insulation value of airspaces. In building practice this often takes the form of bright aluminium foil stuck to the lining or ceiling boards; there are other arrangements involving the use of multiple foil.

Radiation Flux, I

This represents the heat gain from solar and celestial radiation. It is calculated by regarding the roof as a completely enclosed body, small compared with the enclosing body, which in this case is the sky. For this arrangement the factors F_a and F_e in (2.14) are equal to 1 and the emissivity e of the roof surface, respectively. The expression, therefore, for the net radiation flux through the roof surface wholly exposed to the sky vault is

$$H_r = \sigma E T_{so}^4 - I \tag{2.16}$$

where

$$I = \sigma E T_{sp}^4 + \alpha I_s \tag{2.17}$$

T_{sp} is the temperature of the sky or space and I_s is the flux of short-wave radiation at the naturally exposed surface of the roof which absorbs a fraction α of the incident energy. In the present analysis values of I are used that apply to bright clear weather in mid-winter. On clearest days in December, for example, the total flux of solar energy falling on a horizontal surface at Kew was reported by Stagg (1950) as equivalent to 209 W/m^2. If the roof is assumed to radiate as a black body so that $\alpha = e = 1$, and using estimated values of T_{sp} based on radiometer measurements made during clear nights (Dines and Dines, 1927), then I is calculated to have the following mean values

$$205\ \text{W/m}^2\ \text{(night)}; \qquad 414\ \text{W/m}^2\ \text{(day)}$$

Cavity Ventilation Rate, n

There appear to be no published values of the ventilation rates of cavities in roof structures particularly of the sheeted type as considered here. In practice the cavity is usually assumed to be sealed. The IHVE *Guide* proposes reduced values for the resistance of unsealed airspaces but does not specify the relevant ventilation rates. An investigation by Pratt (unpublished work) of heat transmission through sheeted roofs carried out in the Roof Laboratory at the Building Research Station, Garston, included measurements of the ventilation rate of the airspace between the cladding and the insulating lining in structures of this kind with results shown in Fig. 12. At the lower side the roofs faced a heated enclosure, single-storey height, maintained at a constant temperature of 18.3 degC and at the other side were naturally exposed as in a normal practical installation. Although the ventilation rates may seem high, they correspond nevertheless to a volume flow rate of air which in a domestic-size room would give an equivalent rate of about two air changes per hour, a figure that is generally adopted in such cases for design purposes. On the basis of these measurements therefore a ventilation rate of 50 air changes per hour has been adopted in the present analysis for calculating airspace resistance.

Convection Coefficient, h_{co}

The quantity h_{co} denotes the convection coefficient at the outside surface. The ambient conditions at a naturally exposed surface are such that the air flow varies from calm to strong wind. At wind speeds up to 5.03 m/s the value of h_{co} may be calculated according to an empirical relationship obtained by Jurges (1924). Jurges correlated the results of several tests for the flow of air at velocity v parallel to a plane surface with an expression of the form

$$h_c = a + bv \qquad (2.18)$$

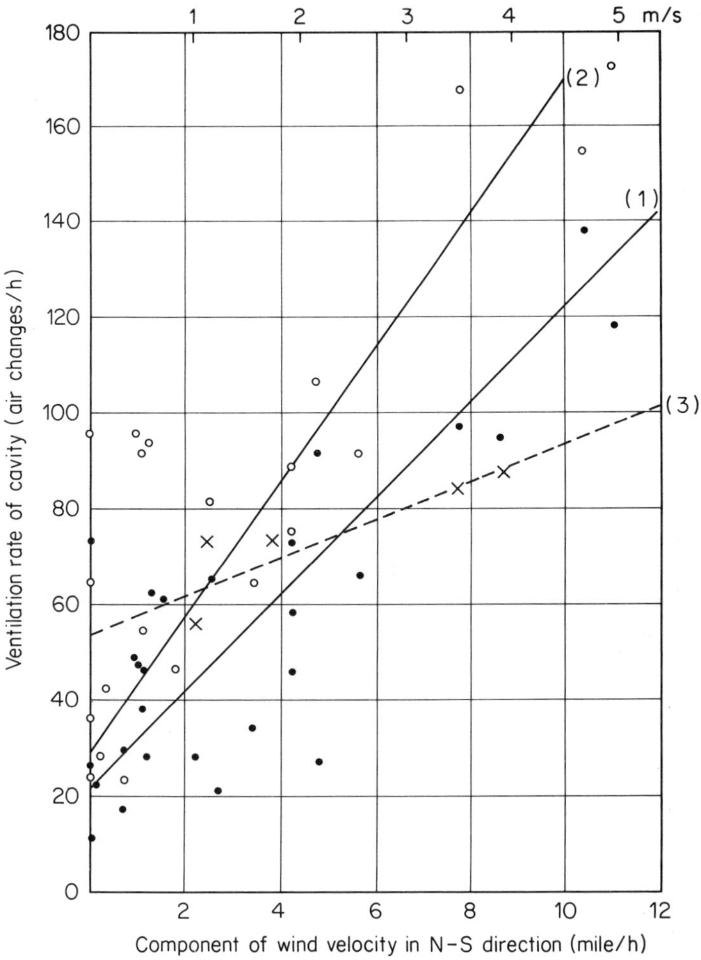

Fig. 12. Cavity ventilation: ventilation rate from catharometer measurements versus north–south component of measured wind vector. ● (1) Metal roof: 86 mm cavity. ○ (2) Metal roof—lap joints open: 86 mm cavity. × (3) Asbestos cement roof: 140 mm cavity

where h_c is the forced convection coefficient and a and b are constants depending on the nature of the surface. At wind speeds in excess of 5.03 m/s a power law of the form

$$h_c = cv^d$$

where c and d are constants, represented the relationship more closely. Although these expressions were derived for a vertical wall, in forced convection they may be applied to a horizontal surface also.

Assuming an air flow of 4.75 m/s, the coefficient for forced convection at the outside surface of the roof is calculated as

$$\begin{aligned} h_c &= 5.8 + 4.1 \times 4.6 \\ &= 24.7\ \mathrm{W/m^2K} \end{aligned} \tag{2.19}$$

The above empirical expressions for h_c were derived from data obtained at room temperature, 21 degC. At other temperatures Schack (1929) recommends a correction to v, based on the gas laws, to adjust the coefficient to the value corresponding to 21 degC. For the present study the correction is small and need not be applied.

It should be noted that Jurges's results are based on measurements made over a relatively small area, actually 0.15 m². It is not known how closely they represent convection transfer over larger surfaces. A less uniform distribution of temperature may be visualized across the air stream over a roof than over a smaller area of the size considered by Jurges. This means that a lower conductance will obtain in practice than the formulae predict, and decrease as the extent of the surface increases. It is estimated (Schack, Goldschmidt, and Partridge, 1933) that the decrease may reach 40 percent for a surface several metres in length.

Again, Jurges's results apply to an air stream moving parallel to the surface. The wind flowing over a roof is more likely to strike the surface at an angle, a condition that leads to higher values of the transfer coefficient.

The probable net effect of the restrictions attached to Jurges's results is such that a coefficient of forced convection at least as high as that calculated according to equation (2.19) may reasonably be accepted.

At very low velocities the heat transfer is largely by natural convection, and for calm weather conditions the following simplified formulae are assumed to apply

$$\left.\begin{aligned} &\text{heat flow upward:} && h_{co} = 2.64(\theta_{so} - \theta_o)^{1/4} \\ &\text{heat flow downward:} && h_{co} = 1.32(\theta_{so} - \theta_o)^{1/4} \end{aligned}\right\} \ \mathrm{W/m^2K} \tag{2.20}$$

Convection Coefficients h_{c1} and h_{c2}

These coefficients refer to convection heat exchange inside the cavity. In the calculations that follow the temperature of the air in the narrow roof space is represented as a mean quantity $\bar{\theta}$ defined by (2.5). In a closed system the coefficients correspond to natural convection, as represented by (2.20). When the cavity is open, forced convection is more likely to occur and here the data are less accurate. For an airspace of practical dimensions ventilated at 50 air changes per hour with $\theta = 5$ degC, the Reynolds number, Re, based on an effective diameter of four times the area divided by the perimeter is about 700 which corresponds to streamline flow. Boundary layer theory (Prandtl, von Karman) applied to streamline flow of gases over a plane surface gives the relationship

$$\mathrm{Nu} = 0.60\ \mathrm{Re}^{0.5} \tag{2.21}$$

Accepting the effective diameter defined above as the characteristic dimension upon which forced convection in a rectangular channel depends, it follows from (2.21) that the convection coefficient in a ventilated cavity for the conditions specified is 2.04 W/m^2K. This value is well below that given by Jurges's formula (2.18) at the equivalent air velocity. At this low rate of air flow the heat transfer will be due largely to natural convection. In such cases McAdams (1951) recommends that the larger of the two values calculated from (2.18) and (2.20) should be used. In the absence therefore of more accurate data, coefficients given by Jurges's formula are used in the present calculations. In a similar situation, at higher velocities corresponding to turbulent flow ($Re > 10\,000$), an appropriate value for the coefficient appears to be no less difficult to determine (Peregrine and Rowland, 1951).

It will be noted that the analysis regards the temperature distribution as being fully developed along the whole length of the cavity. This is a simplification of practical conditions. Following the theoretical treatment by Cess and Schaffer (1959) of heat transfer between parallel plates a thermal entrance length for which the Nusselt number is within 5 percent of the fully developed value may be calculated; for the present case this is about 1.2 m. It is a secondary effect and in view of the complexity and uncertainty of the boundary conditions in a practical installation a correction is thought to be unnecessary.

A summary of the data representing the particular cases considered in the foregoing discussion is given below:

$\theta_o = -1$ degC
$\theta_i = 18.3$ degC
$l = 75$ mm
$L = 3$ m
$n = 0, 50$ air changes per hour
$h_{co} = 22.7\ W/m^2K$ in forced convection

$$\left.\begin{aligned} h_{co} &= 2.64(\theta_{so}-\theta_o)^{1/4} \text{ upward flow, natural convection} \\ h_{co} &= 1.32(\theta_{so}-\theta_o)^{1/4} \text{ downward flow, natural convection} \end{aligned}\right\} W/m^2K$$

$$\left.\begin{aligned} h_{c1} &= \frac{1.32}{2.64}(\theta_1-\bar{\theta}) \\ h_{c2} &= \frac{1.32}{2.64}(\bar{\theta}-\theta_2) \end{aligned}\right\} W/m^2K, \text{ cavity sealed}$$

$h_{c1} = h_{c2} = 5.79 + nL10^{-3}\ W/m^2K$, cavity ventilated

$$\begin{Bmatrix} e_1 \\ e_2 \end{Bmatrix} = \begin{Bmatrix} 0.9 \\ 0.9 \end{Bmatrix}, \begin{Bmatrix} 0.05 \\ 0.9 \end{Bmatrix}$$

$$I = \begin{Bmatrix} 205 \text{ night} \\ 414 \text{ day} \end{Bmatrix} W/m^2$$

$$\begin{Bmatrix} R_{i.1} \\ R_2 \end{Bmatrix} = \begin{Bmatrix} 0.352 \\ 0 \end{Bmatrix}, \begin{Bmatrix} 0.176 \\ 0.493 \end{Bmatrix} m^2K/W$$

$\sigma = 5.67 \times 10^{-8}\ W/m^2K^4$

DISCUSSION OF RESULTS

Table 2 is a summary of the results calculated for different combinations of the various factors representing roof construction and climate. These factors are listed in columns 2–8 of the table. Values of radiation flux and convection coefficients are combined to represent upper and lower limits to the influence of climate in mid-winter. The Table is reproduced from the paper by Pratt and Ball (1966) with the quantities in columns 3–20 as originally given in Imperial units. The discussion that follows is concerned almost entirely with the values of R_c and R'_c and columns 21 and 22 therefore have been added expressing these particular quantities in SI units. In columns 9–12 are values of the temperature distribution through the structure calculated from the set of steady state equations for the conditions defined by the data. These temperatures lead to values of cavity resistance through the defining equations

$$\left.\begin{aligned} R'_c &= (\theta_1 - \theta_2)/Q = R_{i.1}(\theta_1 - \theta_2)/(\theta_i - \theta_1) \\ R_c &= R_{a.a} - (R_{i.1} + R_2 + R_{so}) \\ &= \{R_{i.1}(\theta_i - \theta_o)/(\theta_i - \theta_1)\} - (R_{i.1} + R_2 + R_{so}) \end{aligned}\right\} \tag{2.22}$$

where

$$R_{so} = (\theta_{so} - \theta_o)/\{h_{co}(\theta_{so} - \theta_o) + \sigma T_{so}^4 - I\}$$

or, in cases where $R_2 \neq 0$

$$R_{so} = R_2(\theta_{so} - \theta_o)/(\theta_2 - \theta_{so})$$

According to definition R'_c is a measurable quantity and may be regarded as the formal cavity resistance. The expression for R_c on the other hand shows this as a derived quantity, to be determined experimentally as the term that balances the equation expressing the observed steady transfer of heat between the air on opposite sides of the cavity structure. It may be regarded as an 'effective' resistance satisfying the practical rule for computing air-to-air transmission coefficients as the reciprocal of the sum of resistances. When the cavity is sealed $R_c = R'_c$ as shown in cases 1–16 calculated with $n = 0$ for a variety of conditions.

Cases 1 and 2 serve as a basis for comparing the results of the many different combinations considered. They refer to a sealed cavity in a sheeted roof. Comparable with these are cases 9 and 10 representing an alternative structural arrangement in which the airspace is sealed between insulation as may be found in a cavity roof deck. The day and night values of cavity reistance, R_c, for both structures (0.174, 0.171, 0.167, 0.180, column 21) agree closely with the IHVE recommended value of 0.176 m^{2}K/W for an airspace bounded by high emissivity surfaces. The slight increase by day for the warmer cavity (9 versus 10) has no practical significance. For both positions of the cavity R_c is almost unaffected by changing the convection coefficient, h_{co}, at the outside surface from 4.0 Btu/ft^{2}h (22.7 W/m^{2}K) to a value $\leq$0.86 (4.8 W/m^{2}K) to correspond with natural convection. A similar result applies to a sealed airspace with a low emissivity. With a cavity deck in calm weather the difference between day and night values is more pronounced (15 versus 16) and may be expected to increase with $\bar{\theta}$, the cavity mean temperature. By reducing the emissivity E from 0.82 to 0.05 the value of R_c for a sealed cavity is increased to 0.43 m^{2}K/W or more and the small difference between day and night values is maintained. It is noted that the values for an airspace with a low value of E are about 25 percent higher than the IHVE recommended value of 0.35 m^{2}K/W.

Table 2 has been prepared primarily to evaluate the change in insulating value that results from ventilating the airspace in a naturally exposed cavity structure. Consider first the sheeted roof; results based on the higher value for the coefficients h_{c1} and h_{c2} show that during clear nights R_c for a 75 mm thick airspace is reduced from 0.174 m^{2}K/W when sealed to 0.144 when ventilated at 50 air changes per hour (1 versus 17). The drop in R_c is larger when $E = 0.05$; the corresponding night time values are then 0.433 and 0.306 m^{2}K/W (3 versus 19). By day the influence of ventilation is much greater, the sealed value of 0.171 with high emissivity surfaces reducing to 0.072 and changing by a larger amount from 0.470 to 0.151 when $E = 0.05$. Suppressing the radiation component of the total heat transfer enhances the influence of ventilation and the heat exchange between surfaces when $E = 0.05$ depends almost entirely upon the mechanism of convection. This indicates that the results for foil-lined cavities are likely to be conditioned markedly by the values assigned to h_{c1} and h_{c2}. This is confirmed by repeating the calculations assuming natural convection with values for the coefficients calculated using the formulae in equations (2.20). For a sheeted roof, R_c during the night then has the same value, 0.174, whether the cavity is sealed or ventilated (1 versus 21). The result, though with higher resistance values, is similar for a foil-lined cavity. During the day, and assuming natural convection, R_c is reduced by ventilating the 75 mm thick airspace at 50 air changes per hour from 0.171 to 0.100 when $E = 0.82$ and from 0.470 to 0.208 when $E = 0.05$. Again the reduction in cavity resistance due to ventilation is greater by day than at night, but is less than when calculated with the higher convection coefficient.

The greater daytime reduction is associated with the higher cavity temperatures during this period and suggests that ventilation may be more effective in reducing R_c as θ increases. This leads to an examination of the results for the cavity deck, specified as $R_2 = 0.493$, $R_{i.1} = 0.176$ (column 3). The pattern of R_c-values for a sealed cavity in this position agrees closely with that for a sealed cavity in a sheeted roof. The warmer cavity however is influenced to a greater extent by ventilation. This was anticipated in the discussion earlier of the values assigned to $R_{i.1}$ and R_2, where it was argued that ventilation may be expected to have an increasing effect as the position of the cavity moved towards the higher indoor air temperature. The analysis confirms this. For

Table 2 Summary of numerical results

Case	n	R_2 ft²h degF/Btu	$R_{i.1}$	I Btu/ft²h	E	h_{co} Btu/ft²h degF	h_{c1}	h_{c2}	$\theta_{\delta o}$ °F	θ_2 °F	θ_1 °F	$\bar{\theta}$ °F	θ_e °F	R_{so} ft²h degF/Btu	$R_{\delta 0}(e)$	R_c	$R_c(e)$	R'_c	U Btu/ft²h degF	$U(e)$	R_c m²K/W	R'_c
1	2	3		4	5	6	7	8	9	10	11	12	13	14	15	16	17	18	19	20	21	22
1	0	0	2.0	64.9	0.82	4	0.64	h_{c1}	θ_2	25.4	38.6	32.0	22.2	−0.34	0.25	0.99	0.99	0.99	0.38	0.31	0.174	0.174
2	0			131.3	0.82	4	0.58	h_{c2}	θ_2	38.4	47.1	42.8	38.1	0.94	0.25	0.97	0.97	0.97	0.26	0.31	0.171	0.171
3	0			64.9	0.05	4	0.73	h_{c2}	θ_2	24.6	46.9	35.8	22.4	−0.60	0.25	2.46	2.46	2.46	0.26	0.21	0.433	0.433
4	0			131.3	0.05	4	0.67	h_{c2}	θ_2	37.7	53.3	45.5	36.3	1.32	0.25	2.67	2.67	2.67	0.17	0.205	0.470	0.470
5	0	0	2.0	64.9	0.82	0.40	0.68	h_{c1}	θ_2	14.5	31.6	23.0	−27.4	−0.93	2.52	1.02	1.02	1.02	0.48	0.18	0.180	0.180
6	0			131.3	0.82	0.86	0.50	h_{c1}	θ_2	51.0	55.7	53.3	45.6	4.52	1.17	1.01	1.01	1.01	0.13	0.24	0.178	0.178
7	0			64.9	0.05	0.42	0.78	h_{c1}	θ_2	11.1	40.1	25.6	−18.7	−1.52	2.40	2.33	2.33	2.33	0.36	0.15	0.410	0.410
8	0			131.3	0.05	0.85	0.58	h_{c1}	θ_2	50.2	58.9	54.6	46.6	6.62	1.18	2.85	2.85	2.85	0.09	0.17	0.502	0.502
9	0	2.8	1.0	64.9	0.82	4	0.57	h_{c1}	24.5	48.4	56.5	52.5	22.4	−0.65	0.25	0.95	0.95	0.95	0.24	0.195	0.167	0.167
10	0			131.3	0.82	4	0.48	h_{c1}	37.7	53.5	59.3	56.4	36.3	1.35	0.25	1.02	1.02	1.02	0.16	0.195	0.180	0.180
11	0			64.9	0.05	4	0.69	h_{c1}	24.1	42.0	58.6	50.3	22.5	−0.92	0.25	2.59	2.59	2.59	0.18	0.15	0.456	0.456
12	0			131.3	0.05	4	0.62	h_{c1}	37.4	49.1	60.8	55.0	36.4	1.76	0.25	2.79	2.79	2.79	0.12	0.15	0.497	0.497
13	0	2.8	1.0	64.9	0.82	0.42	0.61	h_{c1}	10.4	42.7	53.5	48.1	−17.0	−1.70	2.38	0.94	0.95	0.94	0.33	0.14	0.165	0.165
14	0			131.3	0.82	0.85	0.44	h_{c1}	50.2	58.9	61.9	60.4	46.6	6.52	1.18	0.97	0.97	0.97	0.09	0.17	0.171	0.171
15	0			64.9	0.05	0.43	0.73	h_{c1}	8.4	33.8	55.9	44.8	−12.7	−2.37	2.32	2.43	2.43	2.43	0.26	0.12	0.428	0.428
16	0			131.3	0.05	0.84	0.55	h_{c1}	49.8	55.9	62.8	59.4	47.1	9.00	1.19	3.14	3.14	3.14	0.06	0.12	0.553	0.553

17	50	0	2.0	64.9	0.82	4	1.05	h_{c1}	θ_2	25.6	37.0	31.2	22.2	−0.32	0.25	0.82	0.81	0.81	0.40	0.33	0.144	0.143
18	50			131.3	0.82	4	1.05	h_{c1}	θ_2	38.1	45.0	40.3	36.2	1.09	0.25	0.41	0.63	0.69	0.29	0.35	0.072	0.126
19	50			64.9	0.05	4	1.05	h_{c1}	θ_2	24.8	43.2	33.6	22.3	−0.52	0.25	1.74	1.67	1.69	0.31	0.255	0.306	0.298
20	50			131.3	0.05	4	1.05	h_{c1}	θ_2	37.6	49.0	41.9	36.3	1.52	0.25	0.86	1.34	1.43	0.23	0.28	0.151	0.252
21	50	0	2.0	64.9	0.82	4	0.65	0.63	θ_2	25.4	38.5	31.7	22.2	−0.38	0.25	0.99	0.98	0.99	0.38	0.31	0.174	0.174
22	50			131.3	0.82	4	0.62	0.49	θ_2	38.0	46.1	40.2	36.2	1.13	0.25	0.57	0.80	0.86	0.27	0.33	0.100	0.151
23	50			64.9	0.05	4	0.74	0.72	θ_2	24.5	46.2	34.7	22.4	−0.65	0.25	2.37	2.28	2.31	0.27	0.22	0.417	0.407
24	50			131.3	0.05	4	0.69	0.60	θ_2	37.3	51.4	42.4	36.4	1.97	0.25	1.18	1.96	2.07	0.19	0.24	0.208	0.365
25	50	2.8	1.0	64.9	0.82	4	1.05	h_{c1}	24.3	45.3	53.2	47.2	22.4	−0.76	0.25	−0.08	−0.44	0.67	0.34	0.28	−0.014	0.118
26	50			131.3	0.82	4	1.05	h_{c1}	37.5	50.0	55.4	50.3	36.3	1.63	0.25	−1.78	−1.06	0.56	0.275	0.33	−0.313	0.099
27	50			64.9	0.05	4	1.05	h_{c1}	24.0	40.0	54.9	46.0	22.5	−0.99	0.25	0.65	0.16	1.39	0.29	0.24	0.114	0.245
28	50			131.3	0.05	4	1.05	h_{c1}	37.2	46.5	56.8	49.3	36.4	2.17	0.25	−1.71	−0.56	1.26	0.23	0.29	−0.301	0.222
29	50	2.8	1.0	64.9	0.82	0.42	1.05	h_{c1}	9.8	40.1	50.7	43.8	−15.7	−1.87	2.36	0.52	−0.52	0.74	0.41	0.18	0.092	0.130
30	50			131.3	0.82	0.84	1.05	h_{c1}	49.5	54.0	57.6	53.1	47.5	11.96	1.19	−11.02	−2.63	0.49	0.21	0.42	−1.94	0.086
31	50			64.9	0.05	0.43	1.05	h_{c1}	8.5	34.0	52.9	42.0	−12.8	−2.36	2.32	1.46	0.30	1.56	0.35	0.16	0.257	0.275
32	50			131.3	0.05	0.84	1.05	h_{c1}	49.1	51.8	58.5	52.4	48.0	19.8	1.20	−19.21	−2.38	1.03	0.19	0.38	−3.38	0.181
33	50	2.8	1.0	64.9	0.82	4	0.66	0.44	24.3	44.8	53.6	46.3	22.4	−0.75	0.25	0.01	0.31	0.77	0.33	0.27	0.002	0.136
34	50			131.3	0.82	4	0.65	0.20	37.45	49.9	55.9	48.8	36.3	1.68	0.25	−1.37	−0.90	0.66	0.26	0.32	0.241	0.116
35	50			64.9	0.05	4	0.73	0.64	23.8	38.3	56.1	45.0	22.5	−1.19	0.25	1.33	0.72	2.00	0.25	0.21	0.234	0.352
36	50			131.3	0.05	4	0.70	0.56	37.1	44.5	57.8	48.3	36.4	2.68	0.25	−1.63	−0.06	1.85	0.21	0.25	−0.287	0.326

example, the value of 0.165 for a sealed airspace (case 13) swings to −0.014 when the cavity is ventilated (case 25); this compares with the much smaller drop from 0.174 to 0.144 for the sheeted roof. Several comparisons giving results of similar order are available in Table 2.

The general pattern of variation in R_c that results from changing the value of the convection transfer coefficient is similar for ventilated cavities in both positions. It is noted that the influence of these changes is more pronounced with the cavity deck.

The analysis shows that the reistance of a sealed cavity in winter is unaffected by changes in the outside climate and independent of its position in the structure. The resistance of an open cavity on the other hand is influenced to a marked degree by these and other factors associated with convection. To each value of R_c for a ventilated cavity there attaches a corresponding value of the outside surface resistance, R_{so}, as demonstrated by the results entered in columns 14 and 16. Furthermore, R_{so} is seen to be no less a variable quantity than R_c over the considered range of climatic factors and thermal conductance.

The interaction between R_c and R_{so} is defined by the set of heat balance equations (2.7) and the result expressed most effectively in the air-to-air heat transmission coefficient of the structure, or U-value. The respective U-values are shown in column 19; they evaluate the insulating property of the structure for the conditions defined and accordingly provide a useful basis for a comparison of results in practical terms.

A comparison of the U-values may be made by an examination similar to that above of R_c. Briefly, cavity ventilation increases the U-value of the cavity deck by about 40 percent at least, rising to a maximum of 60 percent if the cavity has a low effective emissivity. The corresponding range for the sheeted roof is approximately 5–25 percent. In both cases the upper limit refers to daytime values and the trends, though differing in detail, are similar to those in the behaviour of R_c.

Values of surface resistance R_{so} show greater variation in calm weather for the roof structure of lower conductance (29–32). This large swing between the day and night values of R_{so} may be reduced by defining the surface resistance as follows

$$R_{so} = \frac{\theta_{so} - \theta_o}{\sigma T_{so}^4 - \alpha I_s + h_{co}(\theta_{so} - \theta_o)} = \frac{\theta_{so} - \theta_o}{h_{co}(\theta_{so} - \theta_e)} \tag{2.23}$$

where

$$\theta_e = \theta_o + (\alpha I_s - \sigma T_{so}^4)/h_{co} \tag{2.24}$$

Equation (2.24) is an alternative though less comprehensive expression for the equivalent outdoor temperature described in Chapter 1, and is easily recognized as an extension of the formula for the sol-air temperature originated by Mackey and Wright (1942). Although intended to represent the simultaneous influence of radiation and convection on the heat exchange at an exposed surface, the expression for the sol-air temperature in its original form is restricted to daytime conditions. The expression (2.24) extends the use of this concept of a hypothetical equivalent temperature especially for calculating heat flow through naturally exposed building surfaces at night.

A rearrangement of (2.23) gives

$$R_{so}(e) = R_{so}\frac{(\theta_{so} - \theta_e)}{(\theta_{so} - \theta_o)} \tag{2.25}$$

where $R_{so}(e)$ is the reciprocal of h_{co} and defines surface resistance in terms of θ_e. Values of θ_e and $R_{so}(e)$ calculated with the defining expressions above are entered in columns 13 and 15, respectively, in Table 2. The outside surface resistance expressed in this way is less variable than R_{so} calculated according to the conventional definition and, it may be noted, all the values are positive.

A negative value for the surface resistance indicates that the surface is gaining heat by convection and at the same time losing an amount by radiation. It may be imagined that confusion could arise in calculating heat flow through structures with a coefficient that changes sign according to the ambient conditions. Any difficulty in this regard is overcome by using θ_e as the outdoor (hypothetical) temperature and the shade U-value for thermal transmittance as explained in Chapter 1. Briefly, θ_e removes the variation in R_{so} due to radiation by converting it to an equivalent adjustment of the outdoor air temperature.

Further effects of substituting θ_e for θ_o may be assessed by calculating values of thermal transmittance and cavity resistance using the data in Table 2 with the formulae

$$U(e) = Q/(\theta_i - \theta_e)$$

$$R_c(e) = \frac{1}{U(e)} - \{R_{i.1} + R_2 + R_{so}(e)\}$$

Values of $U(e)$ in column 20 of Table 2 are seen not to be affected by the radiation flux I provided the convection coefficient h_{co} remains unchanged. Variation occurs, however, when the cavity is ventilated (see column 20, cases 9–12 versus 25–28).

It may be noted from the figures in columns 16 and 17 that no advantage is gained by substituting θ_e for θ_o in the definition of cavity resistance whether sealed or ventilated.

It is interesting to calculate values of cavity resistance by the method of Robinson, Powlitch, and Dill, and compare these with the corresponding results obtained from solving equations (2.7). The data for cases 1 and 3 in

Table 2 relating to upward heat flow in a horizontal cavity have been selected. The cavity in the former has a high effective emissivity, $E = 0.82$, and in case 3 has the lowest practical value of 0.05; otherwise the cavities are alike and are exposed to the same climatic conditions.

Case 1

Cavity thickness = 76 mm
Orientation = horizontal
Effective emissivity, E = 0.82
Temperature difference $(\theta_1 - \theta_2) = 7.3$ degC
Mean temperature = 0 degC
From Fig. 7, $(h_c)_{7.3}$ = 2.04 W/m^{2}K
From Fig. 8, $(h_r)_0$ = 4.66 W/m^{2}K

$$(h_c)_0 = (h_c)_{7.3}\{1 - 0.0018(0\text{–}7.3)\}$$
$$= 2.04 \times 1.013$$
$$= 2.07$$

Conductance, $C = Eh_r + (h_c)_0$
$$= 0.82 \times 4.66 + 2.07$$
$$= 5.89$$

Resistance, $R_c = 1/5.89 = 0.170$ m^{2}K/W which agrees closely with 0.174 given in Table 2.

Case 3

Cavity thickness = 76 mm
Orientation = horizontal
Effective emissivity, E = 0.05
Temperature difference $(\theta_1 - \theta_2) = 12.4$ degC
Mean temperature = 2.1 degC
From Fig. 7, $(h_c)_{12.4}$ = 2.41 W/m^{2}K
From Fig. 8, $(h_r)_{2.1}$ = 4.77 W/m^{2}K

$$(h_c)_{2.1} = (h_c)_{12.4}\{1 - 0.0018(2.1\text{–}12.4)\}$$
$$= 2.45$$

Conductance, $C = Eh_r + h_c$
$$= 0.05 \times 4.77 + 2.45$$
$$= 2.69$$

Resistance, $R_c = 1/2.69 = 0.372$ m^{2}K/W which differs significantly from the corresponding value of 0.433 in Table 2. With radiation in this example accounting for less than 10 percent of the total heat transfer the result indicates the difference between convection observed experimentally and that calculated using standard formulae considered appropraite to the conditions for the experiment.

The above results have been calculated for an airspace bounded by plane parallel surfaces of indefinite extent. This is an idealization of practical cavity structures. For example, the airspace may be bounded on the weather side by a corrugated surface with the troughs of the corrugation in some cases making uncertain contact with the opposite surface. Often the cavity is bridged with studding that may be of metal, or of material with a low thermal conductivity. Little is known about the transfer of heat through these irregular airspaces which are more common in roofs.

Despite these limitations the data in Table 2 give information of general interest in the calculation of heat transmission through cavity structures and in particular on the effect of airspace ventilation. Theoretical considerations show the effect of ventilation to depend upon climate and structural details and it has been evaluated with data relating to roofs. A roof is the part of a building most severely exposed to the weather and may be constructed in a variety of ways which have been divided for the present purpose between sheeted structures and roof decks. In the former construction the airspace is enclosed between a moisture impermeable cladding and a lining of heat insulating material, whereas the cavity roof deck is usually constructed with a more uniform arrangement of thermal resistance about the cavity. The more important results of the analysis are summarized below.

The thermal resistance of a horizontal or inclined cavity sealed between high emissivity surfaces may be taken as the standard value of 0.176 m^2K/W irrespective of its position in the structure. A slight difference is calculated between day and night values but this is no more than about 4 percent and may be ignored for practical purposes. When the effective emissivity for radiation heat transfer in the cavity is low, normally represented by $E = 0.05$ corresponding to one surface—or both—lined with bright aluminium foil, the resistance of the airspace is 0.43 rising to 0.47 during the daytime for the sheeted roof, and from 0.46 to 0.50 for the roof deck. This increase is due to absorbed solar heat.

For conventional calculations of heat transmission in cavity structures the airspace is generally regarded as sealed; little guidance is available on how to take account of the ventilation that is seemingly inevitable—though sometimes desirable—in normal building practice. Standard resistance values for ventilated airspaces in various cavity structures are given in the IHVE *Guide Book A* (1970) but, to a certain extent, they are indicated to be arbitrary. When the airspace is ventilated, it is necessary to define an 'effective' cavity resistance. This quantity combines linearly with the other component resistances of the structure to conform with the conventional method for calculating air-to-air heat transmission coefficients.

The thermal effect of ventilating a 75 mm thick airspace in a cavity roof structure has been calculated assuming a ventilation rate of 50 air changes per hour. This value is based on experimental results obtained with actual roofs. At night when the roof may be losing heat at a high rate by radiation to an unclouded sky, ventilation causes the resistance to fall from the standard value of 0.176 m^2K/W to an effective value of 0.144. The daytime value is 0.072. These figures apply to airspaces with high emissivity surfaces in sheeted structures; the corresponding night and day values for a ventilated foil-lined cavity are 0.306 and 0.151, respectively. This result indicates the importance of sealing cavities that rely for their insulating value on a low effective emissivity. The effect of ventilation becomes more pronounced as the mean temperature of the airspace increases. For example, with the cavity deck of lower conductance than the sheeted roof, solar heat causes the outside surface resistance to reach abnormally high values and the cavity reistance then takes negative values in order to balance the heat equation by which it is defined. It follows that for each value of effective cavity resistance there exists a corresponding value of the outside surface resistance depending upon the climate and the design of the structure.

For practical design purposes the most important result from the foregoing analysis is that cavity ventilation increases the U-value of the cavity roof deck by about 40 percent, and by 60 percent if the airspace has a low emissivity. The corresponding range for the sheeted roof is 5–25 percent.

SYMBOLS

Temperature

θ_{si}; T_{si}	Inside surface
θ_1, θ_2; T_1, T_2	Cavity surfaces
θ_{so}; T_{so}	Outside surfaces
θ_i	Inside air
θ_o	Outside air
$\bar{\theta}$	Cavity mean
$\theta(0)$	Cavity inlet
$\theta(L)$	Cavity outlet
θ_e	Effective outside air
T_s	Enclosure
T_{sp}	Sky
θ	Difference
h_c	*Convection transfer coefficient*
h_{ci}	Inside surface
h_{c1}; h_{c2}	Cavity surfaces
h_{co}	Outside surface
h_r	Radiation transfer coefficient
h	Surface transfer coefficient
k	Thermal conductivity
C	Thermal conductance
U	*Thermal transmittance*
$U(e)$	Equivalent

Thermal resistance

$R_{a.a}$	Air-to-air
R_{so}	Outside surface
$R_{so}(e)$	Outside surface, equivalent
R_1	Inner leaf
R_2	Outer leaf
R'_c	Cavity, effective
R_c	Cavity, sealed
$R_c(e)$	Cavity, equivalent
$R_{i.1}$	Inside air—cavity surface 1
Q, H	*Heat flux*
H_c	Convection
H_k	Conduction
H_r	Radiation

Radiation flux at outside surface

I_s	Solar
I	Total

Emissivity

e	Outside surface
e_1	Cavity surface 1
e_2	Cavity surface 2
E	Effective
a	Absorptivity
F_a	Configuration factor
F_e	Emissivity factor
σ	Stefan–Boltzmann constant
r	Reflectivity
c	Thermal capacity per unit mass
a	Coefficient of thermal expansion
g	Acceleration due to gravity

Fluid

ρ	Density
μ	Viscosity
v	Forced velocity

Cavity

l	Thickness
L	Length
n	Ventilation rate
A	Area

V	Volume
x, y	Position co-ordinates
t	Time

Dimensionless groups

Nu	Nusselt $Hl/k\theta$
Gr	Grashof $ag\theta l^3\rho^2/\mu^2$
Pr	Prandtl $c\mu/k$
Ra	Rayleigh $agc\theta l^3\rho^2/k\mu$
Re	Reynolds $v\rho l/\mu$

REFERENCES

Bosworth, B. C. L. (1952) *Heat Transfer Phenomena*, Assd. Gen. Publications Ptd. Ltd., Sydney.

Cess, R. D. and Schaffer, E. C. (1959) *Appl. Sci. Res.*, Section A, **8**.

Dines, W. H. and Dines, L. H. G. (1927) *Memoirs of Royal Meterological Society*, **2** (11).

Fishenden, M. and Saunders, O. A. (1932) *Calculation of Heat Transmission*, HMSO, London.

Fishenden, M. and Saunders, O. A. (1959) *An Introduction to Heat Transfer*, Clarendon Press, Oxford.

Griffiths, E. and Davis, A. H. (1922) DSIR, Food Investigation Board, Special Report, no. 9.

Jurges, W. (1924) *Beiheft zum Gesundsheits-Ingenieur*, **1,** 19.

Mackey, C. O. and Wright, L. T. (1942) *Heating, Piping*, **14** (12), 750.

McAdams, W. H. (1951) *Heat Transmission*, McGraw-Hill, New York.

Mull, W. and Reiher, H. (1930) *Beihefte zum Gesundheits-Ingenieur*, **1,** 28.

Peregrine, E. P. and Rowland, G. (1951) *Proc. of the General Discussion on Heat Transfer*, Inst. of Mech. Engrs.

Pratt, A. W. and Ball, E. F. (1966) *IHVE*, **34**.

Robinson, H. E., Powlitch, F. J., and Dill, R. J. (1954) U.S. Housing and Home Finance Agency, *Housing Research Paper*, **32,** Washington.

Rowley, F. B. and Algren, A. B. (1929) *Themal Resistance of Airspaces*, Trans. Am. Soc. of Heating and Ventilating Engineers, vol. 35, pp. 165–175.

Schack, A. (1929) *Der Industrielle Warmerubergang*, Verlag Stahleisen, Dusseldorf.

Schack, A., Goldschmidt, H., and Partridge, E. P. (1933) *Industrial Heat Transfer*, Chapman and Hall, London.

Stagg, J. M. (1950) *Geophysical Memoirs*, no. 86, Air Ministry Meteorological Office.

Wilkes, G. B. and Peterson, Carl M. F. (1937) *Heating, Piping*, **9,** 505.

CHAPTER 4

Heat Flow Through Edges and Corners

Heat transmission coefficients of materials and structures are determined by measuring the steady flow of heat between two parallel faces maintained at different known temperatures. The isothermals are assumed to be parallel to these faces over the test area of the specimen where the heat flux and temperatures are measured. The accuracy of the determination depends upon the validity of this assumption. Almost always a small proportion of the heat flux entering the hot face escapes through the edges of the test specimen; the amount is unknown and usually neglected. It follows that the measured value for the transmission coefficient differs from the true value calculated on the assumption that the direction of heat flow through the specimen is uniform and normal to both faces.

Calculations for practical purposes of heat conduction in buildings likewise assume the flow of heat to be in one direction only, normal to the surfaces, and ignore the distorting effect of edges and corners. These problems in buildings are generally too complex to solve completely with rigorous analysis and approximate results are then sought by examining idealized cases. These may be considered to fall into two groups for which characteristic solutions are readily calculable. One group is represented by the measurement of thermal conductivity using the hot plate method. The other may be identified with the determination of heat transmission coefficients of building structures exposed to the weather. In the former group measurements are made of a specific property and carried out accordingly under prescribed boundary conditions with high accuracy enabling reliable edge heat loss corrections to be calculated and applied to the quantity measured. Measurements in the latter group are of thermal conductance and thermal transmittance. These are properties of a structure and are determined in a specially designed laboratory in which test walls face a heated enclosure on one side and on the other are exposed to the weather, as in a practical installation. An analysis of heat flow through the corners of the heated chamber behind each test wall in this laboratory gives results that apply to buildings generally.

1. ANALYSIS OF ERROR DUE TO THE EDGE HEAT LOSS IN MEASURING THERMAL CONDUCTIVITY BY THE HOT PLATE METHOD

The standard steady state method for the determination of the thermal conductivity of solid material is described in BS 874:1973. For material of low conductivity represented by a k-value not exceeding, say, 0.15 W/mK, a guarded hot plate is used. This is sandwiched between two identical test specimens each nominally measuring $300 \times 300 \times 25$–50 mm thick. A cold plate at the opposite face of each specimen completes the test arrangement. The edges of the test specimens are not perfectly insulated and exchange heat with the surroundings; this may cause an error in the measured value of conductivity calculated on the assumption that the heat flows through the specimen in a single direction normal to both faces over the central test area. For material of higher conductivity an unguarded hot plate may be used and a correction applied to the measured k-value for the passage of heat through the edges. This correction, according to the BS, is determined by a test, performed under similar conditions of thickness and temperature, on a material of known conductivity.

1.1 Edge Loss with a Guard Ring

The purpose of a guard ring is to ensure that the flow of heat through the test volume of the material is normal to the hot and cold faces. The hot plate and surrounding co-planar guard plates are heated electrically and wired so that their temperatures may be varied independently. Thermal guarding is achieved by manual adjustment of the current in the guard plate to maintain the temperatures of the two plates the same on either side of the 3 mm wide gap that separates them. Despite guarding there is still the possibility that distortion of the isothermals generated by heat flow through the edges of the specimens may extend into the central test area and invalidate the assumption of normal flow. Clearly, the conductivity of the specimen and its size relative to the widths of the guard ring and the hot plate are factors that may be expected to influence the resultant error in measured values. Somers and Cyphers (1951) analysed the error due to edge loss in measuring the thermal conductivity of insulating material by the conventional guarded hot plate method and applied the results to evaluate the relationship between the maximum dimensions of the test sample and the minimum dimensions of the square guarded hot plate defined in the American Specification, ASTM: Designation C177–45 (American Society for Testing Materials, 1955). They considered the model defined by the following set of equations

$$
\left.\begin{aligned}
&\frac{\partial^2\theta}{\partial x^2}+\frac{\partial^2\theta}{\partial y^2}+\frac{\partial^2\theta}{\partial z^2}=0 \qquad \begin{array}{l} 0<z<z_0 \\ |x|<x_0 \\ |y|<y_0 \end{array}\\
&\theta(x, y, 0) \qquad\quad = \theta_0\\
&\left.\begin{array}{l}\theta(x, y, z_0)\\ \theta(x_0, y, z)\\ \theta(x, y_0, z)\\ \theta(-x_0, y, z)\\ \theta(x, -y_0, z)\end{array}\right\} = 0
\end{aligned}\right\} \tag{1.1}
$$

This corresponds to the steady flow of heat in a rectangular parallelepiped with length of sides $2x_0$, $2y_0$ and z_0. The hot surface $z=0$ is maintained at a uniform temperature θ_0; all other surfaces are held at a lower temperature which, for convenience, is denoted zero. This is a restrictive case which, the authors explained, was adopted in order to ease the numerical work. Tedious calculation was avoided by putting the temperature of the exposed edges equal to that of the ambient air. The authors point out that these assumptions have the effect of increasing the calculated heat loss to the air and lead to a maximum error.

The solution given by Somers and Cyphers for the temperature distribution in the slab is

$$
\begin{aligned}
\theta(x, y, z)&\\
=\frac{16\theta_0}{\pi^2}\sum_{m,n=0}^{\infty}(-)^{m+n}&\left\{\frac{\sinh(\pi[\{(2m+1)/2x_0\}^2+\{(2n+1)/2y_0\}^2]^{1/2}[z_0-z])}{\sinh(\pi[\{(2m+1)/2x_0\}^2+\{(2n+1)/2y_0\}^2]^{1/2}z_0)}\right\}\\
\times\frac{[\cos\{(2m+1)\pi x/2x_0\}\cos\{(2n+1)\pi y/2y_0\}]}{(2m+1)(2n+1)}&
\end{aligned} \tag{1.2}
$$

Using this expression to evaluate the flux into the test sample ($-x_1 \leqslant x \leqslant x_1$, $-y_1 \leqslant y \leqslant y_1$, $z=0$) and putting $x_1 = y_1$ they expressed their rsults in the form

$$
\frac{k}{k_{\text{exp}}}=\psi\left(\frac{y_1}{y_0}, \frac{z_0}{y_0}\right) \tag{1.3}
$$

which relates the ratio of the true value (k) to the experimentally determined value (k_{exp}) of conductivity to dimensionless groups defining the size of the specimen. The expression enables the error in a thermal conductivity measurement to be calculated provided (a) the isothermal surface condition is fulfilled and (b) that the cold surface and exposed edges of the specimen are at the same temperature and equal to that of the ambient air.

A later analysis due to Woodside (1957) removes the restriction on the

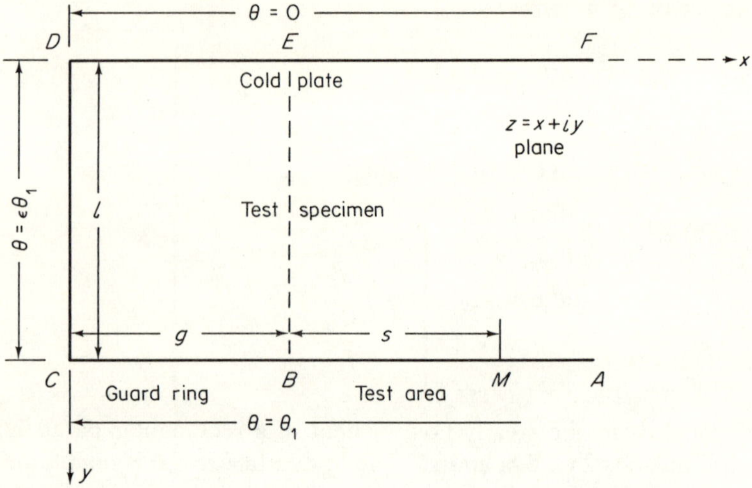

Fig. 1. Diagram representing part of the cross section throught the hot plate–test specimen–cold plate system at the centre plane

edge boundary condition under (b) above and expresses the ratio k/k_{exp} in a simpler and more easily calculable form than the result obtained by Somers and Cyphers. Woodside studied the problem also with particular reference to the determination of thermal conductivity as specified by the ASTM and obtained a solution using the Schwarz–Christoffel transformation theorm. The method of analysis is outlined below in the notation of the original paper.

The polygon in the plane $z = x + iy$ is shown in Fig. 1. It represents part of a cross-section through the assembly at the centre plane. The cold plate DEF is at a temperature $\theta = 0$; the specimen edge CD at temperature $\theta = \varepsilon\theta_1 (0 \leqslant \varepsilon \leqslant 1)$; and the hot plate AMBC at a temperature $\theta = \theta_1$. The Schwarz–Christoffel transformation (see Fig. 2) gives

$$\frac{dz}{d\omega} = \frac{C}{\{\omega(\omega - 1)\}^{1/2}} \tag{1.4}$$

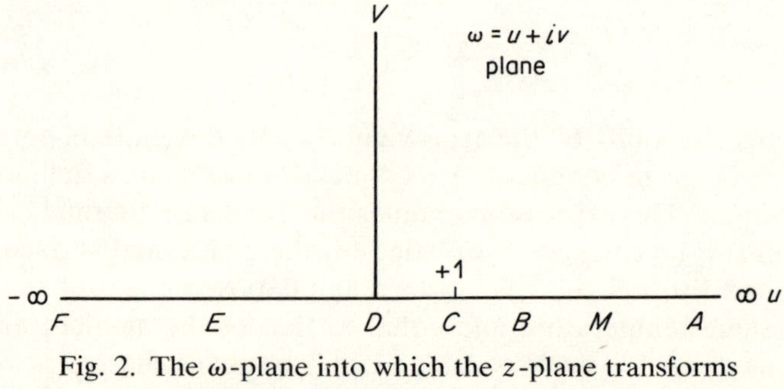

Fig. 2. The ω-plane into which the z-plane transforms

Using the result (Carslaw and Jaeger, 1959)

$$\theta(u, v) = \frac{v}{\pi} \int_{-\infty}^{+\infty} \frac{f(u')\, du'}{v^2 + (u - u')^2}$$

representing the steady state temperature distribution in the half-plane $v \geqslant 0$, with $v = 0$ maintained at temperature $f(u)$, it is shown by substituting the expressions for u and v obtained from integrating (1.4) that the required temperature distribution in the z plane is

$$\theta(x, y) = \frac{\theta_1}{2} + \frac{\theta_1}{\pi}\left[\varepsilon \tan^{-1}\left\{\frac{\cosh(\pi x/l)\cos(\pi y/l) - 1}{\sinh(\pi x/l)\sin(\pi y/l)}\right\} + (1-\varepsilon)\tan^{-1}\left\{\frac{\cosh(\pi x/l)\cos(\pi y/l) + 1}{\sinh(\pi x/l)\sin(\pi y/l)}\right\}\right]$$

The heat flux out of the test area is

$$q = -k\left(\frac{\partial \theta}{\partial y}\right)_{y=l}$$

Let x' be the space variable corresponding to x in a direction normal to the plane of Fig. 1. Then the total flow out of the test area, with length of side $2s$, is

$$Q = 4\int_{x=g}^{s+g}\int_{x=g}^{s+g} q \cdot dx\, dx' = k_{\text{exp}}(2s)^2 \frac{\theta_1}{l}$$

where g denotes the width of the guard ring. Performing the integration leads to the final result

$$\left(\frac{k}{k_{\text{exp}}}\right)^{1/2} = \frac{\pi s}{l}\left[\varepsilon \ln\left\{\frac{\cosh \pi(g+s)/l + 1}{\cosh(\pi g/l) + 1}\right\} + (1-\varepsilon)\ln\left\{\frac{\cosh \pi(g+s)/l - 1}{\cosh(\pi g/l) - 1}\right\}\right]^{-1} \tag{1.5}$$

Woodside concludes from this expression that the error is independent of the conductivity; further analysis however (see equations (1.15) and (1.16)) shows ε to be a function of k.

1.2 Edge Loss without a Guard Ring

The expression (1.2) for the temperature distribution in the slab is probably sufficiently accurate for calculating the error due to edge loss with a guarded hot plate designed in accordance with ASTM requirements. It is doubtful if the simplifying assumptions imposed on the boundary conditions contained in that result may be extended without significant loss of accuracy to a similar analysis of a hot plate without a guard ring. Also, the result in equation (1.5) will be seen to give difficulties with a plate for which $g = 0$.

A more general representation of test conditions is defined by the following set of equations:

$$\frac{\partial^2\theta}{\partial x^2}+\frac{\partial^2\theta}{\partial y^2}+\frac{\partial^2\theta}{\partial z^2}=0; \qquad \left.\begin{array}{r}-a<x<a\\ -b<y<b\\ 0<z<c\end{array}\right\} \tag{1.6}$$

$$\left.\begin{array}{rl}-(\partial\theta/\partial x)+h\theta=0; & x=-a\\ (\partial\theta/\partial x)+h\theta=0; & x=\ \ a\\ -(\partial\theta/\partial y)+h\theta=0; & y=-b\\ (\partial\theta/\partial y)+h\theta=0; & y=\ \ b\\ \theta=\theta_1; & z=\ \ 0\\ \theta=\theta_2; & z=\ \ c\end{array}\right\} \tag{1.7}$$

The hot and cold faces of the specimens are at temperatures θ_1 and θ_2, respectively, and the edges exposed to a medium at zero reference temperature, $h=h_s/k$, where h_s is the surface coefficient for heat transfer at the exposed edges, and k is the thermal conductivity of the specimen. The method of solution is indicated below.

From symmetry about the planes $x=0$ and $y=0$ these equations are replaced by the equivalent set

$$\frac{\partial^2\theta}{\partial x^2}+\frac{\partial^2\theta}{\partial y^2}+\frac{\partial^2\theta}{\partial z^2}=0; \qquad \begin{array}{l}0<x<a\\ 0<y<b\\ 0<z<c\end{array}$$

$$\left.\begin{array}{lll}\dfrac{\partial\theta}{\partial x}=0; & x=0, & \begin{array}{l}0<y<b\\ 0<z<c\end{array}\\ \dfrac{\partial\theta}{\partial y}=0; & y=0, & \begin{array}{l}0<x<a\\ 0<z<c\end{array}\\ \dfrac{\partial\theta}{\partial x}+h\theta=0; & x=a, & \begin{array}{l}0<y<b\\ 0<z<c\end{array}\\ \dfrac{\partial\theta}{\partial y}+h\theta=0; & y=b, & \begin{array}{l}0<x<a\\ 0<z<c\end{array}\end{array}\right\} \tag{1.8}$$

By a repeated use of a finite cosine transform (1.6) becomes

$$\frac{d^2\bar{\theta}}{dz^2}=(\alpha_r^2+\beta_s^2)\bar{\theta}' \tag{1.9}$$

where

$$\bar{\theta}'=\int_0^b \bar{\theta}\cos\beta_s y\,dy=\int_a^b\left(\int_0^a \theta\cos\alpha_r x\,dx\right)\cos\beta_s y\,dy$$

and

$$\alpha_r\tan\alpha_r a=h; \qquad \beta_s\tan\beta_s b=h \tag{1.10}$$

At the boundaries $z=0$ and $z=c$

$$[\bar{\theta}']_{z=0}=\frac{\theta_1}{\alpha_1\beta_s}\sin\alpha_r a\sin\beta_s b$$
$$[\bar{\theta}']_{z=c}=\frac{\theta_2}{\alpha_r\beta_s}\sin\alpha_r a\sin\beta_s b \tag{1.11}$$

The solution of equation (1.9) satisfying (1.11) is

$$\bar{\theta}'=\frac{\sin\alpha_r a\sin\beta_s b}{\alpha_r\beta_s}\{\theta_2\sinh lz+\theta_1\sinh l(c-z)\}\frac{1}{\sinh lc}$$

where $l^2=\alpha_r^2+\beta_s^2$. Inverting $\bar{\theta}^1\to\bar{\theta}\to\theta$ and introducing (1.10) gives

$$\theta(x,y,z)=4h^2\sum_{r=1}^{\infty}\sum_{s=1}^{\infty}\left[\frac{\theta_2\sinh lc+\theta_1\sinh l(c-z)}{\sinh lc\{b(\beta_s^2+h^2)+h\}\{a(\alpha_r^2+h^2)+h\}}\right]\times\frac{\cos\alpha_r x\cos\beta_s y}{\cos\alpha_r a\cos\beta_s b} \tag{1.12}$$

The total flow of heat into the hot face, area $4ab$, is

$$Q=-4\int_0^a\int_0^b k\left(\frac{\partial\theta}{\partial z}\right)_{z=0}\mathrm{d}x\,\mathrm{d}y$$

Also

$$Q=k_{\exp}(\theta_1-\theta_2)\frac{4ab}{c}$$

Hence

$$\frac{k}{k_{\exp}}=\frac{ab}{4h^2c}\left[\sum_{r=1}^{\infty}\sum_{s=1}^{\infty}\frac{\left\{1+\dfrac{\theta_1}{\theta_2-\theta_1}(1-\cosh lc)\right\}}{\{(\alpha_r^2+h^2)a+h\}\{(\beta_s^2+h^2)b+h\}}\left(\frac{l\tan\alpha_r a\tan\beta_s b}{\alpha_r\beta_s\sinh lc}\right)\right]^{-1} \tag{1.13}$$

Similarly, integration between the limits $(0, a_1)$ and $(0, b_1)$ where $0<a_1<a$ and $0<b_1<b$ gives

$$\frac{k}{k_{\exp}}=\frac{a_1b_1}{4h^2c}\times\left[\sum_{r=1}^{\infty}\sum_{s=1}^{\infty}\frac{\left\{1+\dfrac{\theta_1}{\theta_2-\theta_1}(1-\cosh lc)\right\}}{\{(\alpha_r^2+h^2)a+h\}\{(\beta_s^2+h^2)b+h\}}\left(\frac{l\sin\alpha_r a_1\sin\beta_s b_1}{\alpha_r\beta_s\sinh lc\cos\alpha_r a\cos\beta_s b}\right)\right]^{-1} \tag{1.14}$$

corresponding to a guarded hot plate with a sample area measuring $2a_1\times 2b_1$.

1.3 Calculation of Edge Heat Loss

The American Standard for the determination of thermal conductivity specifies edge insulation of such thickness that the resistance to edge losses shall be at least twice and preferably three or more times the thermal resistance of the specimen in the normal direction of heat flow. Woodside applied his solution (1.5) to the American standard and obtained the results shown in Fig. 3 for a 200×200 mm hot plate with $s = g = 50$ mm. The curves show the influence of the factor $\varepsilon (0 \leqslant \varepsilon \leqslant 1)$ which he introduced to represent the edge temperature distribution through the thickness of the specimen. This distribution clearly is determined by the test conditions, but

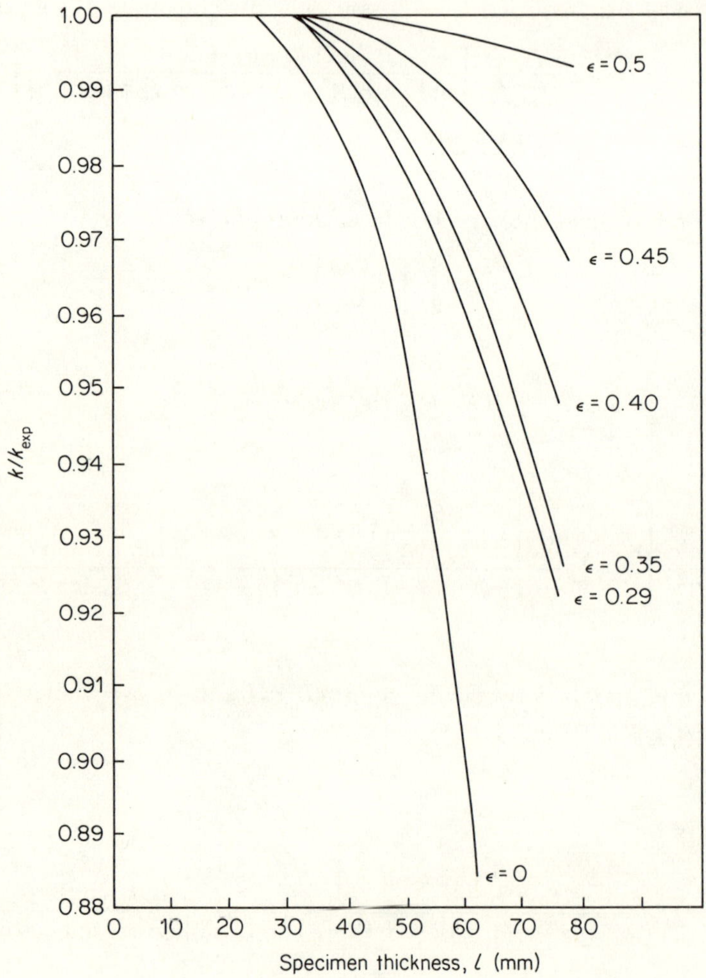

Fig. 3. k/k_{exp} calculated from equation (1.5) plotted against specimen thickness for a 200×200 mm hot plate having $s = g = 50$ mm for different values of ε

these are not known precisely; the choice of a value for ε therefore is arbitrary and from Fig. 3 is seen to be important. For a thick specimen well insulated around its edges Woodside estimated ε to be 0.29 based on measurements of edge temperature due to Pascal.

Where edge insulation is not used, values of ε may be obtained from the expression (1.12) for the temperature distribution in a parallelipiped by calculating the mean temperature over the surface $x=a$, $-b<y<b_1$ and $0<z<c$, giving

$$\varepsilon = 4h^2 \sum_{r=1}^{\infty}\sum_{s=1}^{\infty} \frac{\tan \beta_s b(\cosh lc - 1)}{\sinh lc\{b(\beta_s^2+h^2)+h\}\{a(\alpha_r^2+h^2)+h\}l\beta_s} \tag{1.15}$$

where $l^2=\alpha_r^2+\beta_s^2$, and $a_r \tan \alpha_r a = h$, $\beta_s \tan \beta_s b = h$. Slightly less exact values of ε for similar cases may be obtained more readily by considering the steady flow of heat in a rectangle $-a<x<a$, $0<y<b$ with edges $|x|=a$, $0<y<b$ exposed to a medium at zero temperature. This gives

$$\varepsilon = \frac{2h}{b}\sum_{n=1}^{\infty} \frac{\cosh \alpha_n b - 1}{\alpha_n\{(\alpha_n^2+h^2)a+h\} \sinh \alpha_n b} \tag{1.16}$$

where $\alpha_n \tan \alpha_n a = h$.

Values of ε calculated from (1.15) and (1.16) with the surface transfer coefficient at the exposed edges, h_s, equal to 8.06 W/m^{2}K are compared for different values of conductivity in Table 1. The results are for specimens measuring $300\times300\times50$ mm thick. The difference between the values of ε is about 3 percent when $k=0.062$, falling to approximately 1 percent when k is increased tenfold.

For high values of g/l (see equation (1.5)) the ratio k/k_{exp} rapidly approaches unity irrespective of the value asigned to ε. With a 300 mm square hot plate including a 50 mm wide guard ring, the maximum error obtained by putting $\varepsilon=0$ in (1.5) is calculated to be less than 1 percent.

A solution more representative of practical test conditions is given by equation (1.14). It applies to the steady state method of measurement described in the British Standard using a guarded hot plate. Normally the temperatures of the hot and cold faces of the specimen are arranged to be respectively 10 degC above and below the ambient temperature. In equation

Table 1

k (W/mK)	Rectangle	Parallelipiped
0.062	0.194	0.188
0.137	0.285	0.278
0.247	0.348	0.341
0.619	0.423	0.418

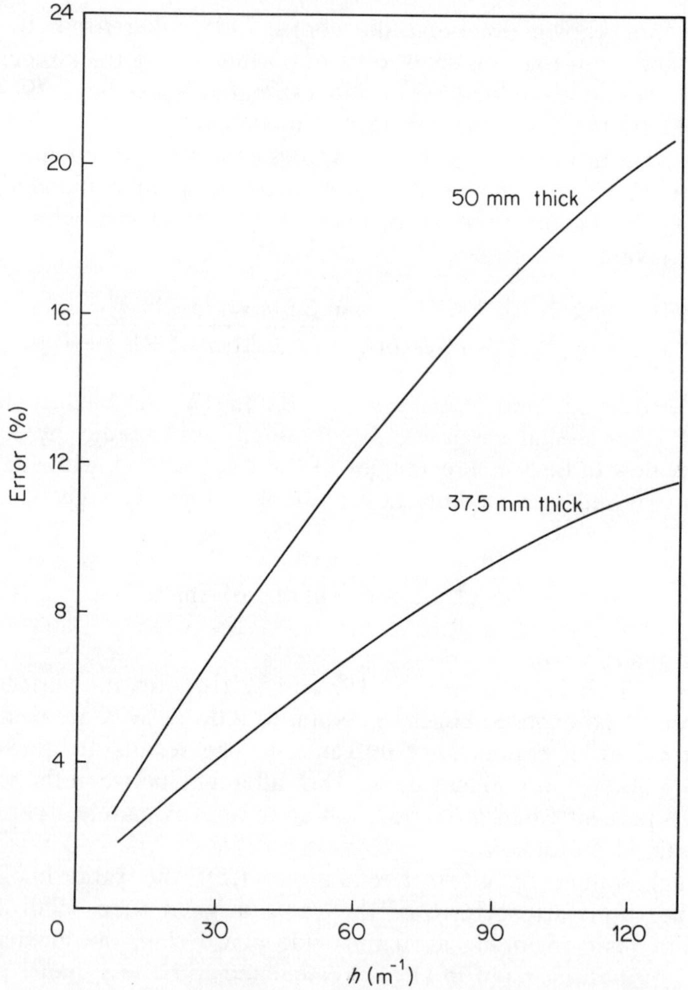

Fig. 4. Errors for 300 mm square unguarded hot plate calculated from temperature distribution in a parallelipiped. Specimen thickness 37.5 and 50 mm

(1.14), therefore, putting $\theta_1 = -\theta_2$ and substituting $a = 150$ mm $= b$, $a_1 =$ 100 mm $= b_1$, the ratio k/k_{exp} for a 50 mm thick specimen of a low conductivity material, $k = 0.062$ W/mK, is calculated to be 1.0075, corresponding to an error in the measured k-value of less than 1 percent.

The error in conductivity measurements using a hot plate without a guard ring may be calculated using equation (1.13). The result is shown in Fig. 4 with the percentage error in k plotted versus $h(= h_s/k)$ for 37.5 and 50 mm thick specimens; these are within the thickness limits recommended in the British Standard. The Standard also recommends the use of an unguarded hot plate with materials whose conductivity is greater than 0.15 W/mK.

These include structural insulating material represented mainly by the wide range of commercial lightweight concretes. These products must satisfy a minimum strength requirement and it is often difficult to prepare test specimens less than 50 mm thick. The graph in Fig. 4 illustrates the advantage of using thinner specimens for reducing edge heat loss. The value of the edge transfer coefficient h_s is not known precisely, but Fig. 4 provides a means of obtaining it by measuring the conductivity of a specimen of known k-value.

The influence of conductivity is shown in Fig. 5. The error falls quickly, though at a diminishing rate, as the conductivity increases, the surface coefficient remaining constant.

Although the above investigation deals specifically with the measurement of thermal conductivity it will be apparent that there are problems of heat

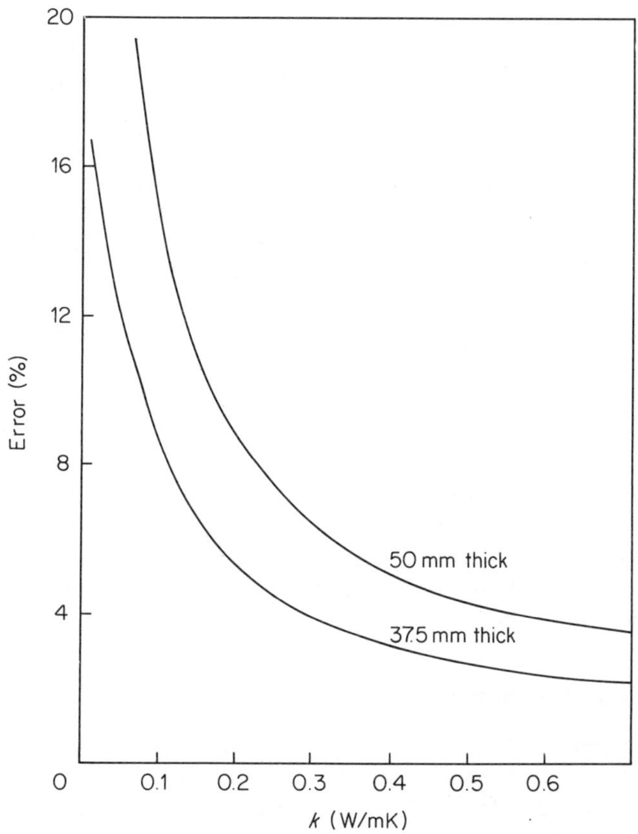

Fig. 5. Errors for 300 mm square unguarded hot plate calculated from temperature distribution in a parallelipiped. Specimen thickness 37.5 and 50 mm. Surface coefficient $h_s = 8.12$ W/m^{2}K

flow in buildings of a similar type to which this method of analysis may be applied.

2. ANALYSIS OF ERROR DUE TO THE EXCESS HEAT FLOW AT A RIGHT-ANGLED CORNER IN THE MEASUREMENT OF THE HEAT TRANSMISSION COEFFICIENTS OF BUILDING STRUCTURES

The analysis relates to the determination of the heat transmission coefficients in the Wall Laboratory that was in use for many years at the Building Research Station of the Department of Scientific and Industrial Research in Britain. It consisted of a row of six cubicles (see Chapter 8, Fig. 1) each enclosed on five sides by well-insulated walls. A test wall measuring 2.4 m square to correspond with storey height in buildings formed the remaining side and was exposed to the weather as in a practical installation. The six test panels were co-planar, faced due north, and had a sheltered to moderate exposure. The laboratory was heated electrically and wholly thermostatically controlled at a uniform temperature of 18.3 degC for a period of about six months extending from October to March. Throughout this time continuous measurements were recorded of the temperature difference between the indoor and the outdoor air, also of the temperature distribution and flow of heat through the test walls. With these data the thermal transmittance and other heat transmission coefficients of the test structures were calculated. The heat loss through the test wall was determined by measuring the heat flow through the floor and roof surfaces and deducting this amount from the heat input to the cubicle. Though indirect, this is the only method for determining the mean heat flow through structures, especially if their surfaces are not plane or isothermal. Since all the rooms in the laboratory were heated to the same temperature, including the corridor that served as a thermal guard on three sides of the building, there was assumed to be no net flow of heat through the partition walls. Although the roof, floor, and all internal walls were well insulated, a small proportion of the heat loss from each cubicle took place through the corners causing an error in the measured coefficients calculated on the assumption of normal heat flow. The source and magnitude of this error is examined below.

2.1 Heat Flow at a Corner

Consider a test panel, thickness a, conductivity λ, with heat flowing in the steady state through a section of unit width extending a distance h at right-angles to its edge from the corner. Let q be the flow normal to the inside surface of the section and continuing through the section without change of direction; then

$$q = \lambda(\theta_1 - \theta_2)\frac{h}{a}$$

where θ_1 and θ_2 are the temperatures of the inside and outside surfaces of the section, respectively. If, however, an amount of heat Δq deviates from this direction and passes through the edge of the section at the corner, then

$$q+\Delta q=\lambda_{\text{exp}}(\theta_1-\theta_2)\frac{h}{a}$$

where λ_{exp} is the measured conductivity. These expressions may be combined to give

$$\frac{\lambda}{\lambda_{\text{exp}}}=\frac{1}{1+\dfrac{a}{h}\dfrac{q'}{\lambda}} \tag{2.1}$$

where q' is the excess heat flow due to the corner per unit temperature difference between the surfaces causing the measured conductivity λ_{exp} to exceed the true value λ.

In the Wall Laboratory two separate problems of heat loss through the edges of the wall need to be considered. One occurs at the corner between the test wall and the roof, the other at the corner between the test wall and the partition wall. These have been studied as two-dimensional problems, assuming in both cases that the inside and outside surfaces are isothermal boundaries. In a practical installation the inside surface temperature will show a progressive fall towards the edges of the wall. Clearly this change in temperature is a consequence of the greater heat flow through the corner and the assumption, therefore, that the boundaries are isothermal does not invalidate the results of the analysis provided the temperature variation along the wall is small compared to $\theta_1-\theta_2$.

2.2 Right-angled Corner between Test Wall and Roof

The flow of heat in steady state through a right-angled corner formed by the test wall and roof will now be considered. Fig. 6 illustrates the system. The problem is defined mathematically as follows

$$\begin{aligned}
&\frac{\partial^2\theta}{\partial x^2}+\frac{\partial^2\theta}{\partial y^2}=0, &&\left|\begin{matrix}0<x<\infty, & 0<y<b\\ 0<x<a, & b<y<\infty\end{matrix}\right|\\
&\theta=0, &&\left|\begin{matrix}x=0, & 0<y<\infty\\ 0<x<\infty, & y=0\end{matrix}\right|\\
&\theta=1, &&\left|\begin{matrix}x=a, & b<y<\infty\\ a<x<\infty, & y=b\end{matrix}\right|
\end{aligned} \tag{2.2}$$

$$\begin{aligned}
&(\theta)_{b-0}=(\theta)_{b+0}\\
&\lambda_1\left(\frac{\partial\theta}{\partial y}\right)_{b-0}=\lambda_2\left(\frac{\partial\theta}{\partial y}\right)_{b+0}, \qquad 0<x<a
\end{aligned} \tag{2.3}$$

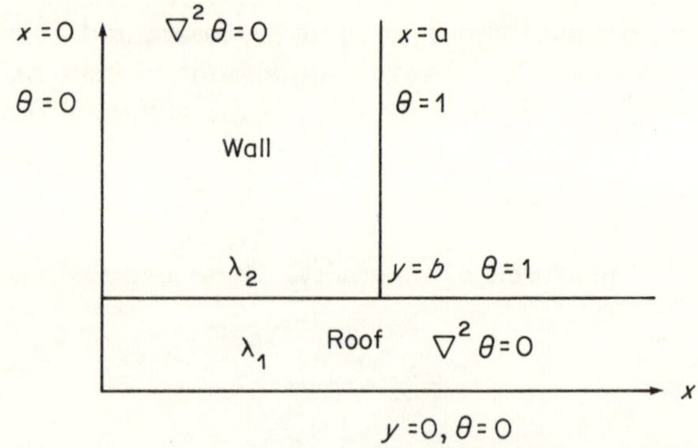

Fig. 6. Right-angled corner between test wall and roof

These equations specify the conditions of heat flow through a corner in which the thermal conductivity and thickness are not necessarily alike for both slabs. Carslaw and Jaeger (1959) consider the more restrictive case in which the thickness only is allowed to differ between the two slabs, the conductivity being uniform throughout.

A solution to these equations has been obtained by Sewell (private communication) who gives the following expression for the excess heat flow through the test panel due to the effect of the corner

$$q' = \frac{2\lambda_1\lambda_2}{\lambda_1+2\lambda_2} \times \left[\frac{\mu}{3} + \frac{1}{\mu}\left\{\frac{1}{6} + \frac{1}{\pi^2}\sum_{m=1}^{\infty}\frac{e^{-2m\pi\mu}}{m^2}\right\} - \frac{1}{\mu^2}\left\{\frac{1}{\pi^3}\sum_{m=1}^{\infty}\frac{1}{m^3} - \frac{1}{\pi^3}\sum_{m=1}^{\infty}\frac{e^{-2m\pi\mu}}{m^3}\right\}\right.$$
$$\left. + \sum_{k=1}^{\infty}\left[A_k\frac{1}{\sqrt{k}}\left\{\frac{\mu}{2k\pi} - \frac{1}{2}\left(\coth\frac{k\pi}{\mu} - 1\right) + \frac{\mu k}{\pi}\sum_{m=1}^{\infty}\frac{e^{-2m\pi\mu}}{\mu^2 m^2 + k^2}\right\} - \frac{\pi}{2}L_k\right]\right]$$
$$\equiv \frac{2\lambda_1\lambda_2}{\lambda_1+2\lambda_2} F\left(\frac{\lambda_2}{\lambda_1}, \mu\right) \tag{2.4}$$

where μ is the thickness ratio a/b, and the coefficients $A_k L_k$ are defined by the set of equations

$$\sum_{n=1}^{\infty} K_{kn} A_n = L_k \tag{2.5}$$

$$K_{kk} = \coth\frac{k\pi}{\mu} + \frac{\lambda_2}{\lambda_1} - k\frac{4}{\pi^2}\sum_{m=1}^{\infty}\frac{\mu^2 m e^{-m\pi\mu}\sinh m\pi\mu}{(m^2\mu^2+k^2)^2}$$
$$K_{kn} = K_{nk} = \sqrt{(kn)}\,\frac{4}{\pi^2}\times\sum_{m=1}^{\infty}\frac{\mu^2 m e^{-m\pi\mu}\sinh m\pi\mu}{(m^2\mu^2+n^2)(m^2\mu^2+k^2)} \qquad (n \neq k) \tag{2.6}$$
$$L_k = \frac{2}{\pi^3}\left\{\frac{\mu}{k^{3/2}} + \sqrt{k}\,\frac{2}{\pi}\sum_{m=1}^{\infty}\frac{e^{-m\pi\mu}\sinh m\pi\mu}{m(m^2\mu^2+k^3)}\right\}$$

As an empirical check on the process of solving the infinite set of equations above for the coefficients A_k, the solution may be compared with that

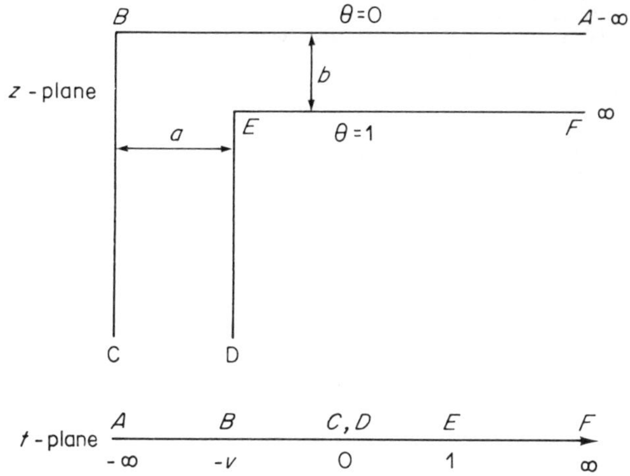

Fig. 7. Corner problem: polygon in z-plane and corresponding points in t-plane

obtained by conformal transformation for the restrictive case where $\lambda_1 = \lambda_2$. The procedure is described briefly below.

The corner, or polygon, in the complex z-plane, and corresponding points in the t-plane, are shown in Fig. 7. The transformation between the z-plane and the t-plane is given by Carslaw and Jaeger; with slight change in notation this is written

$$z = -\frac{2c}{\sqrt{v}} \tan^{-1} \beta\sqrt{v} + c \ln \frac{1+\beta}{1-\beta} \tag{2.7}$$

where

$$\beta = \left(\frac{t-1}{t+v}\right)^{1/2}; \qquad \frac{1}{\sqrt{v}} = \frac{a}{b}$$
$$b = \pi c; \qquad a = \frac{\pi c}{\sqrt{v}} \tag{2.8}$$

It is required to find the flow over DE representing the inside surface of the wall. This is obtained by a method of solution using conjugate functions.

Let $\omega = \theta + i\psi$, where ψ and θ are conjugate functions representing the heat flux and temperature functions, respectively. The heat flux at any point across a surface is $-\lambda\, \partial\theta/\partial n$, where $\partial/\partial n$ denotes differentiation along the outward-drawn normal to the surface. The heat flow across the portion of surface from s_1 to s_2 is

$$-\lambda \int_{s_1}^{s_2} \left(\frac{\partial\theta}{\partial n}\right) ds = -\lambda \int_{s_1}^{s_2} \left(\frac{\partial\psi}{\partial s}\right) ds = -\lambda(\psi_2 - \psi_1)$$

By studying the flux function conjugate to the temperature function for which the expression is already known, the rate of heat flow across any part of one of the boundary surfaces can be obtained.

Now, on DE, t is real and $0<t<1$. Putting

$$\theta+i\psi=1-\frac{1}{\pi i}\ln t,$$

$$z=-\frac{2c}{\sqrt{v}}\tan^{-1}\beta\sqrt{v}+c\ln\frac{1+\beta}{1-\beta}=t\left\{-\frac{c}{\sqrt{v}}\ln\frac{1+(\beta/i)\sqrt{v}}{1-(\beta/i)\sqrt{v}}+2c\tan^{-1}\left(\frac{\beta}{i}\right)\right\} \quad (2.9)$$

or

$$x=0$$

$$y=-\frac{c}{\sqrt{v}}\ln\frac{1+(\beta/i)\sqrt{v}}{1-(\beta/i)\sqrt{v}}+2c\tan^{-1}\left(\frac{\beta}{i}\right) \quad (2.10)$$

To find the flow of heat, per unit length of corner, through a strip extending a distance y (assumed to be large) from the corner, the value of t corresponding to y by (2.10) is required. Noting that when $-y$ is large, $t\to 0$, it can be shown that

$$y=-\frac{c}{\sqrt{v}}\ln\frac{4v}{(1+v)t}+2c\tan^{-1}\frac{1}{\sqrt{v}}$$

Making the substitution $\pi\psi=\ln t$ gives

$$y=a\left\{\psi+\frac{2}{\pi\mu}\tan^{-1}\mu-\frac{1}{\pi}\ln\frac{4}{(\mu^2+1)}\right\}$$

where

$$\mu=a/b=1/\sqrt{v}$$

Hence

$$\psi=\frac{y}{a}-\left\{\frac{2}{\pi\mu}\tan^{-1}\mu-\frac{1}{\pi}\ln\frac{4}{(\mu^2+1)}\right\}\frac{y}{a}\ll-1 \quad (2.11)$$

At E, $t=1$, therefore

$$\psi(z=0,y=0)=\frac{1}{\pi}\ln 1=0$$

The flow therefore through the inner face of a test panel of large height h per unit distance normal to the section is

$$\lambda\left[\psi\begin{pmatrix}x=0\\y=0\end{pmatrix}-\psi\begin{pmatrix}x=\ \ 0\\y=-h\end{pmatrix}\right]=\lambda\left[\frac{h}{a}+\left\{\frac{2}{\pi\mu}\tan^{-1}\mu-\frac{1}{\pi}\ln\frac{4}{(\mu^2+1)}\right\}\right] \quad (2.12)$$

The term $\lambda h/a$ corresponds to the steady flow of heat between parallel planes, distance a apart, when the flow of heat is normal to both faces. The effect of a right-angled bend, therefore, for the case where $\lambda = \lambda_1 = \lambda_2$ is to increase the heat flow through a vertical section of length h, assuming that h is large, in unit time and per unit length along the corner by an amount

$$q' = \lambda\left\{\frac{2}{\pi\mu}\tan^{-1}\mu - \frac{1}{\pi}\ln\frac{4}{(\mu^2+1)}\right\} \tag{2.13}$$

Substituting the excess flow (2.13) into (2.1) gives

$$\frac{\lambda}{\lambda_{exp}} = \frac{1}{1+\frac{a}{h}\left\{\frac{2}{\pi\mu}\tan^{-1}\mu - \frac{1}{\pi}\ln\frac{4}{(\mu^2+1)}\right\}} \tag{2.14}$$

For the case where $a = b$, corresponding to $\mu = 1$

$$\frac{\lambda}{\lambda_{exp}} = \frac{1}{1+0.279a/h} \tag{2.15}$$

2.3 Right-angled Corners between Partition Wall and Adjoining Test Wall and Roof

Since the partition walls face similar rooms heated identically on both sides, the transmission coefficients are calculated on the assumption that no heat escapes from the rooms through these walls. Actually, a small amount of heat enters the surface of the partition wall and flows through the outside edge of this wall where it is exposed to the weather. This amount is now considered.

Assuming $\lambda = \lambda_1 = \lambda_2$, the heat flow over the internal boundary surfaces can be determined by conformal transformation, as described above. The corner is illustrated in Fig. 8, where BC denotes the centre-plane through the partition wall. It can be shown that the total flow over DEF up to a

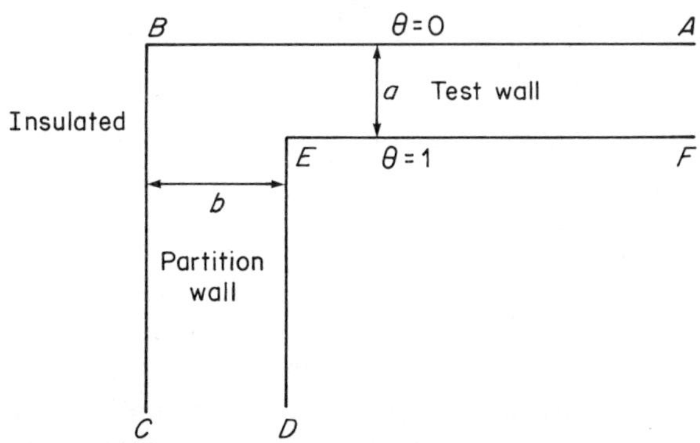

Fig. 8. Corner between test wall and partition wall

Table 2

$\mu = a/b$	$\frac{2}{\mu}\tan^{-1}\mu$	$\ln\frac{(\mu^2+1)}{\mu^2}$	$\cosh^{-1}\frac{(\mu^2+2)}{\mu^2}$
0.5	1.8546	1.6094	2.8873
1.0	1.5708	0.6931	1.7627
2.0	1.1071	0.2231	0.9624
3.0	0.8327	0.1054	0.6549

distance x along EF, where x is large, exceeds $\lambda x/a$ by an amount

$$q' = \lambda \frac{1}{\pi}\left\{\frac{2}{\mu}\tan^{-1}\mu + \ln\frac{(\mu^2+1)}{\mu^2}\right\} \tag{2.16}$$

A similar problem occurs at the junction between the partition wall and the roof. The amount of heat leaking through this corner from the room to outside may be determined by calculating the flow through the surface DE in Fig. 8. This gives

$$q' = \lambda \frac{1}{\pi}\cosh^{-1}\frac{(\mu^2+2)}{\mu^2} \tag{2.17}$$

Values of the terms in (2.16) and (2.17) for an appropriate range of values of μ are given in Table 2.

For the case where $\lambda_1 \neq \lambda_2$ the excess heat flow into the partititon wall over the surface DE may be calculated approximately using the following expression (2.18) obtained by the general analysis illustrated in Fig. 6 for the excess flow into the boundary ($a < x < \infty$, $y = b$)

$$\begin{aligned} q' &= \lambda_1\left[\frac{1}{\mu}\left\{\frac{1}{6} - \frac{1}{\pi^2}\sum_{n=1}^{\infty}\frac{e^{-2n\pi\mu}}{n^2}\right\}\right. \\ &\quad \left. + \sum_{n=1}^{\infty}\frac{A_n}{2\sqrt{n}}\left\{\coth\frac{n\pi}{\mu} - \frac{\mu}{n\pi} - \frac{2\mu n}{\pi}\sum_{m=1}^{\infty}\frac{e^{-2m\pi\mu}}{(m^2\mu^2+n^2)}\right\}\right] \\ &\equiv \lambda_1 G\left(\frac{\lambda_2}{\lambda_1}, \mu\right) \end{aligned} \tag{2.18}$$

The boundary ($0 < x < \infty$, $y = 0$) in Fig. 6 does not correspond exactly with the insulated plane in the partition wall as the temperature of the latter will normally be nearer unity than zero. Equation (2.18) is expected therefore to give too high a value for q' but closer examination indicates that the error is unlikely to be significant and further analysis is considered not to be necessary.

2.4 Computation of Error due to Edge Heat Loss for Particular Cases

A feature of the Wall Laboratory, unique in studies of the thermal properties of structures, was the arrangement whereby test walls were exposed on one side to the weather as in a normal building.

The influence of weather gives rise to boundary conditions that are complex and difficult to interpret analytically in their effect on the measured values of the transmission coefficients, especially if considered over a short time interval, say of a few hours, as required, for example, in the calculation of design loads for air conditioning. These difficulties do not arise with transmission coefficients calculated from mean values of temperature and heat flow measured over the whole winter test period. The mean winter values are assumed to represent the thermal behaviour of the wall in steady state and are used for calculating the space heating requirements of buildings and the sizing of heating plant. The present analysis considers the steady state condition only and assumes (1) that the boundaries are isothermal and (2) that the test walls are sufficiently large (2.4×2.4 m) compared with their thickness (maximum 0.3 m) for the edge heat loss errors to be calculated on the basis of heat flow in a rectangle. The thickness and heat transmission coefficients vary from one test panel to another, corresponding to values of μ ranging approximately from 1 to 3 and the ratio of conductivities, λ_1/λ_2, from 1 to 4 and sometimes higher. As the numerical calculations are lengthy a few cases only have been selected for computation to demonstrate the importance of edge insulation and to give an order of error for pilot scale measurements in the Wall Laboratory.

The amount of heat in unit time entering the inside surface of a square test panel, length of side $2h$ and thickness α, when unit temperature difference exists between opposite faces is

$$\lambda_{\text{exp}}4h^2/\alpha = \lambda 4h^2/\alpha + 2hq'_{\text{w/r}} + 2hq'_{\text{w/f}} + 4hq'_{\text{w/p}} + 4hq'_{\text{p/r}} + 4hq'_{\text{p/f}}$$

where the subscripts w, r, f, and p denote wall, roof, floor, and partition, respectively. For example, $q'_{\text{w/r}}$ signifies the edge heat loss at the corner formed by the wall and roof. It is assumed that the temperature below the floor is equal to the outside temperature.

It is not unreasonable to assume

$$q'_{\text{w/f}} = q'_{\text{w/r}} \quad \text{and} \quad q'_{\text{p/f}} = q'_{\text{p/r}}$$

so that

$$\frac{\lambda}{\lambda_{\text{exp}}} = \left\{1 + \frac{a}{h}\left(\frac{q'_{\text{w/r}}}{\lambda} + \frac{q'_{\text{w/p}}}{\lambda} + \frac{2q'_{\text{p/r}}}{\lambda}\right)\right\}^{-1} \tag{2.19}$$

Case 1. No edge insulation: roof, floor, and partition walls have the same thickness and conductivity as the test wall. The solutions obtained by conformal transformation give the following results

$$\frac{q'_{w/r}}{\lambda}=\left\{\frac{2}{\pi\mu}\tan^{-1}\mu-\frac{1}{\pi}\ln\frac{4}{(\mu^2+1)}\right\}_{\mu=1}=0.279$$

$$\frac{q'_{w/p}}{\lambda}=\left\{\frac{2}{\pi\mu}\tan^{-1}\mu+\frac{1}{\pi}\ln\frac{(\mu^2+1)}{\mu^2}\right\}_{\mu=2}=0.423$$

$$\frac{q'_{p/r}}{\lambda}=\left\{\frac{1}{\pi}\cosh^{-1}\frac{(\mu^2+2)}{\mu^2}\right\}_{\mu=2}=0.306$$

Collecting terms

$$\frac{q'_{w/r}}{\lambda}+\frac{q'_{w/p}}{\lambda}+2\frac{q'_{p/r}}{\lambda}=1.314$$

Putting $\alpha=0.3$ m and $h=1.2$ m yields

$$\frac{\lambda}{\lambda_{exp}}=\frac{1}{1+\frac{1}{4}(1.314)}=0.75$$

Case 2. Test panel insulated along each edge. Let λ_1 be the conductivity of the insulation, thickness b, and λ_2 the conductivity of the test panel, thickness a. For $\lambda_2/\lambda_1=4$, equation (2.4) shows that

$$\frac{q'_{w/r}}{\lambda}=\frac{1}{\lambda_2}\frac{2\lambda_1\lambda_2}{\lambda_1+2\lambda_2}F(4,1)=0.099$$

$$\frac{q'_{p/r}}{\lambda}=\frac{1}{4}\left\{\frac{1}{\pi}\cosh^{-1}\frac{(\mu^2+2)}{\mu^2}\right\}_{\mu=2}=0.306/4$$

Assuming $q'_{w/p}/\lambda=0.099$, then, with $a=0.3$ m and $h=1.2$ m

$$\frac{\lambda}{\lambda_{exp}}=\left[1+\frac{1}{4}\left\{(2\times0.099)+\frac{(2\times0.306)}{4}\right\}\right]^{-1}=0.92$$

This case is not strictly representative of the Wall Laboratory, as will be evident from the assumption concerning $q'_{w/p}$ which implies that the partition wall is exposed to the same external temperature conditions as the roof. This leads to a value that errs on the high side but a comparison of the individual terms in the expression for λ/λ_{exp} indicates that the inaccuracy introduced thereby is relatively small. A new analytical solution would be needed to reduce this small error, involving a considerable amount of further computation which is thought to be unwarranted.

The quantities λ and λ_{exp} represent values of the equivalent conductivity of the structure. Using the above results for λ/λ_{exp}, values may also be calculated for the corresponding ratios of thermal transmittances U/U_{exp}.

2.5 Thermal Transmittance, *U*-Value

Thermal transmittance, or U-value of a structure, is defined as the amount of heat flowing in steady state through unit area of the structure in unit time when the temperature of the air on opposite sides differs by one degree.

Following the expressions leading to equation (2.1) let Q and $Q+\Delta Q$ be the quantities of heat entering the inside surface of the square test panel in unit time, then

$$Q = U4h^2(\theta_i - \theta_o)$$
$$Q+\Delta Q = U_{exp}4h^2(\theta_i - \theta_o)$$

where θ_i and θ_o denote the indoor and outdoor air temperatures, respectively. It follows that

$$\frac{U}{U_{exp}} = 1 - \frac{\Delta Q}{Q+\Delta Q}$$

Now

$$\Delta Q \simeq (2hq'_{w/f} + 2hq'_{w/r} + 4hq'_{p/w} + 4hq'_{p/r} + 4hq'_{p/f})(\theta_{si} - \theta_{so})$$

and

$$Q+\Delta Q = \lambda_{exp}\frac{4h^2}{a}(\theta_{si} - \theta_{so})$$

By substitution

$$\frac{U}{U_{exp}} \simeq 1 - \frac{1}{\lambda_{exp}}\frac{a}{2h}(q'_{w/f} + q'_{w/r} + 2q'_{p/w} + 2q'_{p/r} + 2q'_{p/f}).$$

Case 1. No edge insulation: $\lambda/\lambda_{exp} = 0.75$

$$q'_{p/r} = q'_{p/f} = \frac{\lambda}{\pi}\left\{\cosh^{-1}\frac{(\mu^2+2)}{\mu^2}\right\}_{\mu=2}$$

$$q'_{w/r} = q'_{w/f} = \frac{\lambda}{\pi}\left\{\frac{2}{\mu}\tan^{-1}\mu - \ln\frac{4}{(\mu^2+1)}\right\}_{\mu=1}$$

$$q'_{p/w} = \frac{\lambda}{\pi}\left\{\frac{2}{\mu}\tan^{-1}\mu + \ln\frac{(\mu^2+1)}{\mu^3}\right\}_{\mu=2}$$

Putting $a = 0.3$ m and $h = 1.2$ m it is calculated that

$$U/U_{exp} = 0.74$$

Case 2. Test wall insulated along each edge, $\lambda_2/\lambda_1 = 4$, $\lambda/\lambda_{exp} = 0.92$

$$q'_{p/r} = q'_{p/f} = \frac{\lambda}{4}\frac{1}{\pi}\left\{\cosh^{-1}\frac{(\mu^2+2)}{\mu^2}\right\}_{\mu=2}$$

$$q'_{w/r} = q'_{w/f} = \frac{2(\lambda/4)\lambda}{\lambda/4+2\lambda}F(4, 1) \quad \text{by (2.4)}$$

$$q'_{p/w} = \frac{\lambda}{4}G(4, 2) \qquad \text{by (2.18)}$$

Again, putting $a = 0.3$ m and $h = 1.2$ m it follows that

$$U/U_{exp} = 0.91$$

The experiments in the Wall Laboratory at the Building Research Station described in Chapter 9 necessitated for their interpretation the above detailed examination of heat flow at a corner that would take account of a difference in conductivity as well as thickness between the adjoining sections. The solutions obtained are not rstricted to calculating heat flow through the structure of that particular building; they have a wider application that the reader will readily identify with buildings generally.

REFERENCES

American Society for Testing Materials (1955) ASTM Standard C177-45, *Book of ASTM Standards*, Pt 3.

British Standards Institution (1973) *Methods for determining Thermal Insulating Properties with Definitions of Thermal Insulating Terms*, BS 874:1973.

Carslaw, H. S. and Jaeger, J. C. (1959) *Conduction of Heat in Solids*, Clarendon Press, Oxford.

Somers, E. V. and Cyphers, J. A. (1951) *Rev. Sci. Instrum.*, **22,** 583.,

Woodside W. (1957). *Symposium on Thermal Insulating Materials*, ASTM, Philadelphia, p. 49.

CHAPTER 5

Transient Temperature Change in Buildings

In Chapter 1 expressions were obtained for the temperature in a large homogeneous slab representing a uniform wall or roof exposed to a variable ambient temperature. The analysis will now be extended to a space enclosed by a uniform shell representing a simple building to obtain solutions that relate the temperature of the air within it to the thermal properties of the enclosing and internal structures, and with ventilation. Calculations with these solutions then follow to evaluate the interplay and relative importance of these factors for different building types.

1. The situation considered is illustrated in Fig. 1. A solution is required giving the variation with time of the air temperature inside an enclosure following a sudden drop through θ_o in the outdoor temperature. This and the indoor temperature are assumed to be uniform over the respective surfaces. In the initial steady state condition, and at all subsequent values of time, heat is added to the air inside the enclosure from a source independent of the structure at a constant rate q. Heat is lost from the enclosure by conduction and ventilation at the respective rates H and v per unit difference of temperature between the inside and outside air. Compared with the structure the thermal capacity of the enclosed air is assumed to be negligibly small.

The enclosure contains internal mass of thermal capacity, c, whose temperature at any instant is uniform throughout its volume and initially equal to that of the indoor air. The quantities q, c, and v are defined per unit area of the enclosure wall. For all values of time the surfaces of the internal mass at temperature $\theta_{is}(t)$, and of the enclosure wall at temperature $\theta(l, t)$ (inside surface), and $\theta(0, t)$ (outside surface), exchange heat with the contiguous air at h_s, h_i, and h_o times their respective temperature differences per unit area of the wall and per unit time. The enclosure walls are regarded as large slabs, thickness l, of isotropic material, conductivity k, diffusivity K, and conducting heat in one direction only along the x axis perpendicular to the surfaces.

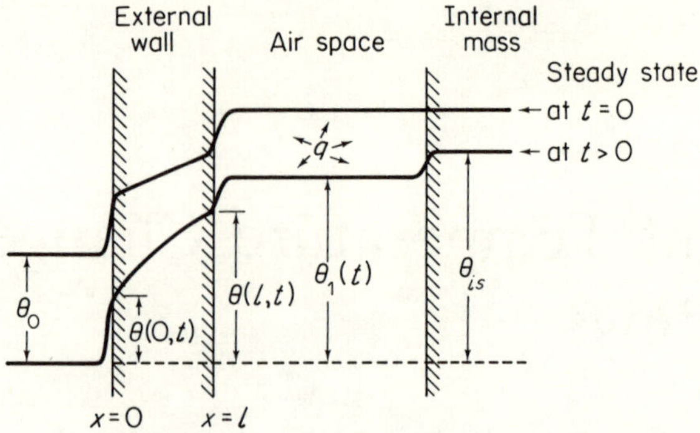

Fig. 1. Schematic representation of temperature change in a heated enclosure

The transfer of heat within this arrangement is represented by the following set of equations which will be solved to obtain an expression for $\theta_i(t)$, denoting the variation of the indoor temperature with time

$$\kappa \frac{\partial^2 \theta(x, t)}{\partial x^2} = \frac{\partial \theta(x, t)}{\partial t}, \qquad 0<x<l, \quad t>0 \tag{1.1}$$

$$k \frac{\partial \theta(x, t)}{\partial x} = h_o \theta(x, t), \qquad x=0, \quad t>0 \tag{1.2}$$

$$k \frac{\partial \theta(x, t)}{\partial x} = h_i\{\theta_i(t) - \theta(x, t)\}, \qquad x=l, \quad t>0 \tag{1.3}$$

$$q = h_i\{\theta_i(t) - \theta(x, t)\} + v\theta_i(t) - h_s\{\theta_{is}(t) - \theta_i(t)\}, \qquad x=l, \quad t>0 \tag{1.4}$$

$$c \frac{d\theta_{is}(t)}{dt} = -h_s\{\theta_{is}(t) - \theta_i(t)\}, \qquad t>0 \tag{1.5}$$

Initially the system is in steady state so that

$$\theta(x, 0) = \theta_o + \frac{(1/h_o + x/k)(\theta_i - \theta_o)}{1/h_o + l/k + 1/h_i}$$

and

$$\theta_i - \theta_o = \frac{q}{v + (1/h_o + l/k + 1/h_i)^{-1}}$$

By substitution

$$\theta(x, 0) = \theta_o + \frac{(1/h_o + x/k)}{1/h_o + l/k + 1/h_i} \times \frac{q}{v + (1/h_o + l/k + 1/h_i)^{-1}}$$

$$= \theta_o + \frac{(1/h_o + x/k)q}{1 + v(1/h_o + l/k + 1/h_i)}, \qquad 0<x<l, \quad t=0 \tag{1.6}$$

Also

$$\theta_{is} = \theta_i, \qquad t = 0$$

Making the substitutions

$$\xi = x/l; \qquad \tau = \kappa t/l^2; \qquad B_i = h_i l/k; \qquad B_o = h_o l/k$$
$$B_s = h_s l/k; \quad Q = ql/k; \qquad C = c\kappa/lk; \qquad V = vl/k$$

equations (1.1)–(1.6) are written more conveniently as

$$\frac{\partial^2\theta(\xi,\tau)}{\partial\xi^2} = \frac{\partial\theta(\xi,\tau)}{\partial\tau}, \qquad 0<\xi<1, \quad \tau>0 \tag{1.7}$$

$$\frac{\partial\theta(\xi,\tau)}{\partial\xi} = B_o\theta(\xi,\tau), \qquad \xi=0, \quad \tau>0 \tag{1.8}$$

$$\frac{\partial\theta(\xi,\tau)}{\partial\xi} = B_i\{\theta_i(\tau)-\theta(\xi,\tau)\}, \qquad \xi=1, \quad \tau>0 \tag{1.9}$$

$$Q = B_i\{\theta_i(\tau)-\theta(\xi,\tau)\} + V\theta_i(\tau) - B_s\{\theta_{is}(\tau)-\theta_i(\tau)\}, \qquad \xi=1, \quad \tau>0 \tag{1.10}$$

$$C\frac{d\theta_{is}(\tau)}{d\tau} = -B_s\{\theta_{is}(\tau)-\theta_i(\tau)\}, \qquad \tau>0 \tag{1.11}$$

$$\theta(\xi,0) = \theta_o + \frac{Q(\xi+1/B_o)}{1+VA} \tag{1.12}$$

where $A = 1+1/B_i+1/B_o$.

The equations (1.7)–(1.11), expressed with dimensionless quantities, are solved by a routine application of the Laplace transform. Multiplying throughout by $e^{-p\tau}$ and integrating with respect to τ between the limits 0 and ∞ gives the following

$$\frac{d^2\bar{\theta}(\xi,p)}{d\xi^2} - p\bar{\theta}(\xi,p) + \frac{Q(\xi+1/B_o)}{1+VA} + \theta_o = 0, \qquad 0<\xi<1 \quad \tau>0 \tag{1.13}$$

$$\frac{d\bar{\theta}(\xi,p)}{d\xi} = B_o\bar{\theta}(\xi,p), \qquad \xi=0, \quad \tau>0 \tag{1.14}$$

$$\frac{d\bar{\theta}(\xi,p)}{d\xi} = B_i\{\bar{\theta}_i(p)-\bar{\theta}(\xi,p)\}, \qquad \xi=1, \quad \tau>0 \tag{1.15}$$

$$\frac{Q}{p} = B_i\{\bar{\theta}_i(p)-\bar{\theta}(\xi,p)\} + V\bar{\theta}_i(p) - B_s\{\bar{\theta}_{is}(p)-\bar{\theta}_i(p)\}, \qquad \xi=1, \quad \tau>0 \tag{1.16}$$

$$Cp\bar{\theta}_{is}(p) = C\left(\theta_o + \frac{QA}{1+VA}\right) - B_s\{\bar{\theta}_{is}(p)-\bar{\theta}_i(p)\}, \qquad \tau>0 \tag{1.17}$$

where

$$\bar{\theta}(\xi, p) = \int_0^\infty e^{-p\tau}\theta(\xi, \tau)\,d\tau$$

The solution of (1.13) satisfying (1.14)–(1.17) is

$$\begin{aligned}\bar{\theta}(\xi, p) = {} & \frac{1}{p}\left[\theta_o + Q\frac{(\xi + 1/B_o)}{1+VA}\right] - \frac{1}{\sqrt{p}}\frac{\theta_o}{\phi(p)} \\ & \times\left(\frac{VB_iB_o}{p}(Cp+B_s) - \frac{B_o}{p}\{B_i(Cp+B_s)\sqrt{p}\sinh\sqrt{p}\right. \\ & \left. + [V(Cp+B_s)+CpB_s][B_i\cosh\sqrt{p} + \sqrt{p}\sinh\sqrt{p}]\}\right) \\ & \times\sinh\sqrt{p}\xi) - \frac{1}{\sqrt{p}}\frac{\theta_o}{\phi(p)}\left(\frac{VB_i}{\sqrt{p}}(Cp+B_s) + \frac{B_o}{p}\right. \\ & \times\{B_i(Cp+B_s)\sqrt{p}\sinh\sqrt{p} + [V(Cp+B_s)+CpB_s] \\ & \left.\times[B_i\sinh\sqrt{p} + \sqrt{p}\cosh\sqrt{p}]\}\right)\cosh\sqrt{p}\xi \end{aligned} \tag{1.18}$$

where

$$\begin{aligned}\phi(p) = {} & [V(Cp+B_s)(B_i+B_o) + CpB_s(B_i+B_o) + B_iB_o(Cp+B_s)] \\ & \times\cosh\sqrt{p} + [(Cp+B_s) + B_sCp + CB_iB_oB_s]\sinh\sqrt{p} \\ & + V(Cp+B_s)(p - B_iB_o)\sinh\sqrt{p}/\sqrt{p}\end{aligned} \tag{1.19}$$

Equation (1.18) may be inverted to give the temperature distribution $\theta(\xi, \tau)$ within the enclosure wall: attention however will be confined to calculating a solution for the inside air temperature $\theta_i(\tau)$. From (1.17)

$$\bar{\theta}_{is}(p) = \frac{1}{(Cp+B_s)}\left\{C\left(\theta_o + \frac{QA}{1+VA}\right) + B_s\bar{\theta}_i(p)\right\} \tag{1.20}$$

Substituting the expression (1.20) for $\bar{\theta}_{is}(p)$ into (1.16) gives

$$\begin{aligned}\bar{\theta}_i(p) = {} & \frac{1}{\left(B_i + V + \dfrac{B_sCp}{Cp+B_s}\right)}\left[\frac{Q}{p} + \frac{B_sC}{Cp+B_s}\left(\theta_o + \frac{QA}{1+VA}\right)\right] \\ & + \frac{B_i\bar{\theta}(\xi, p)}{\left(B_i + V + \dfrac{B_sCp}{Cp+B_s}\right)}, \qquad (\xi = 1)\end{aligned} \tag{1.21}$$

Substituting for $\{\bar{\theta}(\xi, p)\}_{\xi=1}$ from (1.18) into (1.21) leads, after rearrangement, to

$$\begin{aligned}\bar{\theta}_i(p) = {} & \frac{QA}{p(1+VA)} + \frac{\theta_o}{p}\left[\frac{B_i + CpB_s/(Cp+B_s)}{B_i + V + CpB_s/(Cp+B_s)}\right] - \frac{B_iB_o\theta}{p\phi(p)}(Cp+B_s) \\ & - \frac{B_iV\theta_o(Cp+B_s)}{p\phi(p)\left(B_i + V + \dfrac{B_sCp}{Cp+B_s}\right)}\left[B_i\cosh\sqrt{(p)} + \frac{B_iB_o}{\sqrt{(p)}}\sinh\sqrt{(p)}\right]\end{aligned} \tag{1.22}$$

The air temperature $\theta_i(\tau)$ is found from its Laplace transform (1.22) by use of the inversion theorem and contour integration. Thus

$$\theta_i(\tau)=\frac{1}{2\pi i}\int_{\gamma-i\infty}^{\gamma+i\infty} e^{p\tau}\bar{\theta}_i(p)\,dp, \qquad \gamma>0$$

The expression for $\bar{\theta}_i(p)$ is an analytic function of p with a simple pole at the origin of the p-plane and the other singularities simple poles at points located on the negative real axis. Carrying out the integration indicated, the solution for the air temperature inside the enclosure is

$$\begin{aligned}\theta_i(\tau)=&\frac{QA}{1+VA}+2B_iB_o\theta_o\sum_{n=1}^{\infty}\frac{(B_s-C\alpha_n^2)}{\alpha_n^2F(\alpha_n)}\exp(-\alpha_n^2\tau)\\ &+V\left\{\frac{B_s\theta_o e^{-\beta\tau}}{(B_i+B_s+V)(B_i+V)}+B_iB_oB_s^3\theta_o\right.\\ &\times\left.\frac{[(B_i/B_o)\cos\sqrt{\beta}+(B_i/\sqrt{\beta})\sin\sqrt{\beta}]e^{-\beta\tau}}{[\phi(p)]_{p=-\beta}(B_i+B_s+V)^2(B_i+V)}\right\}-2VB_iB_o\theta_o\\ &\times\sum_{n=1}^{\infty}\frac{(B_s-C\alpha_n^2)^2[(B_i/B_o)\cos\alpha_n+(B_i/\alpha_n)\sin\alpha_n]\exp(-\alpha_n^2\tau)}{\alpha_n^2F(\alpha_n)(B_i+B_s+V)C(\alpha_n^2-\beta)}\end{aligned} \tag{1.23}$$

where

$$\beta=\frac{B_s(B_i+V)}{C(B_i+B_s+V)}$$

$$\begin{aligned}F(\alpha_n)=2\left[\frac{d\phi(p)}{dp}\right]_{p=-\alpha_n^2}&=[B_sC(2B_i+2B_o+B_iB_o)+B_i(B_s+2CB_o)\\ &-C(B_i+B_s)\alpha_n^2]\cos\alpha_n+\{B_iB_s[1+B_o(1+C)]-[B_s(B_i+B_o+3)\\ &+B_i(B_o+3)]C\alpha_n^2\}(\sin\alpha_n)/a_n+V\{[B_s+C(2B_i+2B_o+B_iB_o)\\ &-C\alpha_n^2-B_iB_oB_s/\alpha_n^2]\cos\alpha_n+[CB_iB_o+B_s(1+B_i+B_o)\\ &-C(B_i+B_o+3)\alpha_n^2+B_iB_oB_s/a_n^2](\sin\alpha_n)/\alpha_n\}\end{aligned}$$

and the summation is taken over the positive roots α_n, $n=1, 2, 3$, etc. of

$$\alpha\tan\alpha=\frac{C\alpha^4[(V+B_s)(B_i+B_o)+B_iB_o]-B_s\alpha^2[V(B_i+B_o)+B_iB_o]}{C\alpha^4(B_i+B_s+V)-\alpha^2[CB_iB_o(V+B_s)+B_s(V+B_i)]+VB_iB_oB_s} \tag{1.24}$$

Evaluation of the complete solution (1.23) is tedious largely because of the labour involved in calculating the zeros of $\phi(p)$ using (1.24). Except for small values of time, however, roots higher than the second do not contribute significantly to the computed result and may be neglected for numerical purposes.

Approximate values of the zeros α_1^2 and α_2^2 may be calculated more readily by expressing $\phi(p)$ as a polynomial in p; thus

$$\begin{aligned}\phi(p) = {} & B_s[B_iB_o + V(B_i+B_o+B_iB_o)]\\ & + \left[B_s\left(B_i+\frac{B_iB_o}{2!}\right)+C[B_s(B_i+B_o+B_iB_o)+B_iB_o]\right.\\ & \left. + CV(B_i+B_o+B_iB_o)+VB_s\left(1+\frac{B_i+B_o}{2!}+\frac{B_iB_o}{3!}\right)\right]p\\ & + \left\{B_s\left(\frac{B_i}{3!}+\frac{B_iB_o}{4!}\right)+C\left[B_s\left(1+\frac{B_i+B_o}{2!}+\frac{B_iB_o}{3!}\right)+B_i\left(1+\frac{B_o}{2!}\right)\right]\right.\\ & \left. + CV\left(1+\frac{B_i+B_o}{2!}+\frac{B_iB_o}{3!}\right)+VB_s\left(\frac{1}{3!}+\frac{B_i+B_o}{4!}+\frac{B_iB_o}{5!}\right)\right\}p^2+\cdots \end{aligned} \tag{1.25}$$

The polynomial (1.25) has zeros at $p=-\alpha_1^2, -\alpha_2^2, \ldots$ and may be expressed as $\phi(p)=a(1+p/\alpha_1^2)(1+p/\alpha_2^2)\ldots$, where a is a constant which, by putting $p=0$ in (1.25), is recognized as $(1+VA)B_iB_oB_s$. Writing therefore

$$\phi(p)=(1+VA)B_iB_oB_s\prod_{r=1}^{\infty}(1+p/\alpha_r^2)$$

the calculation of the residues now takes the form

$$L^{-1}\left[\frac{1}{\phi(p)}\right]=\sum_{r=1}^{\infty}\frac{\exp -(\alpha_r^2\tau)}{(1+VA)B_iB_oB_s\prod_{s=1}^{\infty}{}'(1+p/\alpha_s^2)_{p=-\alpha_r^2}} \tag{1.26}$$

where L^{-1} denotes the Laplace inversion and the prime in $\prod'$ denotes that the term given by $s=r$ is omitted from the infinite product.

As indicated above, residues contributed at $r=1, 2$ only need be calculated in the inversion integrand. Assuming therefore that the zeros of $\phi(p)$ may be determined approximately from (1.25) as the roots of a quadratic in p, and writing these as $[\alpha_1^2]$ and $[\alpha_2^2]$, the transform solution for $\bar{\theta}_i(p)$ may be inverted to give

$$\begin{aligned}\theta_i(\tau) \simeq {} & \frac{QA}{1+VA}+\frac{\theta_o(B_s-C[\alpha_1^2])[\alpha_2^2]}{(1+VA)B_s([\alpha_2^2]-[\alpha_1^2])}\\ & \times\left\{1+\frac{B_iV(B_s-C[\alpha_1^2])(\cos[\alpha_1]+(B_o/[\alpha_1])\sin[\alpha_1])}{B_oC(B_i+B_s+V)(\beta-[\alpha_1^2])}\right\}\\ & \times\exp(-[\alpha_1^2]\tau)+\frac{\theta_o(B_s-C[\alpha_2^2])[\alpha_1^2]}{(1+VA)B_s([\alpha_1^2]-[\alpha_2^2])}\\ & \times\left\{1+\frac{B_iV(B_s-C[\alpha_2^2])(\cos[\alpha_2]+(B_o/[\alpha_2])\sin[\alpha_2])}{B_oC(B_i+B_s+V)(\beta-[\alpha_2^2])}\right\}\\ & \times\exp(-[\alpha_2^2]\tau)+\frac{\theta_oVB_s^2}{C\beta(B_i+B_s+V)^2}\\ & \times\left\{1+\frac{B_i(B_s-C\beta)^2(B_i+B_s+V)[\cos\sqrt{(\beta)}+(B_o/\sqrt{\beta})\sin\sqrt{\beta}]}{B_oB_s^3(1+VA)(1-\beta/[\alpha_1^2])(1-\beta/[\alpha_2^2])}\right\}\\ & \times\exp(-\beta\tau)\end{aligned} \tag{1.27}$$

In either form of solution, (1.23) or (1.27), the steady state term $QA/(1+VA)$ is equivalent to $q(H+v)$, where H and v denote, respectively, the initial conduction and ventilation rates of heat loss from the enclosure per unit difference between the inside and outside temperatures. The first term or constant therefore represents the initial inside temperature excess over the outside temperature. Writing this as $\{\theta_i(0)-\theta_o\}$ and substituting into (1.27), the approximate solution passes into the more convenient form

$$\begin{aligned}\frac{\theta_i(0)-\theta_i(\tau)}{\theta_o} \simeq 1 &- \frac{(B_s-C[\alpha_1^2])[\alpha_2^2]}{B_s(1+VA)([\alpha_2^2]-[\alpha_1^2])}\\ &\times\left\{1+\frac{VB_i(B_s-C[\alpha_1^2])(\cos[\alpha_1]+(B_o/[\alpha_1])\sin[\alpha_1])}{CB_o(B_i+B_s+V)(\beta-[\alpha_1^2])}\right\}\\ &\times\exp(-[\alpha_1^2]\tau)+\frac{(B_s-C[\alpha_2^2])[\alpha_1^2]}{B_s(1+VA)([\alpha_2^2]-[\alpha_1^2])}\\ &\times\left\{1+\frac{VB_i(B_s-C[\alpha_2^2])(\cos[\alpha_2]+(B_o/[\alpha_2])\sin[\alpha_2])}{CB_o(B_i+B_s+V)(\beta-[\alpha_2^2])}\right\}\\ &\times\exp(-[\alpha_2^2]\tau)-\frac{VB_s^2}{C\beta(B_i+B_s+V)^2}\\ &\times\left\{1+\frac{B_i(B_s-C\beta)^2(B_i+B_s+V)[\cos\sqrt{\beta}+(B_o/\sqrt{\beta})\sin\sqrt{\beta}]}{B_oB_s^3(1+VA)(1-\beta/[\alpha_1^2])(1-\beta/[\alpha_2^2])}\right\}\\ &\times\exp(-\beta\tau) \qquad (1.28)\end{aligned}$$

Particular cases follow by taking limiting values of the parameters; for example, putting $V=0$ to correspond with an unventilated enclosure having internal heat storage, (1.28) reduces to

$$\begin{aligned}\frac{\theta_i(0)-\theta_i(\tau)}{\theta_o}=1&-\frac{(B_s-C[\alpha_1^2])[\alpha_2^2]\exp(-[\alpha_1^2]\tau)}{B_s([\alpha_2^2]-[\alpha_1^2])}\\ &+\frac{(B_s-C[\alpha_2^2])[\alpha_1^2]\exp(-[\alpha_2^2]\tau)}{B_s([\alpha_2^2]-[\alpha_1^2])} \qquad (1.29)\end{aligned}$$

The simplest solution applies to the case $C=0$, $V=0$, when

$$\frac{\theta_i(0)-\theta_i(\tau)}{\theta_o}=1-\frac{[\alpha_2^2]}{[\alpha_2^2]-[\alpha_1^2]}\exp(-[\alpha_1^2]\tau)+\frac{[\alpha_1^2]}{[\alpha_2^2]-[\alpha_1^2]}\exp(-[\alpha_2^2]\tau) \qquad (1.30)$$

Further examination shows $[\alpha_1^2]$ to be a satisfactory approximation to α_1^2 but less so for the second root.

The results of numerical calculation given below indicate that in general $[\alpha_2^2]\gg[\alpha_1^2]$. In the region where $|p|$ is small, therefore, only the first root is required and the solution, for example, for the fractional temperature change inside a sealed enclosure with no internal heat storage approximates satisfactorily to the simple decay expression of familiar form

$$\frac{\theta_i(0)-\theta_i(\tau)}{\theta_o}=1-\exp(-[\alpha_1^2]\tau) \qquad (1.31)$$

Assuming further that $[\alpha_1^2]$ in (1.31) may be calculated by writing $\phi(p)$ as a linear function in p it follows that with $C=0$ and $V=0$

$$[\alpha_1^2] \simeq \frac{1}{1/B_o + \frac{1}{2}} \tag{1.32}$$

The approximate solution (1.31) will be recognized as defining the exponential decay in the air temperature that follows from representing the cooling process simply as

$$\frac{d\theta_i(\tau)}{dt} + [\alpha_1^2]\theta_i(\tau) = 0 \tag{1.33}$$

From (1.31) when $\tau = 1/[\alpha_1^2]$, $\{\theta_i(0) - \theta_i(\tau)\}/\theta_o = 1 - e^{-1}$ showing that $(1/[\alpha_1^2])$ is the value of τ for which the air temperature inside the unventilated enclosure has fallen by about 63 percent of the temperature change applied externally. By analogy with electric current flow in a capacitative circuit consisting of a condenser with capacitance C, discharging through a series resistance R, $l^2/K[\alpha_1^2]$ may be identified as the thermal time constant of the cooling system represented by (1.33). In electric current flow the time constant is well known as the product RC: a similar analogous expression may be shown to apply in the present approximate calculation of heat flow. Relative to θ_o as base temperature the initial heat content per unit area of the wall of thickness l and volumetric specific heat ρs is

$$W = \rho s l \left\{ \frac{\theta(l, 0) + \theta(0, 0)}{2} - \theta_o \right\} \tag{1.34}$$

Now

$$\theta(l, 0) = \theta_o + (\theta_i - \theta_o) \frac{(1/h_o + l/k)}{1/h_i + 1/h_o + l/k}$$

$$\theta(0, 0) = \theta_o + (\theta_i - \theta_o) \frac{1/h_o}{1/h_i + 1/h_o + l/k}$$

so that

$$\frac{\theta(l, 0) + \theta(0, 0)}{2} - \theta_o = (\theta_i - \theta_o) \frac{(1/h_o + l/k)}{1/h_i + 1/h_o + l/k}$$

Noting also that

$$q = \frac{\theta_i - \theta_o}{1/h_i + 1/h_o + l/k} \tag{1.35}$$

representing heat transmitted through the enclosure walls in steady state, then by substitution in (1.34)

$$W = \rho s l q (1/h_o + l/2k)$$

or

$$\frac{W}{q}=\frac{\rho s l^2}{k}(1/B_o+\tfrac{1}{2})$$

which from (1.32) is

$$\frac{W}{q}=\frac{l^2}{K}\cdot\frac{1}{[\alpha_1^2]} \tag{1.36}$$

Hence, from (1.31)

$$\frac{\theta_i(0)-\theta_i(\tau)}{\theta_o}=1-\exp\left\{\frac{-[\alpha_1^2]Kt}{l^2}\right\}$$

$$=1-\exp\left\{-t\Big/\left(\frac{W}{q}\right)\right\} \tag{1.37}$$

The expression (1.37) identifies the time constant as the ratio (heat stored/heat transmitted) in the steady state. If the enclosure is unventilated and without internal thermal capacity the quantity *heat transmitted* in the above ratio is the thermal transmittance or U-value of the enclosing structure.

The cooling curves of six enclosures of different construction have been calculated to illustrate the application of the above solutions. A ventilation rate of two air changes per hour has been assumed. By repeating the calculations for zero ventilation the transient response of the enclosure structure alone is evaluated. The cases considered cover a wide range of building types and are described briefly in Table 1. The constructional details and heat transfer data in these examples are reproduced from an examination made by Jamieson (1954) of the effect of structural thermal capacity on the assessment of basic design temperatures for space heating installations in buildings.

The structure is described as heavy or light according to the weight per unit area of the external walls. Three different pairs of building types are considered: 1 and 2, 3 and 4, 5 and 6. Cases 1 and 2, 3 and 4 differ only in the weight of the external wall; in all other respects they are identical. The space enclosed in cases 1–4 is four times that for cases 5 and 6; the ratio (volume/wall area) is 10.5 for cases 1–4 and 5.5 for cases 5 and 6. The examples have been selected and grouped to demonstrate within each pair the influence of the thermal capacity of the external wall.

In all cases the calculations assume the following values

$$h_i=1.4\ \text{Btu/ft}^2\text{h degF}$$

$$h_o=3.3\ \text{Btu/ft}^2\text{h degF}$$

$$v=0.02\ \text{Btu/ft}^3\text{h degF}$$

$$\theta_i(0)=65\ \text{degF}$$

$$\theta_o=30\ \text{degF}$$

Table 1

Case	Building type
1	Office building: heavy construction. Four storeys with internal partitions. Ventilation two air changes per hour.
1a	Building as case 1; no ventilation.
2	Factory building: heavy construction. Same dimensions as case 1 and constructed in the same way, except that there are no internal floors or internal partitions. Ventilation two air changes per hour.
2a	Building as case 2; no ventilation.
3	Office building: light construction. Same dimensions as case 1 though walls and roof are of lighter construction. Intermediate floors and partitions as case 1. Ventilation two air changes per hour.
3a	Building as case 3; no ventilation.
4	Factory building: light construction. Same dimensions as case 1 and constructed in the same way as case 3 though without internal partitions and intermediate floors. Ventilation two air changes per hour.
4a	Building as case 4; no ventilation.
5	Office building: heavy construction. Building has half the height and half the width of case 1, with more widely spaced partitions, but length is the same and construction similar. Ventilation two air changes per hour.
5a	Building as case 5; no ventilation.
6	Factory building: heavy construction. Same dimensions as case 5, and constructed in the same way though without internal partitions and intermediate floors. Ventilation two air changes per hour.
6a	Building as case 6; no ventilation.

and a ground temperature of 50 degF for calculating heat loss through the floor.

The calculations have been made with the variables converted to dimensionless quantities using the multiplying factors l/k and l^2/K. The factor l/k is obtained from

$$\bar{q} = (\theta_i - \theta_o)/(1/h_i + 1/h_o + l/k)$$

where $\bar{q}$ is the rate of total heat loss by conduction from the building. This expression defines l/k as an equivalent thermal resistance for the whole external structure, i.e. walls, roof, and ground floor.

The method for obtaining l^2/K is less direct. Let W be the initial heat content of the complete structure above θ_o per unit area of the external structure surface, then neglecting the relatively small heat content of the air

$$W = c\theta_i(0) + \frac{\rho s l}{2}\{\theta(0, 0) + \theta(l, 0)\} \tag{1.39}$$

Now

$$\theta_i(0) = q/(U + v)$$
$$\theta(0, 0) = \bar{q}/h_o$$
$$\theta(l, 0) = \bar{q}(l/k + 1/h_o)$$

and

$$\bar{q} = qU/(U+v)$$

By substitution in (1.39)

$$W = q\left\{\frac{c}{U} + \frac{\rho s l}{2}\left(\frac{2}{h_o} + \frac{l}{k}\right)\right\}\frac{U}{U+v}$$

and putting

$$\frac{1}{U} = \frac{l}{k}\left(1 + \frac{1}{B_i} + \frac{1}{B_o}\right)$$

gives

$$\frac{W}{q} = \left\{\frac{cl}{k}\left(1 + \frac{1}{B_i} + \frac{1}{B_o}\right) + \frac{l^2}{2K}\left(1 + \frac{2}{B_o}\right)\right\}\frac{1}{1 + v/U}$$

which provides an expression for calculating the factor l^2/K. It may be simplified slightly by putting $v = 0$ when $\bar{q}$, the conduction heat loss, is substituted for the total heat loss to give

$$\frac{W}{\bar{q}} = \frac{cl}{k}\left(1 + \frac{1}{B_i} + \frac{1}{B_o}\right) + \frac{l^2}{2K}\left(1 + \frac{2}{B_o}\right) \quad (1.40)$$

Preparation of the data given in Table 2 for calculating the cooling curves is shown in detail later in the Chapter.

In discussing the results of the calculations it is helpful to distinguish between the various cases by referring to Table 3.

Table 2 Data used in numerical examples

Case	k/l	K/l^2	c	v	h_i	h_o	h_s
1	0.750	0.053	19.2	0.398	1.43	3.33	1.43
1a	0.750	0.053	19.2	—	1.43	3.33	1.43
2	0.750	0.053	—	0.398	1.43	3.33	—
2a	0.750	0.053	—	—	1.43	3.33	—
3	0.750	0.367	19.2	0.398	1.43	3.33	1.43
3a	0.750	0.367	19.2	—	1.43	3.33	1.43
4	0.750	0.367	—	0.398	1.43	3.33	—
4a	0.750	0.367	—	—	1.43	3.33	—
5	0.723	0.053	7.34	0.210	1.43	3.33	1.43
5a	0.723	0.053	7.34	—	1.43	3.33	1.43
6	0.723	0.053	—	0.210	1.43	3.33	—
6a	0.723	0.053	—	—	1.43	3.33	—

Notes:

(i) The value selected for h_s implies that the surface areas of the enclosing wall and the enclosed mass are equal.

(ii) The values selected for v, being based on a ventilation rate of two air changes per hour, imply that the ratio of the enclosed volume to the area of the enclosing wall takes the following values:

Cases 1, 2, 3, 4 = 10.48,

Cases 5, 6 = 5.53.

Table 3 Reference code of numerical examples

Structure type	Ventilated $V=0$	Not ventilated $V\neq 0$
Heavy, $C\neq 0$	1	1a
Heavy, $C=0$	2	2a
Light, $C\neq 0$	3	3a
Light, $C=0$	4	4a
Heavy, $C\neq 0$	5	5a
Heavy, $C=0$	6	6a

The curves for cases shown in Figs. 2 and 3 have been calculated using the exact form of solution and are plotted to show the change with time in the inside air temperature expressed as a fraction of the causal drop in the ambient outside temperature. In Fig. 2 the curves refer to enclosures ventilated at two air changes per hour; the corresponding curves for zero ventilation are shown in Fig. 3. A horizontal line drawn through the vertical scale at $\{\theta_i(0)-\theta_i(t)\}/\theta_o=1-e^{-1}$, or 0.63 approximately, intersects each curve at the corresponding value of the thermal time constant, T, which is read off along the time scale. A summary of the time constants obtained from the curves is given in Table 4.

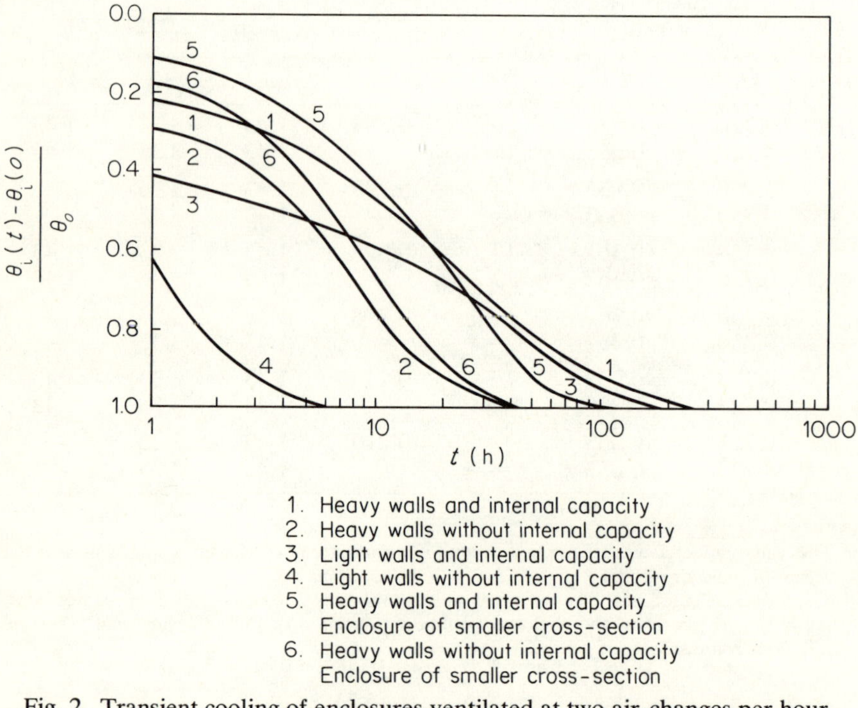

Fig. 2. Transient cooling of enclosures ventilated at two air-changes per hour

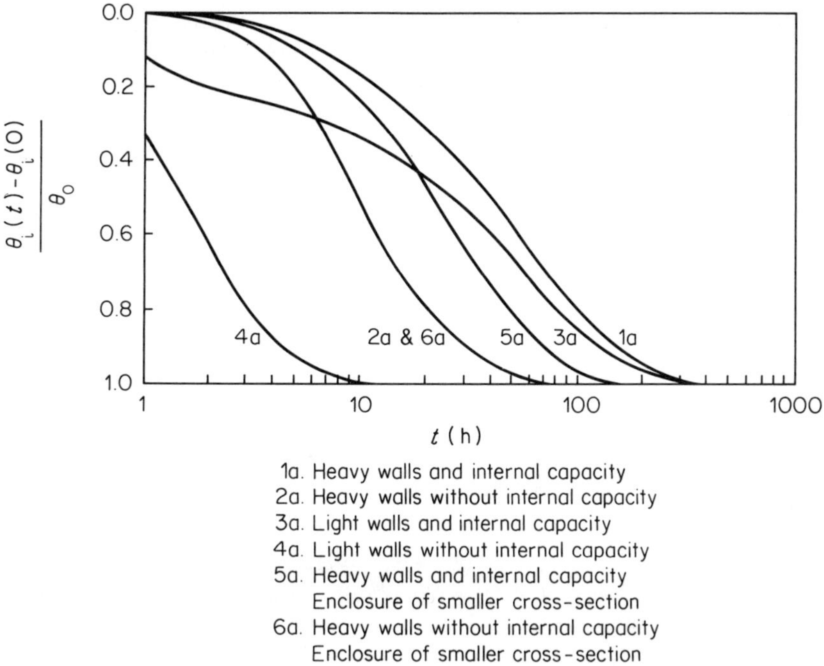

Fig. 3. Transient cooling of unventilated enclosures

An indication of the influence of ventilation on the rate at which an enclosure cools can be obtained from a comparison of the corresponding curves in Figs. 2 and 3 or of the respective values of the thermal time constant. In cases 1–4 the time constant of the sealed enclosure is roughly two to three times that of an identical one ventilated at two air changes per hour. For a smaller structure, represented by cases 5 and 6, the corresponding increase is less, being about 50 percent for the same air change rate. The

Table 4 Summary of values of time constant, T, for all cases determined by the exact solution

	Time constant (h)			
	Ventilated		Not ventilated	
Structure	Case	$V \neq 0$ T(h)	Case	$V = 0$ T(h)
Heavy, $C \neq 0$	1	22.5	1a	57
Heavy, $C = 0$	2	7.2	2a	13.7
Light, $C \neq 0$	3	14.5	3a	44.3
Light, $C = 0$	4	1.0	4a	2.0
Heavy, $C \neq 0$	5	20.0	5a	30.5
Heavy, $C = 0$	6	9.1	6a	13.7

curves in Fig. 2 indicate an instantaneous drop in the inside air temperature. In these cases the applied step-function drop in the outside air temperature is transmitted directly by the ventilating air to the inside air, and θ_i responds accordingly in a manner influenced by the thermal capacity of the structure. In those cases with $C \neq 0$, the heat stored internally transfers to the inside air as its temperature begins to fall, thereby helping to offset the cooling influence of the cooler outside air as it enters the heated enclosure. In all cases the effect of the internal mass is to increase the time constant; with the values chosen for the illustrative examples, T is increased by several times the corresponding value for the enclosure when empty.

The curves demonstrate the comparative influence of the thermal capacities of the external structure and internal mass. The air enclosed behind a lightweight wall cools more quickly than in a similar enclosure with a heavy cladding, as indicated by the appropriate comparision of values of T in Table 4. For the cases considered this effect, though large, is less than that of the internal heat storage in its influence on the cooling rate of the inside air.

The effect of internal heat storage will become even greater as the value of h_s increases. The present numerical results assume $h_s = h_i$, and that the surface area of the internal mass is the same as that of the enclosure wall; if it were possible to arrange circumstances so that the value of $1/h_s$ were negligibly small, which is equivalent to writing $\theta_i(t) = \theta_{is}(t)$ for all t, the cooling interval from initial time to T would be lengthened reaching maximum values of 24 and 46 h in the lightweight structures, cases 3 and 3a, respectively. With the exception of case 1, which gives an increase in T of about 40 percent, the heavily-clad structures are found to be little affected by such an extreme change in the value of the interior surface transfer coefficient, h_s.

The temperature of the internal mass may be of interest. Following the statement in the formulation of the problem above that the internal mass is at a uniform temperature θ_{is} throughout its bulk, an expression for the time variation of θ_{is} is obtained from (1.17) by convolution, giving

$$\theta_{is}(\tau) = \theta_{is}(0) \exp(-[B_s/C]\tau) + \frac{B_s}{C}\int_0^{\tau} \exp(-[B_s/C][\tau - \tau'])\theta_i(\tau')\,d\tau'$$

The exact solution, equation (1.23), for $\theta_i(\tau)$ is lengthy and the calculation of the thermal time constant T using this form of expression involves considerable numerical work. For certain of the cases considered simple expressions for the time variation of θ_i follow from using the first and second roots calculated by the approximate method indicated by (1.25). Equation (1.29), for example, is an approximation solution for the cases 1a, 3a, and 5a ($C \neq 0$, $V = 0$) and the results in Table 5 demonstrate the efficiency of the method. For relatively small values of time the error is large, but it diminishes thereafter and is negligible when T is reached. Similar results are obtained with the approximate solution (1.31) applied to cases 2a, 4a, and

Table 5 Comparison of results calculated by exact and approximate* methods: cases 1a, 3a, and 5a ($C \neq 0$, $V = 0$)

Time (h)	$\dfrac{(B_s - C[\alpha_1^2])[\alpha_2^2]\exp(-[\alpha_1^2]\tau)}{B_s([\alpha_2^2]-[\alpha_1^2])}$	$\dfrac{-(B_s - C[\alpha_2^2])[\alpha_1^2]\exp(-[\alpha_2^2]\tau)}{B_s([\alpha_2^2]-[\alpha_1^2])}$	$[\theta_i(0)-\theta_i(\tau)]/\theta_0$; Approx.	Exact
Case 1a				
10	0.7753	0.0278	0.1969	0.1666
20	0.6622	0.0084	0.3294	0.3067
40	0.4831	0.0008	0.5161	0.4971
100	0.1875	—	0.8125	0.8041
200	0.0387	—	0.9613	0.9592
Case 3a				
1	0.7756	0.0943	0.1301	0.1219
2	0.7626	0.0421	0.1953	0.2056
4	0.7371	0.0084	0.2545	0.2598
10	0.6655	0.0001	0.3344	0.3377
20	0.5615	—	0.4385	0.4410
40	0.3995	—	0.6005	0.6018
100	0.1440	—	0.8560	0.8561
Case 5a				
4	0.9794	−0.0758	0.0964	0.0609
10	0.7837	−0.0350	0.2413	0.2380
20	0.5591	−0.0097	0.4506	0.4601
40	0.2773	−0.0007	0.7234	0.7283
100	0.0339	—	0.9661	0.9677

* See equation (1.29)

6a ($C = 0$, $V = 0$). Cooling curves calculated with the approximate result (1.31) are shown in Fig. 4. The quantity $l^2/K[\alpha_1^2]$ had been recognized in the analysis (1.36) as the ratio (heat stored/rate of heat loss from the enclosure) in the steady state and, by definition, is the time constant of the system $C = 0$, $V = 0$, as represented by (1.31). Values of this ratio, denoted by W/q, are found to agree exactly with the corresponding values of the thermal time constant T calculated from the exact solution for these particular cases.

The approximate form of solution (1.31) had been extended to the cases 2, 4, and 6 ($C = 0$, $V \neq 0$) with q in the ratio W/q representing the total heat loss including the amount due to ventilation, and gives the curves shown in Fig. 5. The pattern of results is similar to that of the previous group with $C = 0$, $V = 0$, and again the time constants are found to agree closely with the exact values (see Table 6). For the most general ($C \neq 0$, $V \neq 0$) cases 1, 3, and 5, the approximate solution (1.28) is unsatisfactory.

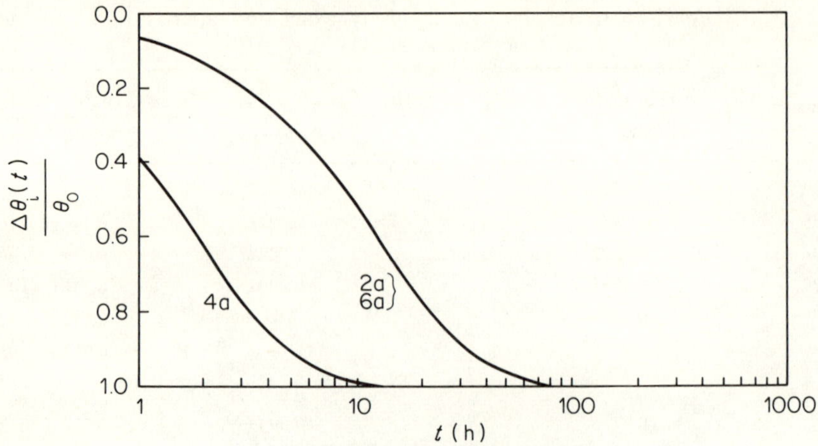

2a. Enclosure with heavy walls
4a. Enclosure with light walls
6a. Enclosure with heavy walls (smaller size)

Fig. 4. Transient cooling curves based on W/q, i.e. (heat stored/heat loss) for unventilated enclosure

The fundamental significance of the ratio W/q in transient heat flow appears to have been recognized first by Reiher (1928). Esser and Krischer (1930) described the cooling of a plane wall with an equation of the form

$$\theta(x, t) = \theta(x, t_u) \exp (t - t_u)/\psi(W/q)$$

where t_u is defined as the time taken for the cooling process to spread

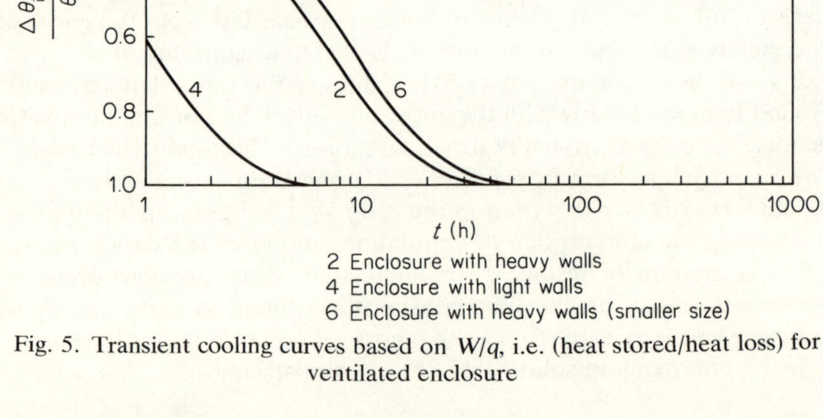

2 Enclosure with heavy walls
4 Enclosure with light walls
6 Enclosure with heavy walls (smaller size)

Fig. 5. Transient cooling curves based on W/q, i.e. (heat stored/heat loss) for ventilated enclosure

Table 6 W/q, ratio of heat stored to heat transmitted compared with T, the exact value of the time constant

	Ventilated $V \neq 0$			Not ventilated $V = 0$		
Structure type	Case	W/q (h)	T (h)	Case	W/q (h)	T (h)
Heavy, $C = 0$	2	7.1	7.2	2a	13.7	13.7
Light, $C = 0$	4	1.0	1.0	4a	2.0	2.0
Heavy, $C = 0$	6	9.0	9.1	6a	13.7	13.7

through the whole wall, and ψ is a numerical quantity depending on the value of t_u. When the cooling process is rapid, ψ approaches unity. The application of this simple ratio of steady state terms to the cooling and warming of buildings is discussed by Bruckmeyer (1951).

2. Mention has been made of the substantial amount of computation required for calculating the transient response curves using the exact solution (1.23). This may be reduced, though possibly with a loss in the accuracy of the model for representing practical conditions, by assuming the internal structure of partition walls and intermediate floors to have the same temperature as the inside air for all values of time. A mathematical statement of this problem follows by putting $\theta_{is}(t) = \theta_i(t)$, $t \geq 0$, in the equations (1.1)–(1.5) which then become

$$
\left.\begin{array}{l}
\left.\begin{array}{ll}
K\dfrac{\partial^2\theta(x,t)}{\partial x^2} = \dfrac{\partial\theta(x,t)}{\partial t}, & 0<x<l \\[2ex]
k\dfrac{\partial\theta(x,t)}{\partial x} = h_o\theta(x,t), & x=0 \\[2ex]
k\dfrac{\partial\theta(x,t)}{\partial x} = h_i\{\theta_i(t)-\theta(x,t)\}, & x=l \\[2ex]
q = c\dfrac{d\theta_i(t)}{dt} + h_i\{\theta_i(t)-\theta(x,t)\} + v\theta_i(t), & x=l
\end{array}\right] t>0 \\[2ex]
\left.\begin{array}{ll}
q = h_i\{\theta_i(t)-\theta(x,t)\} + v\{\theta_i(x,t)-\theta_o\}, & x=l \\[1ex]
\theta(x,t) = \theta_o + \dfrac{(1/h_o + x/k)q}{1+v(1/h_o + l/k + 1/h_i)} &
\end{array}\right] t=0
\end{array}\right] \qquad (2.1)
$$

The assumption in the above statement that the air and internal mass are at the same temperature implies that $1/h_s = 0$. Introducing therefore this condition into the more general solution (1.23) by putting $1/B_s (= k/lh_s) = 0$ gives directly the required result

$$\theta_i(\tau)=\frac{QA}{1+VA}+2B_iB_o\theta_o\sum_{n=1}^{\infty}\frac{e^{-\alpha_n^2\tau}}{\alpha_n^2F(\alpha_n)}$$
$$+V\left[\frac{\theta_o}{V+B_i}\left\{1+\frac{B_iB_o}{[\phi(p)]_{p=-\beta}}\left[\frac{B_i\cos\sqrt{\beta}}{B_o}+\frac{B_i\sin\sqrt{\beta}}{\sqrt{\beta}}\right]\right\}e^{-\beta\tau}\right.$$
$$\left.-2B_iB_o\theta_o\sum_{n=1}^{\infty}\left\{\frac{(B_i\cos\alpha_n)/B_o+(B_i\sin\alpha_n)/\alpha_n}{\alpha_n^2(\alpha_n^2-\beta)F(\alpha_n)}\right\}e^{-\alpha_n^2\tau}\right] \tag{2.2}$$

where

$$\phi(p)=\{(B_i+B_o)Cp+B_iB_o\}\cosh\sqrt{p}+\{Cp+B_i(1+B_oC)\}\sqrt{p}\sinh\sqrt{p}$$
$$+V\{(B_i+B_o)\cosh\sqrt{p}+(B_iB_o+p)\sinh\sqrt{p}/\sqrt{p}\} \tag{2.3}$$
$$p=-\alpha_n^2,\quad n=1,2,\ldots,\ \text{zeros of }\phi(p)$$
$$\beta=(V+B_i)/C$$

and

$$F(\alpha_n)=2\left\{\frac{d\phi(p)}{dp}\right\}_{p=-\alpha_n^2}$$
$$=\{2C(B_i+B_o)+B_i(1+B_oC)-\alpha_n^2C\}\cos\alpha_n$$
$$+\{B_i(1+B_oC)+B_iB_o-(3+B_i+B_o)C\alpha_n^2\}(\sin\alpha_n)/\alpha_n$$
$$+V\{(1+B_i+B_o+B_iB_o/\alpha_n^2)(\sin\alpha_n)/\alpha_n+(1-B_iB_o/\alpha_n^2)\cos\alpha_n\}$$

Also, following (1.25) by writing the expression (2.3) as a polynomial in p, thus

$$\phi(p)=B_iB_o+V(B_i+B_o+B_iB_o)$$
$$+\left\{B_i+\frac{B_iB_o}{2!}+C(B_i+B_o+B_iB_o)+V\left(1+\frac{B_i+B_o}{2!}+\frac{B_iB_o}{3!}\right)\right\}p$$
$$+\left[\frac{B_i}{3!}+\frac{B_iB_o}{4!}+C\left\{1+\frac{B_i+B_o}{2!}+\frac{B_iB_o}{3!}\right\}\right.$$
$$\left.+V\left\{\frac{1}{3!}+\frac{B_i+B_o}{4!}+\frac{B_iB_o}{5!}\right\}\right]p^2+\cdots \tag{2.4}$$

then, corresponding to (1.27) the alternative form of solution is

$$\theta_i(\tau)\simeq\frac{QA}{1+VA}+\frac{\theta_o}{1+VA}\left[1+\frac{VB_i}{CB_o}\left\{\frac{\cos[\alpha_1]+(B_o\sin[\alpha_1])/[\alpha_1]}{\beta-[\alpha_1^2]}\right\}\right]\frac{[\alpha_2^2]e^{-[\alpha_1^2]\tau}}{[\alpha_2^2]-[\alpha_1^2]}$$
$$-\left[1+\frac{VB_i}{CB_o}\left\{\frac{\cos[\alpha_2]+(B_o\sin[\alpha_2]/[\alpha_2]}{\beta-[\alpha_2^2]}\right\}\right]\frac{[\alpha_1^2]e^{-[\alpha_2^2]\tau}}{[\alpha_2^2]-[\alpha_1^2]}$$
$$+V\theta_o\left[1+\frac{B_i}{B_o}\left\{\frac{\cos\sqrt{\beta}+(B_o\sin\sqrt{\beta})/\sqrt{\beta}}{(1+VA)(1-\beta/[\alpha_1^2])(1-\beta/[\alpha_2^2])}\right\}\right. \tag{2.5}$$

where $[\alpha_1^2]$ and $[\alpha_2^2]$ are the roots of the limited expansion of $\phi(p)$ to $\mathcal{O}(p^3)$ as written in (2.4) and have the following expressions

$$[\alpha_1^2]=\frac{B_iB_o+V(B_i+B_o+B_iB_o)}{B_i+\dfrac{B_iB_o}{2!}+C(B_i+B_o+B_iB_o)+V\left(1+\dfrac{B_i+B_o}{2!}+\dfrac{B_iB_o}{3!}\right)} \tag{2.6}$$

$$[\alpha_2^2]=\frac{\begin{array}{l}B_i+B_iB_o/2!+C(B_i+B_o+B_iB_o)\\ \qquad +V\{1+(B_i+B_o)/2!+B_iB_o/3!\}\end{array}}{\begin{array}{l}B_i/3!+B_iB_o/4!+C\{1+(B_i+B_o)/2!+B_iB_o/3!\}\\ \qquad +V\{1/3!+(B_i+B_o)/4!+B_iB_o/5!\}\end{array}}-[\alpha_1^2] \tag{2.7}$$

Thermal Time Constant, *T*

The thermal time constant is accepted as the time taken for the temperature of the air inside the enclosure to change by approximately 63 percent of the causative step-function change in the ambient temperature applied externally. In the present notation it is the value of t, given the symbol T, for which

$$\theta_i(0)-\theta_i(t)=\theta_o(1-e^{-1})$$

It will be recognized that the thermal time constant is a convenient quantity for evaluating, and hence comparing, the transient heat response of buildings, although to obtain it by constructing the temperature response curve involves a considerable amount of numerical work. However, this may be avoided by accepting that equation (1.31) describes the temperature change after an interval of time equal to the time constant which is $l^2/K[\alpha_1^2]$, where $[\alpha_1^2]=-p$, obtained by truncating the series in (1.25) to include the first root only. It then follows that

$$T=\frac{l^2}{K}\left\{\frac{\begin{array}{l}\left(\dfrac{1}{2}+\dfrac{1}{B_o}+C\left(1+\dfrac{1}{B_i}+\dfrac{1}{B_o}+\dfrac{1}{B_s}\right)+V\left(\dfrac{1}{3!}+\dfrac{1}{2B_i}+\dfrac{1}{2B_o}+\dfrac{1}{B_iB_o}\right)\right.\\ \qquad\qquad \left.+CV\left(1+\dfrac{1}{B_i}+\dfrac{1}{B_o}\right)\right/B_s\end{array}}{1+V\left(1+\dfrac{1}{B_i}+\dfrac{1}{B_o}\right)}\right\} \tag{2.8}$$

Substituting

$$B_i=\frac{h_il}{k};\quad B_o=\frac{h_ol}{k};\quad B_s=\frac{h_sl}{k};\quad C=\frac{cK}{lk};\quad V=\frac{vl}{k};\quad \frac{1}{2}+\frac{1}{B_o}=\frac{K}{l^2}\frac{W}{U}$$

(see (1.32) *et seq.*) the expression (2.8) for the time constant reduces to

$$T=\frac{W+c}{U+v}+\frac{c}{h_s}+\Phi \tag{2.9}$$

where

$$\Phi = \frac{l^2}{K}\left\{\frac{Uv\left(\frac{1}{2U} - \frac{R_{ss}}{3} + \frac{R_{si}R_{so}}{R_{ss}}\right)}{U+v}\right\},$$

R_{ss} is the surface-to-surface resistance of the external wall, and W is the heat stored in the wall per unit air-to-air temperature difference. For buildings generally Φ is relatively small and may be neglected for heavy structures.

Expressions for the time constant of other enclosures follow from (2.9) and are listed below for the conditions specified

$\frac{W}{U}$	$c=0,$	$v=0$	
$\frac{W}{U+v}+\Phi$	$c=0,$	$v\neq 0$	
$\frac{W+c}{U}$	$c\neq 0,$	$v=0,$	$\theta_i(t)=\theta_{is}(t)$
$\frac{W+c}{U+v}+\Phi$	$c\neq 0,$	$v\neq 0,$	$\theta_i(t)=\theta_{is}(t)$
$\frac{W+c}{U+v}+\frac{c}{h_s}+\Phi,$	$c\neq 0,$	$v\neq 0,$	$\theta_i(t)\neq\theta_{is}(t)$

Preparation of Data for Calculation of Cooling Curves

Case 1. *Office Building—Heavy Construction*

The building is 200 ft long × 50 ft wide and has four storeys giving a total height of 44 ft. On each floor there is a central corridor running the whole length of the building and flanked with partitions every 10 ft. The constructional details of the main elements of the buildings are as follows:

Element	Construction	Thermal Transmittance Btu/ft²h degF
Walls	13½ in. brick, plastered internally	0.31
Roof	Asphalt, 6 in. concrete, ½ in. cork	0.31
Windows	Single glass ⅛ in.	1.0
Ground floor	6 in. concrete on hard-core	0.2
Intermediate floors	8 in. solid concrete	—
Partitions	3 in. hollow clay blocks	—

The area of glass is 40 percent of the total area of the external walls.

The conduction heat loss per degF difference in temperature between the inside and the outside of the building is:

Walls—$0.6 \times 2 \times 44(200+50) \times 0.31$		4 100 Btu/h degF
Glass—$0.4 \times 2 \times 44(200+50) \times 1$		8 800 Btu/h degF
Roof—$200 \times 50 \times 0.31$		3 100 Btu/h degF
Floor—$200 \times 50 \times 0.2$		2 000 Btu/h degF
	Total	18 000 $= \bar{q}$

Ventilation heat loss at two air changes per hour:

$$2 \times 200 \times 44 \times 50 \times 0.019 = 16\,700 \text{ Btu/h degF.}$$

Heat Stored in Building

Assume: Internal temperature, 65 degF
External temperature, 30 degF
Temperature of ground under floor, 50 degF

Element	Density lb/ft^3	Weight lb	Specific heat Btu/lb degF	Mean temperature. degF	Heat content Btu
Roof: Cork	9	3 750	0.43	49.5	31 400
Concrete	150	750 000	0.21	37.2	1 140 000
Asphalt	140	87 500	0.2	33.4	59 500
Walls	110	1 635 000	0.25	45.4	6 300 000
Glass	157	14 400	0.2	40.5	30 300
Ground Floor	150	750 000	0.21	56.9	1 085 000
Total for external structure					8 646 200
Intermediate floors	150	3 000 000	0.21	65.0	22 050 000
Partitions	55	702 000	0.25	65.0	6 120 000
Total for internal structure					28 170 000
Total for building					36 816 200

Heat stored in building for 1 degF temperature difference between inside and outside: 1 050 000 Btu.

Surface area of external walls, roof and floors:

$$2(200\times 50+200\times 44+50\times 44)=42\,000\ \text{ft}^2.$$

Mean Thermal Transmittance

$$\frac{18\,000}{42\,000}=\frac{3}{7}=\frac{1}{R_{si}+R_{so}+l/k}$$

$$R_{si}=0.7;\qquad R_{so}=0.3;\qquad \therefore\ l/k=1.333$$

$$B_{si}=h_i l/k=1.905;\qquad B_o=h_o l/k=4.444$$

Heat stored in the internal structure per unit area of external surface per degF temperature difference between inside and outside

$$c=\frac{28\,170\,000}{35\times 42\,000}=19.16$$

Heat stored/heat transmitted by conduction

$$\frac{W}{\bar{q}}=\frac{1\,050\,000}{18\,000}=58.33$$

Substituting in equation (1.40)

$$\frac{W}{\bar{q}}=\frac{cl}{k}\left(1+\frac{1}{B_i}+\frac{1}{B_o}\right)+\frac{1}{2}\frac{l^2}{K}\left(1+\frac{2}{B_o}\right)$$

gives

$$\frac{l^2}{K}=18.81$$

$$\therefore C=\frac{cl}{k}\frac{K}{l^2}=\frac{19.16\times 1.333}{18.81}=1.358$$

Ventilation heat loss at two air changes per hour

$$v=(2\times 44\times 50\times 0.019)/42\,000$$

$$=0.3981\ \text{Btu/h ft}^2\ \text{external surface}$$

$$\therefore V=\frac{vl}{k}=0.5307$$

Case 2. Factory Building—Heavy Construction

This building is identical with the office building of case 1 except that it has no intermediate floors or partitions. It follows therefore that: $B_i=1.905$, $B_o=4.444$, $l/k=1.333$, $l^2/K=18.81$, $C=0$, as in case 1, and $v=0$.

Case 3. Office Building—Light Construction

The dimensions of this building are the same as those of the office building of case 1. The constructional details of the main elements are:

Element	Construction	Thermal transmittance Btu/ft^{2}h degF
Walls	Thin sheet cladding—cavity—0.49 in. fibreboard.	0.31
Roof	Metal deck—cavity—0.55 in. fibreboard	0.31
Windows	Single glass $\frac{1}{8}$ in.	1.0
Ground floor	6 in. concrete on hardcore	0.2
Intermediate floors	8 in. solid concrete	—
Partitions	3 in. hollow clay blocks	—

The area of glass is 40 percent of the total area of the external walls.

$$\text{Heat loss by conduction} = 18\,000 \text{ Btu/h} = \bar{q}$$

$$\text{Heat loss by ventilation} = 16\,700 \text{ Btu/h}$$

Heat Stored in Building

Internal temperature, 65 degF
External temperature, 30 degF
Temperature of ground under floor, 50 degF

Element	Density (lb/ft^3)	Weight (lb)	Specific heat (Btu/lb degF)	Mean temperature (degF)	Heat content (Btu)
Roof	20	9 170	0.38	51.0	73 300
Walls	20	10 800	0.38	50.7	85 000
Glass	157	14 400	0.2	40.5	30 300
Ground floor	150	750 000	0.21	56.9	1 085 000
Total for external structure					1 273 600
Intermediate floors	150	3 000 000	0.21	65.0	22 050 000
Partitions	55	702 000	0.25	65.0	6 120 000
Total for internal structure					28 170 000
Total for building					29 443 600

Heat stored in building for 1 degF temperature difference between inside and outside: 840 000 Btu = W.

As for case 1, $l/k = 1.333$, $B_i = 1.905$, $B_o = 4.444$, and $c = 19.16$.

Heat stored/heat transmitted by conduction

$$\frac{W}{\bar{q}} = \frac{840\,000}{18\,000} = 46.67$$

Hence by substitution in equation (1.40)

$$\frac{l^2}{K} = 2.726$$

$$\therefore C = \frac{cl}{k}\frac{K}{l^2} = 9.369$$

and

$$V = \frac{vl}{k} = 0.5307$$

Case 4. *Factory Building—Light Construction*

This building is identical with the office building of case 3 except that it has no intermediate floors or partitions. It follows therefore that $B_i = 1.905$, $B_o = 4.444$, $l/k = 1.333$, $l^2/K = 2.726$, $V = 0.5307$, as in case 3, and $C = 0$.

Case 5. *Office Building—Heavy Construction*

This building is half the height and half the width of the building in case 1. The construction and internal arrangement is the same as for the larger building except that the partitions are spaced 20 ft apart instead of 10 ft. Conduction heat loss per degF temperature difference between inside and outside:

Walls—$0.6 \times 2 \times 22(200+25) \times 0.31$	1840 Btu/h degF.
Glass—$0.4 \times 2 \times 22(200+25) \times 1$	3960 Btu/h degF
Roof—$200 \times 25 \times 0.31$	1550 Btu/h degF
Floor—$200 \times 25 \times 0.2$	1000 Btu/h degF
Total	8350 Btu/h degF
	$= \bar{q}$

Ventilation heat loss at two air changes per hour:

$$2 \times 22 \times 25 \times 200 \times 0.019 = 4180 \text{ Btu/h degF.}$$

Heat Stored in Building

Internal temperature, 65 degF
External temperature, 30 degF
Temperature of ground under floor, 50 degF

Element	Density (lb/ft^3)	Weight (lb)	Specific heat (Btu/lb degF)	Mean temperature (degF)	Heat content (Btu)
Roof: Cork.	9	1 875	0.43	49.5	15 700
Concrete	150	375 000	0.21	37.2	570 000
Asphalt	140	43 700	0.2	33.4	29 800
Walls	110	735 000	0.25	45.4	2 820 000
Glass	157	6 450	0.2	40.5	13 600
Ground floor	150	375 000	0.21	56.9	543 000
Total for external structure					3 992 100
Intermediate floors.	150	500 000	0.21	65.0	3 670 000
Partitions	55	164 500	0.25	65.0	1 445 000
Total for internal structure					5 115 000
Total for building					9 107 000

Heat stored in building for 1 degF temperature difference between inside and outside: 260 000 Btu $= W$.

Surface area of external walls, roof, and floor:

$$2(200\times25+200\times22+22\times25)=19\,900\ \text{ft}^2$$

Mean thermal transmittance

$$\frac{8350}{19\,900}=\frac{1}{R_{si}+R_{so}+l/k}$$

$$\therefore l/k=1.383;\qquad B_i=1.976;\qquad B_o=4.610$$

Heat stored in internal structure per unit area of external surface per 1 degF temperature difference between inside and outside

$$c=\frac{5\,115\,000}{35\times 19\,900}=7.344$$

Heat stored/heat transmitted by conduction:

$$\frac{W}{\bar{q}}=\frac{260\,000}{8350}=31.14$$

Substituting in equation (1.40) gives

$$\frac{l^2}{K}=19.02$$

$$\therefore C=\frac{cl}{k}\frac{K}{l^2}=0.5340$$

and

$$V=\frac{vl}{k}=0.2906$$

Case 6. *Factory Building—Heavy Construction*

This building is identical with the office building of case 5 except that it has no intermediate floors or partitions. It follows therefore that: $B_i=1.976$, $B_o=4.610$, $l/k=1.383$, $l^2/\kappa=19.04$, $V=0.2906$, as in case 5, and $C=0$

Calculation of Cooling Curves by the Exact Method

To illustrate the method of calculation cases 1 and 1(a) are worked out in detail below using equation (2.2) which for convenience is written in a modified form as

$$\frac{\theta_i(0)-\theta_i(\tau)}{\theta_o}=1-2B_iB_o\sum_{n=1}^{\infty}\frac{e^{-\alpha_n^2\tau}}{\alpha_n^2F(\alpha_n)}$$
$$+\frac{2VB_iB_o}{C}\sum_{n=1}^{\infty}\frac{\{(B_i\cos\alpha_n)/B_o+(B_i\sin\alpha_n)/\alpha_n\}}{\alpha_n^2(\alpha_n^2-\beta)F(\alpha_n)}e^{-\alpha_n^2\tau}$$
$$-\frac{V}{C\beta}\left[1+\frac{B_iB_o}{\{\phi(p)\}_{p=-\beta}}\left\{\frac{(B_i\cos\sqrt{\beta}}{B_o}+\frac{(B_i\sin\sqrt{\beta})}{\sqrt{\beta}}\right\}\right]e^{-\beta\tau} \quad (2.7)$$

Case 1

Office building of heavy construction ventilated at two air changes per hour.

The calculations for this example (see above) give $B_i=1.905$, $B_o=4.444$, $C=1.358$, $V=0.5307$, $l^2/K=18.81$, $1/k=1.333$.

The Term in $e^{-\beta\tau}$

The coefficient of $e^{-\beta\tau}$ is

$$\frac{V}{C\beta}\left[1+\frac{B_iB_o}{\{\phi(p)\}_{p=-\beta}}\left\{\frac{(B_i\cos\sqrt{\beta})}{B_o}+\frac{(B_i\sin\sqrt{\beta})}{\sqrt{\beta}}\right\}\right]$$

$$\beta=(V+B_i)/C=1.794$$

$$\sqrt{\beta}=1.339;\qquad \cos\sqrt{\beta}=0.2297;\qquad \sin\sqrt{\beta}=0.9733$$

$$\begin{aligned}\{\phi(p)\}_{p=-\beta}&=\{B_iB_o-(B_i+B_o)C\beta\}\cos\sqrt{\beta}-\{B_i(1+B_oC)-C\beta\}\sqrt{\beta}\sin\sqrt{\beta}\\&\quad+V\{(B_i+B_o)\cos\sqrt{\beta}+(B_iB_o-\beta)(\sin\sqrt{\beta}/\sqrt{\beta}\}\\&=\{8.466-(6.349\times1.358\times1.794)\}0.2297\\&\quad-\{(1.905\times7.035)-(1.358\times1.794)\}1.339\times0.9733\\&\quad+\{(6.349\times0.2297)+(6.672\times0.7468\times0.9733)\}0.5307\\&=-12.69\end{aligned}$$

$$\begin{aligned}B_iB_o\{(B_i\cos\sqrt{\beta})/B_o+(B_i\sin\sqrt{\beta})/\sqrt{\beta}\}&=8.466(0.4376/4.444+1.854/1.339)\\&=12.55\end{aligned}$$

Hence the coefficient of $e^{-\beta\tau}$ is $(0.5307/2.436)\times(1-0.9893)=0.0023$. The term is small and may be neglected.

The Terms in $e^{-\alpha_n^2\tau}$

The summations are rapidly convergent so that the first and second terms only in each series will normally be required. Putting $p=-\alpha_n^2$ in the expression (2.3) for $\phi(p)$ and equating to zero gives, after rearrangement

$$\alpha_n\frac{\{C(B_i+B_o)\alpha_n^2-V(B_i+B_o)-B_iB_o\}}{C\alpha_n^4-\{B_i(1+B_oC)+V\}\alpha_n^2+VB_iB_o}$$

Values of α_n, $n=1,2,\ldots,$ are determined from this expression using a graphical method. By substitution

$$\frac{\alpha_n(8.622\alpha_n^2-11.835)}{1.358\alpha_n^4-13.93\alpha_n^2+4.493}=\tan\alpha_n$$

Using the same axes to plot the functions

$$y=\tan x$$

$$y=\frac{x(8.622x-11.835)}{1.358x^4-13.93x^2+4.493}$$

the values of x at the points of intersection of the curves are the required values of α_n. This is shown in Fig. 6 where from the enlarged plots of the

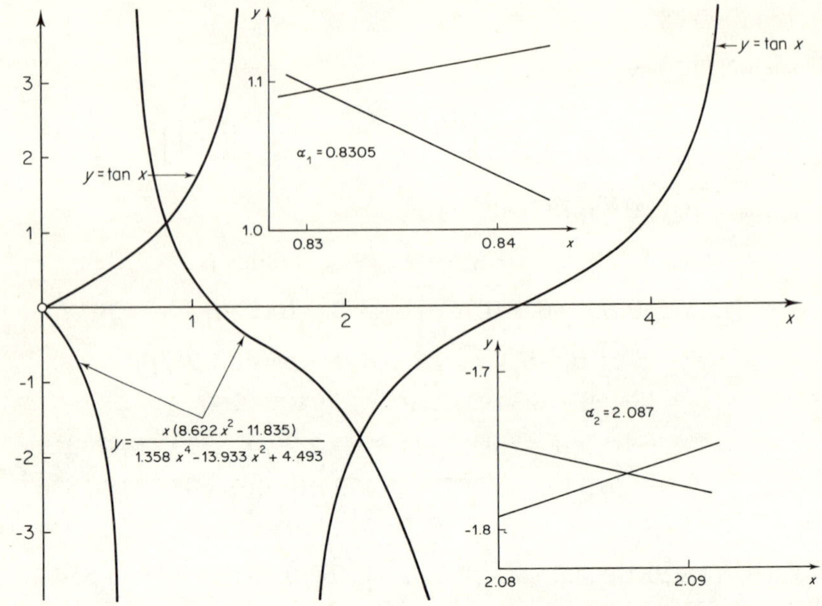

Fig. 6 Graphical solution of transcendental equations

intersections it will be seen that $\alpha_1 = 0.8305$ and $\alpha_2 = 2.087$. Hence

$$\alpha_1^2 = 0.6897; \quad \cos\alpha_1 = \ \ 0.6745; \quad (\sin\alpha_1)/\alpha_1 = 0.8890$$
$$\alpha_2^2 = 4.356; \quad \cos\alpha_1 = -0.4936; \quad (\sin\alpha_2)/\alpha_2 = 0.4167$$

Substituting numerical values gives

$$\begin{aligned} F(\alpha_1) &= [31.18 - (1.358\times 0.6897) - 4.493/0.6897]0.6745 \\ &\quad + [25.77 - (12.70\times 0.6897) + 4.493/0.6897]0.8890 \\ &= 36.92 \end{aligned}$$

Similarly

$$F(\alpha_2) = -23.84$$

The equation to the cooling curve for case 1 is therefore

$$\begin{aligned} \frac{\theta_i(0) - \theta_i(t)}{\theta_o} &= -16.93\left[\frac{\exp\left(-\dfrac{0.6897}{18.81}t\right)}{0.6897\times 36.92} + \frac{\exp\left(-\dfrac{4.356}{18.81}t\right)}{4.356\times(-23.84)}\right] \\ &\quad + 6.617\left[\frac{1.983\exp\left(-\dfrac{0.6897}{18.81}t\right)}{(1-1.104)\times 0.6897\times 36.92} + \frac{0.5822\exp\left(-\dfrac{4.356}{18.81}t\right)}{2.562\times 4.356\times(-23.84)}\right] \\ &\quad - 0.0023\exp\left(-\frac{1.794}{18.81}t\right) \\ &= 1 - 1.132e^{-0.0367t} + 0.1486e^{-0.2315t} - 0.0023e^{-0.0954t} \end{aligned}$$

Case 1(a)

This building is the same as for case 1 but without ventilation. The equation to the cooling curve follows from (2.7) by putting $V=0$, giving

$$\frac{\theta_i(0)-\theta_i(\tau)}{\theta_o}=1-2B_iB_o\sum_{n=1}^{\infty}\frac{e^{-\alpha_n^2\tau}}{\alpha_n^2F(\alpha_n)}$$

and the equation defining the roots

$$\frac{C(B_i+B_o)\alpha_n^2-B_iB_o}{\alpha_n\{C\alpha_n^2-B_i(1+B_oC)\}}=\tan\alpha_n$$

Solving graphically as above

$$\alpha_1=0.6020; \qquad \alpha_2=2.073$$

In this case

$$F(\alpha_n)=\{2C(B_i+B_o)+B_i(1+B_oC)-C\alpha_n^2\}\cos\alpha_n$$
$$+\{B_i(1+B_oC)+B_iB_o-(3+B_i+B_o)C\alpha_n^2\}(\sin\alpha_n)/\alpha_n$$

Substituting numerical values gives

$$F(\alpha_1)=41.10; \qquad F(\alpha_2)=-25.76$$

and finally

$$\frac{\theta_i(0)-\theta_i(\tau)}{\theta_o}=1-1.137e^{-0.0193t}+0.153e^{-0.2284t}$$

A summary of the equations to the cooling curves for all the cases calculated with the solution (2.2) is given in Table 7 where they are signified as exact. Simpler expressions, shown in the Table as approximate, have been calculated following the inversion method defined by (1.26) and using the first root only given by (2.6). For small values of time the approximate expressions give results that differ appreciably from those calculated with the exact solutions. This is demonstrated in Table 8 with $\Delta\theta_i(t)/\theta_o$ calculated for a series of values of t/T. It will be seen that the disagreement becomes steadily less with time increasing and that both solutions give the same value for the time constant. Thereafter the agreement is exact.

3. The analysis so far has considered the envelope of the enclosure as a uniformly thick slab of homogeneous material. This represents the most elementary form of construction which, with parallel simplicity, has a time constant for transient temperature change that is easily determined as the ratio of heat stored in the slab to heat transmitted in the steady state. It may be enquired how multi-layer walls, which are by far the more common form of construction, respond to the same pattern of ambient temperature change. This is now examined by repeating the analysis with the external

Table 7 Transient cooling of buildings; summary of calculations

Case		$\frac{W}{q}$	$\frac{l}{k}$	$\frac{l^2}{K}$	c	C	V	B_i	B_o	α_1^2	α_2^2	$\frac{\Delta\theta_i(t)}{\theta_o}$
1	Exact	30.3	1.333	18.81	19.16	1.358	0.5307	1.905	4.444	0.6897	4.356	$1-1.132e^{-0.0367t}+0.149e^{-0.232t}$
	Approx.											$1-e^{-0.0330t}$
1a	Exact	58.3	1.333	18.81	19.16	1.358	0	1.905	4.444	0.3624	4.297	$1-1.137e^{-0.0193t}+0.153e^{-0.228t}$
	Approx.											$1-e^{-0.0171t}$
2	Exact	7.1	1.383	18.92	0	0	0.5307	1.905	4.444	2.283	16.58	$1-0.882e^{-0.121t}+0.212e^{-0.876t}$
	Approx.											$1-e^{-0.141t}$
2a	Exact	13.7	1.333	18.92	0	0	0	1.905	4.444	1.661	15.85	$1-1.234e^{-0.0878t}+0.333e^{-0.838t}$
	Approx.											$1-e^{-0.0730t}$
3	Exact	24.2	1.333	2.726	19.16	9.369	0.5307	1.905	4.444	0.1151	3.826	$1-1.026e^{-0.0423t}+0.0251e^{-1.403t}$
	Approx.											$1-e^{-0.0413t}$
3a	Exact	46.7	1.333	2.726	19.16	9.369	0	1.905	4.444	0.0598	3.826	$1-1.021e^{-0.0219t}+0.0249e^{-1.403t}$
	Approx.											$1-e^{-0.0214t}$
4	Exact	1.05	1.333	2.797	0	0	0.5307	1.905	4.444	2.283	16.58	$1-0.882e^{-0.816t}+0.212e^{-5.93t}$
	Approx.											$1-e^{-0.952t}$
4a	Exact	2.03	1.333	2.797	0	0	0	1.905	4.444	1.661	15.85	$1-1.234e^{-0.594t}+0.333e^{-5.67t}$
	Approx.											$1-e^{-0.493t}$
5	Exact	20.7	1.383	19.02	7.344	0.534	0.2906	1.976	4.610	1.079	5.577	$1-1.227e^{-0.0567t}+0.273e^{-0.293t}$
	Approx.											$1-e^{-0.0483t}$
5a	Exact	31.1	1.383	19.02	7.344	0.534	0	1.976	4.610	0.7448	5.392	$1-1.258e^{-0.0392t}+0.302e^{-0.284t}$
	Approx.											$1-e^{-0.0321t}$
6	Exact	9.1	1.383	19.04	0	0	0.2906	1.976	4.610	2.079	16.43	$1-1.021e^{-0.109t}+0.260e^{-0.863t}$
	Approx.											$1-e^{-0.110t}$
6a	Exact	13.6	1.383	19.04	0	0	0	1.976	4.610	1.681	15.98	$1-1.236e^{-0.0883t}+0.336e^{-0.839t}$
	Approx.											$1-e^{-0.0730t}$

Table 8 Comparison of results calculated from exact and approximate equations to cooling curves

Case	$\frac{\Delta\theta_i(t)}{\theta_o}$ \ t/T	0.1	0.2	0.3	0.4	0.5	0.6	0.7	0.8	0.9	1.0	1.1	1.2	1.3	1.4	1.5
1	Exact	6	14	21	28	35	42	48	54	59	63	67	70	73	76	
	Approx.	10	18	26	33	40	46	51	55	60	64	67	70	73	76	78
1a	Exact	2	10	19	27	35	42	48	54	59	63	67	71	74	77	79
	Approx.	11	18	26	33	39	45	50	55	60	64	67	70	73	76	78
2	Exact	28	31	35	40	44	48	52	56	60	63	66	69	71	73	75
	Approx.	10	18	26	33	39	45	51	56	60	64	67	70	73	76	78
2a	Exact	1	6	15	24	32	40	47	53	58	63	68	71	75	78	80
	Approx.	9	18	26	33	39	45	50	55	59	63	67	70	73	76	78
3	Exact	8	16	24	32	38	44	50	55	59	63	67	70	73	76	78
	Approx.	9	18	26	33	39	45	50	55	59	63	67	70	73	76	78
3a	Exact	8	17	25	33	39	45	50	55	59	63	67	71	74	77	
	Approx.	10	18	26	33	39	45	50	55	59	63	67	70	73	76	78
4	Exact	21	31	35	39	44	48	52	56	60	63	66	69	71	74	76
	Approx.	10	18	26	33	39	45	50	55	60	64	67	70	73	76	78
4a	Exact	1	6	15	24	33	40	47	53	58	63	67	71	74	77	80
	Approx.	10	18	26	33	39	45	50	55	59	63	67	70	73	76	78
5	Exact	5	11	19	26	33	40	47	53	58	63	67	71	74	77	79
	Approx.	9	18	27	34	40	46	51	56	61	65	68	71	74	77	79
5a	Exact	1	7	15	24	32	40	47	53	58	63	68	72	75	78	80
	Approx.	10	19	26	33	39	45	50	55	59	63	67	70	73	76	78
6	Exact	18	22	27	33	39	45	50	54	59	63	66	69	73	76	78
	Approx.	10	18	26	33	39	45	51	56	60	64	67	71	74	77	79
6a	Exact	1	6	15	24	32	40	47	53	58	63	68	71	75	78	80
	Approx.	9	18	26	33	39	45	50	55	59	63	67	70	73	76	78

T denotes the *exact* value of the time constant. $\frac{\Delta\theta_i(t)}{\theta_o}$ is expressed as a percentage.

wall consisting of two homogeneous slabs of different material and thickness, and an airspace, unventilated and of uniform thickness, sandwiched between them. The two slabs, denoted regions 1 and 2, have properties

$$(l_1, k_1, K_1), \{(l_2-l_1), k_2, K_2\}$$

and temperatures

$$\theta_1(x, t), \quad 0<x<l_1; \qquad \theta_2(x, t), \quad l_1<x<l_2$$

At $x=l$, the surfaces exchange heat across the cavity at a rate h times their temperature difference per unit area, x being invariate in the cavity. The heated enclosure is ventilated but has no internal heat storage. A mathematical model of heat transfer in this arrangement is formulated as follows

$$K_1 \frac{\partial^2 \theta_1(x,t)}{\partial x^2} = \frac{\partial \theta_1(x,t)}{\partial t} \qquad (0<x<l_1,\ t>0) \tag{3.1}$$

$$-k_1 \frac{\partial \theta_1(x,t)}{\partial x} = -h_o \theta_1(x,t), \qquad (x=0,\ t>0) \tag{3.2}$$

$$-k_1 \frac{\partial \theta_1(x,t)}{\partial x} = h[\theta_1(x,t) - \theta_2(x,t)] \qquad (x=l_1,\ t>0) \tag{3.3}$$

$$K_2 \frac{\partial^2 \theta_2(x,t)}{\partial x^2} = \frac{\partial \theta_2(x,t)}{\partial t} \qquad (l_1<x<l_2,\ t>0) \tag{3.4}$$

$$-k_2 \frac{\partial \theta_2(x,t)}{\partial x} = h[\theta_1(x,t) - \theta_2(x,t)] \qquad (x=l_1,\ t>0) \tag{3.5}$$

$$-k_2 \frac{\partial \theta_2(x,t)}{\partial x} = h_i[\theta_2(x,t) - \theta_i(t)] \qquad (x=l_2,\ t>0) \tag{3.6}$$

$$q = h_i[\theta_i(t) - \theta_2(x,t)] + v\theta_i(t) \qquad (x=l_2,\ t>0) \tag{3.7}$$

$$\theta_1(x,0) = \theta_o + \frac{(1/h_o + x/k_1)}{1+vR_{i.o}} q \qquad (0<x<l_1,\ t=0)$$

$$\theta_2(x,0) = \theta_o \frac{(1/h_o + l_1/k_1 + 1/h + (x-l_1)/k_2)}{1+vR_{i.o}} \qquad (l_1<x<l_2,\ t=0)$$

with

$$R_{i.o} = 1/h_o + l_1/k_1 + 1/h + (l_2 - l_1)/k_2 + l/h_i$$

representing the air-to-air thermal resistance of the wall. Applying the Laplace transform to (3.1) through (3.7) gives the following subsidiary equations

$$K_1 \frac{d^2\bar{\theta}_1(x,p)}{dx^2} - p\bar{\theta}_1(x,p) + \frac{(1/h_o + x/k_1)}{1+vR_{i.o}} q + \theta_o = 0 \qquad (0<x<l_1,\ t>0) \tag{3.8}$$

$$k_1 \frac{d\bar{\theta}_1(x,p)}{dx} = h_o \bar{\theta}_1(x,p) \qquad (x=0,\ t>0) \tag{3.9}$$

$$k_1 \frac{d\bar{\theta}_1(x,p)}{dx} = h[\bar{\theta}_2(x,p) - \bar{\theta}_1(x,p)] \qquad (x=l_1,\ t>0) \tag{3.10}$$

$$K_2 \frac{d^2\bar{\theta}_2(x,p)}{dx^2} - p\bar{\theta}_2(x,p) + \frac{(1/h_o + l_1/k_1 + 1/h + (x-l_1)/k_2)}{1+vR_{i.o}} q + \theta_o = 0 \qquad (l_1<x<l_2,\ t>0) \tag{3.11}$$

$$k_2 \frac{d\bar{\theta}_2(x, p)}{dx} = h[\bar{\theta}_2(x, p) - \bar{\theta}_1(x, p)] \qquad (x = l_1, t > 0) \tag{3.12}$$

$$k_2 \frac{d\bar{\theta}_2(x, p)}{dx} = h_i[\bar{\theta}_1(p) - \bar{\theta}_2(x, p)] \qquad (x = l_2, t > 0) \tag{3.13}$$

$$\frac{q}{p} = h_i[\bar{\theta}_i(p) - \bar{\theta}_2(x, p)] + v\bar{\theta}_i(p) \qquad (x = l_2, t > 0) \tag{3.14}$$

The solution of (3.8) and (3.11) subject to (3.9), (3.10) and (3.12), (3.13), respectively, is

$$\left.\begin{aligned}
\bar{\theta}_1(x, p) = {} & \frac{\theta_o}{p} + \frac{R_1(x)}{(1+vR_{i.o})}\frac{q}{p} - \frac{a_0\theta_o}{pf(p)}\left[\surd(p)\cosh\sqrt{\left(\frac{p}{K_1}\right)}(l_1 - x)\right.\\
& \left. + a\sinh\sqrt{\left(\frac{p}{K_1}\right)}(l_1 - x)\right] + \frac{a_0 a a_1\theta_o}{pf(p)g(p)}\\
& \times\left[\surd(p)\cosh\sqrt{\left(\frac{p}{K_2}\right)}(l_2 - l_1) + \frac{a_2 v}{h_i + v}\sinh\sqrt{\left(\frac{p}{K_2}\right)}(l_2 - l_1)\right]\\
& \times\left[\surd(p)\cosh\sqrt{\left(\frac{p}{K_1}\right)}x + a_0\sinh\sqrt{\left(\frac{p}{K_1}\right)}x\right]\\
& + \frac{a a_2\theta_o}{pf(p)g(p)}\frac{v}{(h_i + v)}\\
& \times\left[(a_o + a)\surd(p)\cosh\sqrt{\left(\frac{p}{K_1}\right)}l_1 + (p + a_0 a)\sinh\sqrt{\left(\frac{p}{K_1}\right)}l_1\right]\\
& \times\left[p\cosh\sqrt{\left(\frac{p}{K_1}\right)}x + a_o\sinh\sqrt{\left(\frac{p}{K_1}\right)}x\right] \qquad (0 \leqslant x \leqslant l_1)
\end{aligned}\right\} \tag{3.15}$$

$$\left.\begin{aligned}
\bar{\theta}_2(x, p) = {} & \frac{\theta_o}{p} + \frac{R_2(x)}{(1+vR_{i.o})}\frac{q}{p} + \frac{a_0 a_1\theta_o}{pg(p)}\left[\surd(p)\cosh\sqrt{\left(\frac{p}{K_2}\right)}(l_2 - x)\right.\\
& \left. + \frac{a_2 v}{h_i + v}\sinh\sqrt{\left(\frac{p}{K_2}\right)}(l_2 - x)\right] + \frac{a_2\theta_o}{pg(p)}\frac{v}{(h_i + v)}\\
& \times\left\{\left[(a_0 + a)\surd(p)\cosh\sqrt{\left(\frac{p}{K_1}\right)}l_1 + (p + a_0 a)\sinh\sqrt{\left(\frac{p}{K_2}\right)}l_1\right]\right.\\
& \times\cosh\sqrt{\left(\frac{p}{K_2}\right)}(x - l_1)\\
& + \left[a_1\surd(p)\sinh\sqrt{\left(\frac{p}{K_1}\right)}l_1 + a_0 a_1\cosh\sqrt{\left(\frac{p}{K_1}\right)}l_1\right]\\
& \left.\times\sinh\sqrt{\left(\frac{p}{K_1}\right)}(x - l_1)\right\} \qquad (l_1 \leqslant x \leqslant l_2)
\end{aligned}\right\} \tag{3.16}$$

where

$$f(p)=(a_0+a)\sqrt{(p)}\cosh\sqrt{\left(\frac{p}{K_1}\right)}l_1+(p+a_0a)\sinh\sqrt{\left(\frac{p}{K_1}\right)}l_1 \tag{3.17}$$

$$\left.\begin{aligned}
g(p)=-\Bigg\{&p^{3/2}\sinh\sqrt{\left(\frac{p}{K_1}\right)}l_1\sinh\sqrt{\left(\frac{p}{K_2}\right)}(l_2-l_1)\\
&+p\Bigg[\left(a_1+\frac{a_2v}{h_\mathrm{i}+v}\right)\sinh\sqrt{\left(\frac{p}{K_2}\right)}l_1\cosh\sqrt{\left(\frac{p}{K_2}\right)}(l_2-l_1)\\
&+(a_0+a)\cosh\sqrt{\left(\frac{p}{K_1}\right)}l_1\sinh\sqrt{\left(\frac{p}{K_2}\right)}(l_2-l_1)\Bigg]\\
&+\sqrt{(p)}\Bigg\{\left(a_0a+\frac{a_1a_2v}{h_\mathrm{i}+v}\right)\sinh\sqrt{\left(\frac{p}{K_1}\right)}l_1\sinh\sqrt{\left(\frac{p}{K_2}\right)}(l_2-l_1)\\
&+\left[a_0a_1+(a_0+a)\frac{a_2v}{h_\mathrm{i}+v}\right]\cosh\sqrt{\left(\frac{p}{K_1}\right)}l_1\\
&\times\cosh\sqrt{\left(\frac{p}{K_2}\right)}(l_2-l_1)\Bigg\}+\frac{a_0a_2v}{h_\mathrm{i}+v}\\
&\times\left[a\sinh\sqrt{\left(\frac{p}{K_1}\right)}l_1\cosh\sqrt{\left(\frac{p}{K_2}\right)}(l_2-l_1)a_1\right.\\
&\left.\times\cosh\sqrt{\left(\frac{p}{K_1}\right)}l_1\sinh\sqrt{\left(\frac{p}{K_2}\right)}(l_2-l_1)\right]\Bigg\}
\end{aligned}\right\} \tag{3.18}$$

$$a_0=h_\mathrm{o}\sqrt{(K_1)}/k_1;\qquad a=h\sqrt{(K_1)}/k_1;$$

$$a_1=h\sqrt{(K_2)}/k_2;\qquad a_2=h_\mathrm{i}\sqrt{(K_2)}/k_2$$

$$R_1(x)=1/h_\mathrm{o}+x/k_1\qquad(0\leqslant x\leqslant l_1)$$

$$R_2(x)=1/h_\mathrm{o}+l_1/k_1+1/h+(x-l_1)/k_2\qquad(l_1\leqslant x\leqslant l_2)$$

$$R_\mathrm{i.o}=1/h_\mathrm{o}+l_1/k_1+1/h+(l_2-l_1)/k_2+1/h_\mathrm{i} \tag{3.19}$$

From (3.14)

$$\bar{\theta}_\mathrm{i}(p)=[h_\mathrm{i}\bar{\theta}_2(l_2,p)+q/p]/(h_\mathrm{i}+v) \tag{3.20}$$

Substituting for $\bar{\theta}_2(l_2, p)$ from (3.16) in (3.20) and applying the inversion theorem to the resulting expression for $\bar{\theta}_\mathrm{i}(p)$ gives the following solution for the temperature of the air inside the enclosure

$$\begin{aligned}
\theta_\mathrm{t}(t)=&\frac{(1/h_\mathrm{o}+l_1/k_1+1/h+(l_2-l_1)/k_2)}{1+vR_\mathrm{i.o}}q+\frac{q}{h_\mathrm{i}+v}+\frac{2h_\mathrm{i}\theta_\mathrm{o}}{h_\mathrm{i}+v}\sum_{n=1}^{\infty}\frac{\exp(-\alpha_n^2t)}{\alpha_nG(\alpha_n)}\\
&\times\left\{a_0a_1+\frac{a_2v}{h_\mathrm{i}+v}\left[(a_0+a)\cos b_1\alpha_n+\left(\frac{a_0a}{\alpha_n}-\alpha_n\right)\sin b_1\alpha_n\right]\cos b_2\alpha_n\right.\\
&\left.+(a_0\cos b_1\alpha_n-\alpha_n\sin b_1\alpha_n)\frac{a_1}{\alpha_n}\sin b_2\alpha_n\right\}
\end{aligned} \tag{3.21}$$

where the summation extends over the positive roots, α_n, of

$$\left(\alpha^2-a_0a-\frac{a_1a_2v}{h_i+v}\right)\tan b_1\alpha\tan b_2\alpha-\left[\alpha\left(a_1+\frac{a_2v}{h_i+v}\right)-\frac{a_0aa_2v}{(h_i+v)\alpha}\right]\tan b_1\alpha$$
$$-\left[\alpha(a_o+a)-\frac{a_oa_1a_2v}{(h_i+v)\alpha}\right]\tan b_2\alpha+a_oa_1+\frac{(a_1+a)a_2v}{h_i+v}=0$$

and

$$G(\alpha_n)=\left\{\frac{a_0aa_2v}{(h_i+v)}+\frac{1}{\alpha_n}\,a_1+\frac{a_2v}{h_i+v}+\frac{b_1}{\alpha_n}\left[a_0a_1+\frac{(a_0+a)a_2v}{h_i+v}\right]\right.$$
$$\left.+\frac{b_2}{\alpha_n}\,a_0a+\frac{a_1a_2v}{h_i+v}-\alpha_nb_2\right\}\sin b_1\alpha_n\cos b_2\alpha_n$$
$$+\left\{\frac{a_0a_1a_2v}{(h_i+v)}+\frac{b_1}{\alpha_n}\left(aa_0+\frac{a_1a_2v}{h_i+v}\right)\right.$$
$$\left.+\frac{b_2}{\alpha_n}\left[a_0a_1+\frac{(a_0+a)a_2v}{h_i+v}\right]+\frac{(a_0+a)}{\alpha_n}-b_1\alpha_n\right\}$$
$$\times\cos b_1\alpha_n\sin b_2\alpha_n-\left\{2+\left[\frac{a_1h_i+(a_1+a_2)v}{(h_i+v)}\right]b_2\right.$$
$$\left.+(a_0+a)b_1-\frac{a_0aa_2v}{\alpha_n^2(h_i+v)}\,b_2-\frac{a_0aa_2v}{\alpha_n^2(h_i+v)}\,b_1\right\}$$
$$\times\sin b_1\alpha_n\sin b_2\alpha_n+\left\{\left[\frac{a_1h_i+(a_1+a_2)v}{h_i+v}\right]b_1\right.$$
$$\left.+(a_0+a)b_2-\frac{a_0aa_2v}{\alpha_n^2(h_i+v)}\,b_1-\frac{a_0aa_2v}{\alpha_n^2(h_i+v)}\,b_2\right\}\cos b_1\alpha_n\cos b_2\alpha_n$$

with $b_1=l_1/\sqrt{K_1}$ and $b_2=(l_2-l_1)/\sqrt{K_2}$.

Application of the inversion procedure indicated in (1.25) through (1.27) yields, for the present problem, an approximate form of solution similar to that obtained above for the enclosure with a homogeneous slab representing the wall. For, expanding $g(p)/\sqrt{p}$ from (3.18) to a linear term in p and putting $v=0$ it may be shown that

$$\frac{\theta_i(0)-\theta_i(t)}{\theta_o}\simeq 1-\exp(-\alpha_1^2t)\qquad (t\text{ large})$$

where

$$\alpha_1^2=a_0a_1\{a_1b_1+(a_0+a)b_2+a_0ab_1b_2+(a_0a_1/2)[b_1^2+b_2^2]\}^{-1}\tag{3.22}$$

The heat stored in the wall per unit area in the initial steady state relative to θ_o as base temperature is

$$W = \rho_1 s_1 l_1 \left[\frac{\theta_1(0,0)+\theta_1(l_1,0)}{2} - \theta_o\right] + \rho_2 s_2 (l_2 - l_1)\left[\frac{\theta_2(l_1,0)+\theta_2(l_2,0)}{2} - \theta_o\right]$$

$$= \rho_1 s_1 l_1 (\theta_i - \theta_o)\left(\frac{1/h_o + l_1/2k_1}{R_{i.0}}\right) + \rho_2 s_2 (l_2 - l_1)(\theta_i - \theta_o)$$

$$\times \left[\frac{1/h_o + l_1/k_1 + 1/h + (l_2 - l_1)/2k_2}{R_{i.o}}\right]$$

Assuming no ventilation, heat transmission is by conduction only through the wall so that in the initial steady state

$$q = (\theta_i - \theta_o)/R_{i.o}$$

Composing the ratio (heat stored/rate of heat transmission) in the steady state gives

$$\frac{W}{q} = \rho_1 s_1 l_1 \left(\frac{1}{h_o} + \frac{l_1}{2k_1}\right) + \rho_2 s_2 (l_2 - l_1)\left(\frac{1}{h_o} + \frac{l_1}{k_1} + \frac{1}{h} + \frac{l_2 - l_1}{2k_2}\right) \tag{3.23}$$

and introducing the relationships (3.19) into (3.23) leads to the result

$$\frac{W}{q} = \frac{1}{a_1 a_0}\left\{\frac{a_1 l_1}{\sqrt{K_1}} + (a_0 + a)\frac{(l_2 - l_1)}{\sqrt{K_2}} + a_0 a \frac{l_1}{\sqrt{K_1}}\frac{(l_2 - l_1)}{\sqrt{K_2}} + \frac{a_0 a_1}{2}\left[\frac{l_1^2}{K_1} + \frac{(l_2 - l_1)^2}{K_2}\right]\right\} \tag{3.24}$$

where the quantity on the r.h.s. will be recognized as $1/[\alpha_1^2]$, (3.22).

From this result it may be accepted, without further analysis, that the ratio (heat stored/heat transmitted) in the steady state is the thermal time constant for single- and multi-layer walls that compares the rates at which a transient change in the ambient temperature at one surface passes through structures enclosing a space that is heated at a constant rate of energy input.

SYMBOLS

α_n	Successive roots of relevant function, $n = 1, 2, 3, \ldots$
$[\alpha_1], [\alpha_2]$	Approximations to
c	Thermal capacity of internal mass per unit area of enclosure wall
h	Coefficient of heat transfer across cavity
h_i	Inside wall surface heat transfer coefficient
h_o	Outside wall surface heat transfer coefficient
h_s	Heat transfer coefficient at surface of internal mass expressed per unit area of enclosure wall
H	Conduction heat flux per unit temperature difference
k	Thermal conductivity of single leaf wall
k_1	Thermal conductivity of outer leaf of cavity wall
k_2	Thermal conductivity of inner leaf of cavity wall

K	Thermal diffusivity of single-leaf wall
K_1	Thermal diffusivity of outer leaf of cavity wall
K_2	Thermal diffusivity of inner leaf of cavity wall
l	Thickness of single-leaf wall
l_1	Thickness of outer leaf of cavity wall
l_2-l_1	Thickness of inner leaf of cavity wall
q	Heat input per unit area of enclosure wall
ρ	Density of single-leaf wall
ρ_1	Density of outer leaf of cavity wall
ρ_2	Density of inner leaf of cavity wall
$R_{i.o}$	Air-to-air thermal resistance of cavity wall
s	Specific heat of single-leaf wall
s_1	Specific heat of outer leaf of cavity wall
s_2	Specific heat of inner leaf of cavity wall
t	Time
T	Thermal time constant
v	Ventilation heat loss rate per unit temperature difference and per unit area of enclosure wall
W	Heat stored per unit area of wall
x	Position in wall
a	$h\sqrt{(K_1)}/k_1$
a_0	$h_o\sqrt{(K_1)}/k_1$
a_1	$h\sqrt{(K_2)}/k_2$
a_2	$h_i\sqrt{(K_2)}/k$
A	$1+1/B_i+1/B_o$
b_1	$l_1/\sqrt{K_1}$
b_2	$(l_2-l_1)/\sqrt{K_2}$
B_i	h_il/k
B_o	h_ol/k
B_s	h_sl/k
β	$[B_s(B_i+V)]/[C(B_i+B_s+V)]$
C	cK/lk
Q	ql/k
$R_1(x)$	$1/h_o+x/k_1$, $(0<x<l_1)$
$R_2(x)$	$1/h_o+l_1/k_1+1/h+(x-l_1)/k_2$, $(l_1<x<l_2)$
V	vl/k

REFERENCES

Bruckmayer, F. (1951) Building Research Congress, Division 3, pp. 66–74.

Esser, W. and Krischer, O. (1930) *The Computation of the Warming up and Cooling of Plane and Cylindrical Walls*, Springer, Berlin.

Jamieson, H. C. (1954) *Inst. Htg. and Vent. Engrs. Jnl.*, **22,** 465–495.

Reiher, H. (1928) *Ges. Ing.*, **51,** 737.

CHAPTER 6

Variable Heat Flow through Walls of Cavity Construction

Boundary conditions qualifying the flow of heat through the naturally exposed surfaces of a building assume a variety of different forms; rarely is the heat flux steady. A situation of practical interest, for example in the air conditioning of buildings, is the daily variation of the outside temperature combined with an indoor temperature maintained at a constant value. For rigorous analysis the complex boundary conditions created by natural exposure have, perforce, to be simplified and a solution of the practical problem sought by examining an idealized case having as close a similarity to the actual conditions as the analysis will allow. In settled weather the diurnal change in the outside air temperature approximates to a regular pattern that is roughly sinusoidal and may be described as steady periodic. At other times it is more appropriate to consider the heat flow as transient. There is an extensive literature giving methods of solution for both forms of variation in the external temperature applied to an infinite plane slab of isotropic material (see Crank, 1956; Carslaw and Jaeger, 1959). Little attention has been given to obtaining the corresponding analytical solutions defining heat flow in a cavity structure although less exact methods of calculation have been devised. For example, Mackey and Wright (1946), following Shklover (1945), apply the simplifying concept of an equivalent homogeneous wall for calculating steady periodic heat flow through composite structures. The properties of the (hypothetical) isotropic solid equivalent to the composite assembly of different conducting media, airspaces included, are derived empirically from a correlative examination of results calculated by more exact methods. A similar concept ('brick wall thickness of equivalent thermal capacity') is described by Bruckmayer (1944) based on a more fundamental treatment due to Esser and Krischer (1930).

Matrix methods commonly used in electric circuit theory are described in their application to thermal problems by Carslaw and Jaeger and considered by them to be the most suitable for calculating steady periodic temperatures in composite slabs. Even so, analytical methods are manageable for almost all walls in conventional building construction.

Two problems in heat conduction with important practical applications are solved below, using the Laplace transform method, to give the temperatue in each layer of a single-cavity construction. Solutions for certain other arrangements follow by taking limiting values of the parameters. The first deals with periodic heat flow applied as a sinusoidal variation in the ambient temperature at the outside surface. The second examines transient flow in which the applied boundary condition is a single step-function change of temperature. It is assumed that the heat flow is one-dimensional and that the thermal properties, conductivity k and diffusivity K of each slab are independent of temperature θ, position x, and time t.

The boundary conditions are expressed in air temperature but this may be replaced by equivalent temperature (see Chapter 1) for defining surface heat exchange more conveniently in situations where the mean radiant temperature viewed by the surface is considered to be significantly different from the air temperature.

1. SINUSOIDAL HEAT FLOW THROUGH A CAVITY WALL

Consider a plane slab of thickness l_1, thermal properties k_1 (conductivity), K_1 (diffusivity) representing the outer leaf, and thickness $(l_2 - l_1)$ of another, properties k_2, K_2 representing the inner leaf. The two slabs are separated by a closed airspace denoted by a thermal conductance h and with no thermal capacity. At the positions $x=0$ and $x=l_2$ heat is transferred between the wall surface and air at the respective rates h_o and h_i times their temperature difference. Initially the wall and the ambient air are at the same temperature. Subsequently the air temperature at $x=0$ varies repeatedly with time as $\theta_o \sin \omega t$ while at $x=l_2$ the temperature of the air is raised by a step-change through θ_i and thereafter held constant.

The flow of heat through the three-layer wall, the middle layer of negligible thermal capacity, is formulated mathematically as follows

Differential equations

$$K_1 \frac{\partial^2 \theta_1}{\partial x^2}(x, t) = \frac{\partial \theta_1}{\partial t}(x, t), \qquad 0 < x < l_i, \quad t > 0 \tag{1.1}$$

$$K_2 \frac{\partial^2 \theta_2}{\partial x^2}(x, t) = \frac{\partial \theta_2}{\partial t}(x, t) \qquad l_1 < x < l_2, \quad t > 0 \tag{1.2}$$

Boundary conditions, $t > 0$

$$-k_1 \frac{\partial \theta_1}{\partial x}(x, t) = h_o\{\theta_o e^{i\omega t} - \theta_1(x, t)\}, \qquad x = 0 \tag{1.3}$$

$$-k_1 \frac{\partial \theta_1}{\partial x}(x, t) = h\{\theta_1(x, t) - \theta_2(x, t)\}, \qquad x = l_1 \tag{1.4}$$

$$-k_2 \frac{\partial \theta_2}{\partial x}(x, t) = h\{\theta_1(x, t) - \theta_2(x, t)\}, \qquad x = l_1 \tag{1.5}$$

$$-k_2 \frac{\partial \theta_2}{\partial x}(x, t) = h_i\{\theta_2(x, t) - \theta_i(t)\}, \qquad x = l_2 \tag{1.6}$$

Initial conditions

$$\left.\begin{aligned} \theta_1(x, 0) &= 0, & 0 \leqslant x \leqslant l_1 \\ \theta_2(x, 0) &= 0, & l_1 \leqslant x \leqslant l_2 \\ \theta_o(0) &= 0, & x = l_1 \\ \theta_i(0) &= 0, & x = l_2 \end{aligned}\right\} \tag{1.7}$$

Applied temperature functions

$$\begin{aligned} \theta_o &= 0, \quad x = 0, \quad t \leqslant 0 \\ &= \theta_o \sin \omega t, \quad x = 0, \quad t > 0 \\ \theta_i &= 0, \quad x = l_2, \quad t \leqslant 0 \\ &= \theta_i, \quad x = l_2, \quad t > 0 \end{aligned}$$

Let $\bar{\theta}_r(x, p)$, $(r = 1, 2)$, denote the Laplace transform of $\bar{\theta}_r(x, p)$, defined as

$$\bar{\theta}_r(x, p) = \int_0^\infty e^{-pt} \theta_r(x, t)\, dt, \qquad R(p) > 0$$

Applying the Laplace transform to equations (1.1) through (1.6) gives the following subsidiary equations

Transformed differential equations

$$\frac{d^2\bar{\theta}_1}{dx^2}(x, p) - \frac{p\bar{\theta}_1}{K_1}(x, p) = 0, \qquad 0 < x < l_1 \tag{1.8}$$

$$\frac{d^2\bar{\theta}_2}{dx^2}(x, p) - \frac{p\bar{\theta}_2}{K_2}(x, p) = 0, \qquad l_1 < x < l_2 \tag{1.9}$$

Transformed boundary conditions

$$\frac{d\bar{\theta}_1}{dx}(x, p) + \frac{h_o}{k_1}\left\{\frac{\theta_o}{(p - i\omega)} - \bar{\theta}_1(x, p)\right\} = 0, \qquad x = 0 \tag{1.10}$$

$$\frac{d\bar{\theta}_1}{dx}(x, p) + \frac{h}{k_1}\{\bar{\theta}_1(x, p) - \bar{\theta}_2(x, p)\} = 0, \qquad x = l_1 \tag{1.11}$$

$$\frac{d\bar{\theta}_2}{dx}(x, p) + \frac{h}{k_2}\{\bar{\theta}_1(x, p) - \bar{\theta}_2(x, p)\} = 0, \qquad x = l_1 \tag{1.12}$$

$$\frac{d\bar{\theta}_2}{dx}(x, p) + \frac{h_i}{k_2}\left\{\bar{\theta}_2(x, p) - \frac{\theta_i}{p}\right\} = 0, \qquad x = l_2 \tag{1.13}$$

Attention will be confined to obtaining a solution for the temperature at the position $x = l_2$ after a long time when the sum of transients involving the initial condition has died away.

The system of equations (1.8) through (1.13) yields the result

$$\bar{\theta}_2(l_2, p) = \frac{h_i\sqrt{K_2}\theta_i}{k_2 p f_2(p)}\left\{\sqrt{p}\cosh\sqrt{\left(\frac{p}{K_2}\right)}(l_2-l_1) + \frac{h\sqrt{K_2}}{k_2}\sinh\sqrt{\left(\frac{p}{K_2}\right)}(l_2-l_1)\right\}$$
$$+\frac{h_o h\sqrt{(K_1K_2)}p\theta_o}{k_1k_2(p-\mathrm{i}\omega)F(p)} + \frac{h^2 h_i\sqrt{(K_1)}K_2\theta_i}{k_1k_2^2F(p)f_2(p)}$$
$$\times\left\{\sqrt{p}\cosh\sqrt{\left(\frac{p}{K_1}\right)}l_1 + \frac{h_o\sqrt{K_1}}{k_1}\sinh\sqrt{\left(\frac{p}{K_1}\right)}l_1\right\} \tag{1.14}$$

where

$$f_2(p) = 2\left(p + \frac{hh_iK_2}{k_2^2}\right)\sinh\sqrt{\left(\frac{p}{K_2}\right)}(l_2-l_1)$$
$$+2(h+h_i)\sqrt{p}\frac{\sqrt{K_2}}{k_2}\cosh\sqrt{\left(\frac{p}{K_2}\right)}(l_2-l_1) \tag{1.15}$$

and

$$F(p) = (h_oh + hh_i + h_oh_i)\frac{\sqrt{(K_1K_2)}}{k_1k_2}p\cosh\sqrt{\left(\frac{p}{K_1}\right)}l_1\cosh\sqrt{\left(\frac{p}{K_2}\right)}(l_2-l_1)$$
$$+p\left(p + \frac{h_ohK_1}{k_1^2} + \frac{hh_iK_2}{k_2^2}\right)\sinh\sqrt{\left(\frac{p}{K_1}\right)}l_1\sinh\sqrt{\left(\frac{p}{K_2}\right)}(l_2-l_1)$$
$$+\left\{p(h+h_i) + \frac{h_ohh_iK_1}{k_1^2}\right\}\frac{\sqrt{K_2}}{k_2}\sqrt{p}\sinh\sqrt{\left(\frac{p}{K_1}\right)}l_1\cosh\sqrt{\left(\frac{p}{K_2}\right)}(l_2-l_1)$$
$$+\left\{p(h_o+h) + \frac{h_ohh_iK_2}{k_2^2}\right\}\frac{\sqrt{K_1}}{k_1}\sqrt{p}\cosh\sqrt{\left(\frac{p}{K_1}\right)}l_1\sinh\sqrt{\left(\frac{p}{K_2}\right)}(l_2-l_1) \tag{1.16}$$

The Laplace inversion of the expression for $\bar{\theta}_2(l_2, p)$ gives the required solution as the sum of the residues at the simple poles represented by $p=0$ and $p=\mathrm{i}\omega$. The form of the result is

$$\theta_2(l_2, t) = \theta_2(l_2, t)_\omega + \theta_i\left\{1 - \frac{1}{h_i}\left(\frac{1}{h_o} + \frac{1}{h} + \frac{1}{h_i} + \frac{l_1}{k_1} + \frac{l_2-l_1}{k_2}\right)^{-1}\right\} \tag{1.17}$$

defining a steady-state temperature distribution through the wall upon which is superimposed a steady periodic temperature function defined by

$$\theta_2(l_2, t)_\omega = \frac{hh_o\sqrt{(K_1K_2)}\theta_o\mathrm{e}^{\mathrm{i}\omega t}}{k_1k_2\{F(p)/p\}_{p=\mathrm{i}\omega}} \tag{1.18}$$

In manipulating equation (1.18) it is found convenient to rewrite the expression (1.16) for $F(p)$ as

$$\frac{F(p)}{p} = \frac{1}{4\sqrt{p}}\{(\sqrt{p}+a_0)(\sqrt{p}+a_2)(\sqrt{p}+\overline{a+a_1})\mathrm{e}^{\sqrt{p}\alpha}$$
$$+(\sqrt{p}-a_0)(\sqrt{p}-a_2)(\sqrt{p}-\overline{a+a_1})\mathrm{e}^{-\sqrt{p}\alpha}$$
$$-(\sqrt{p}+a_0)(\sqrt{p}-a_2)(\sqrt{p}+\overline{a-a_1})\mathrm{e}^{\sqrt{p}\beta}$$
$$-(\sqrt{p}-a_0)(\sqrt{p}+a_2)(\sqrt{p}-\overline{a-a_1})\mathrm{e}^{-\sqrt{p}\beta} \tag{1.19}$$

in which

$$a_0 = h_o \frac{\sqrt{K_1}}{k_1}; \qquad a = h\frac{\sqrt{K_1}}{k_1}$$

$$a_1 = h\frac{\sqrt{K_2}}{k_2}; \qquad a_2 = h_i \frac{\sqrt{K_2}}{k_2}$$

$$\alpha = \frac{l_1}{\sqrt{K_1}} + \frac{(l_2 - l_1)}{\sqrt{K_2}}, \qquad \beta = \frac{l_1}{\sqrt{K_1}} - \frac{(l_2 - l_1)}{\sqrt{K_2}}$$

In (1.19), dropping the negligibly small term involving $e^{-\sqrt{p}\alpha}$, and making the substitutions as indicated, leads finally to a solution of the form

$$\theta_2(l_2, t)_\omega = \frac{4a_0 a_1 \theta_o \sqrt{\omega} \sin(\omega t - \psi)}{\sqrt{(G^2 + H^2)}} \tag{1.20}$$

where

$$\tan \psi = (H - G)/(H + G) \tag{1.21}$$

with

$$\left.\begin{aligned} G = {} & [A \cos \alpha\sqrt{(\omega/2)} - B \sin \alpha\sqrt{(\omega/2)}]e^{\alpha\sqrt{(\omega/2)}} \\ & -[C \cos \beta\sqrt{(\omega/2)} - D \sin \beta\sqrt{(\omega/2)}]e^{\beta\sqrt{(\omega/2)}} \\ & -[E \cos \beta\sqrt{(\omega/2)} + F \sin \beta\sqrt{(\omega/2)}]e^{-\beta\sqrt{(\omega/2)}} \\ H = {} & [A \sin \alpha\sqrt{(\omega/2)} + B \cos \alpha\sqrt{(\omega/2)}]e^{\alpha\sqrt{(\omega/2)}} \\ & -[C \sin \beta\sqrt{(\omega/2)} + D \cos \beta\sqrt{(\omega/2)}]e^{\beta\sqrt{(\omega/2)}} \\ & +[E \sin \beta\sqrt{(\omega/2)} - F \cos \beta\sqrt{(\omega/2)}]e^{-\beta\sqrt{(\omega/2)}} \end{aligned}\right\} \tag{1.22}$$

and

$$\begin{aligned} A &= -\omega\sqrt{(\omega/2)} + \{(a + a_1)(a_0 + a_2) + a_0 a_2\}\sqrt{(\omega/2)} + (a + a_1)a_0 a_2 \\ B &= \omega\sqrt{(\omega/2)} + (a_0 + a + a_1 + a_2)\omega + \{(a + a_1)(a_0 + a_2) + a_0 a_2\}\sqrt{(\omega/2)} \\ C &= -\omega(\omega/2) + \{(a - a_1)(a_0 - a_2) - a_0 a_2\}\sqrt{(\omega/2)} - (a - a_1)a_0 a_2 \\ D &= \omega\sqrt{(\omega/2)} + (a_0 + a - a_1 - a_2)\omega + \{(a - a_1)(a_0 - a_2) - a_0 a_2\}\sqrt{(\omega/2)} \\ E &= -\omega\sqrt{(\omega/2)} + \{(a - a_1)(a_0 - a_2) - a_0 a_2\}\sqrt{(\omega/2)} + (a_0 - a_1)a_0 a_2 \\ F &= \omega\sqrt{(\omega/2)} - (a_0 + a - a_1 - a_2)\omega + \{(a - a_1)(a_0 - a_2) - a_0 a_2\}\sqrt{(\omega/2)} \end{aligned}$$

Two-layer Wall

Equation (1.20) defines the steady cyclic temperature of the inside surface of a cavity wall facing a constant indoor temperature and exposed to a regular sinusoidal variation in the ambient temperature at the outside surface. The corresponding solution for a solid wall consisting of two layers (l_1, k_1, K_1) and ($l_2 - l_1$, k_2, K_2) follows immediately by putting $1/h = 0$ in (1.20), giving

$$\theta_2(l_2, t)_\omega = \frac{4a_0\sqrt{\omega}\theta_o}{\sqrt{(G^2 + H^2)}} \sin(\omega t - \psi) \tag{1.23}$$

where ψ, G, and H are defined by the expressions in (1.21) and (1.22), and the terms $A, \ldots, F$ now read

$$A = (a_0 + a_2)(1+b)\sqrt{(\omega/2)} + a_0 a_2(1+b)$$
$$B = (a_0 + a_2)(1+b)\sqrt{(\omega/2)} + \omega(1+b)$$
$$C = (a_2 - a_0)(1-b)\sqrt{(\omega/2)} + a_0 a_2(1+b)$$
$$D = (a_2 - a_0)(1-b)\sqrt{(\omega/2)} - \omega(1-b)$$
$$E = (a_2 - a_0)(1-b)\sqrt{(\omega/2)} - a_0 a_2(1-b)$$
$$F = (a_2 - a_0)(1-b)\sqrt{(\omega/2)} + \omega(1-b)$$

where $b = (k_2/k_1)\sqrt{(K_1/K_2)}$, and a_0, a_2 are as defined above.

Numerical Results

Calculated values of the time-lag and amplitude ratio are given in Table 1 for a solid brick wall and cavity walls incorporating brick and other material. The time-lag is defined as the interval between a sinusoidal variation in the ambient temperature at the outside surface and the resulting variation in the temperature of the inside surface. The amplitude ratio measures the extent to which the simple harmonic variation in the outside ambient temperature is attenuated in its passage through the wall to the inside surface. It is interesting to note that for the thicknesses specified, the respective properties of the two alternative arrangements of such dissimilar materials as brick and timber differ by rather less than 20 percent. An attempt to reduce the amount of computational work involved has been made by repeating the calculations and omitting the two terms that include the β quantity. In the few cases examined the amplitude ratio is not changed appreciably by this approximation but the time-lag is underestimated by a significant amount.

Table 1 Calculated values of time-lag and amplitude ratio of solid and cavity walls

	Complete solution		Ignoring β terms	
Wall (outside to inside)	Time-lag (h)	Amplitude ratio	Time-lag (h)	Amplitude ratio
$4\frac{1}{2}$ in. brick	3.7	0.324		
$4\frac{1}{2}$ in. brick—cavity—1 in. timber	5.0	0.114	4.4	0.112
1 in. timber—cavity—$4\frac{1}{2}$ in. brick	5.7	0.093	4.8	0.081
$4\frac{1}{2}$ in. brick—cavity—3 in. aerated concrete	6.7	0.073	5.5	0.071

2. TRANSIENT HEAT FLOW THROUGH A CAVITY WALL

The wall structure considered is that described above, with one set of thermal properties (subscript 1) in the region $0<x<l_1$ denoting the outer leaf, and a second set (subscript 2) in the region $l_1<x<l_2$ denoting the inner leaf. A closed airspace separates the two leaves. At $x=0$, $x=l_2$ heat is transferred between the environment and the outside and inside surfaces of the wall at the respective rates, h_o and h_i, times their temperature difference. For all values of time the ambient temperature at the inside surface, $x=l_2$, is held constant at θ_i. Initially the system is in steady state. Subsequently the ambient temperature at the outside surface, $x=0$, undergoes a single step-function change represented as a drop through θ_o. It is required to calculate the temperature $\theta_r(x, t)$, $(r=1, 2)$ that satisfies the following mathematical model of the above.

Differential equations

$$K_1 \frac{\partial^2 \theta_1}{\partial x^2}(x, t) = \frac{\partial \theta_1}{\partial t}(x, t), \qquad 0<x<l_1, \quad t>0 \tag{2.1}$$

$$K_2 \frac{\partial^2 \theta_2}{\partial x^2}(x, t) = \frac{\partial \theta_2}{\partial t}(x, t), \qquad l_1<x<l_2, \quad t>0 \tag{2.2}$$

Boundary conditions

$$k_1 \frac{\partial \theta_1}{\partial x}(x, t) = h_o \theta_1(x, t), \qquad x=0, \quad t>0 \tag{2.3}$$

$$k_1 \frac{\partial \theta_1}{\partial x}(x, t) = h\{\theta_2(x, t) - \theta_1(x, t)\}, \qquad x=l_1, \quad t>0 \tag{2.4}$$

$$k_2 \frac{\partial \theta_2}{\partial x}(x, t) = h\{\theta_2(x, t) - \theta_1(x, t)\}, \qquad x=l_1, \quad t>0 \tag{2.5}$$

$$k_2 \frac{\partial \theta_2}{\partial x}(x, t) = h_i\{\theta_i - \theta_2(x, t)\}, \qquad x=l_2, \quad t>0 \tag{2.6}$$

Initial conditions

$$\left.\begin{aligned} \theta_1(x, 0) &= \theta_o + \frac{(1/h_o + x/k_1)}{(1/h_o + l_1/k_1 + 1/h + \overline{l_2 - l_1}/k_2 + 1/h_i)}(\theta_i - \theta_o) \\ \theta_2(x, 0) &= \theta_o + \frac{(1/h_o + l_1/k_1 + 1/h + \overline{x - l_1}/k_2)}{(1/h_o + l_1/k_1 + 1/h + \overline{l_2 - l_1}/k_2 + 1/h_i)}(\theta_i - \theta_o) \end{aligned}\right\} \tag{2.7}$$

Applied temperature functions

$$\begin{aligned} \theta_o &= \theta_o, & x&=0, & t&\leqslant 0 \\ &= 0, & x&=0, & t&>0 \\ \theta_i &= \theta_i, & x&=l_2, & &\text{all } t \end{aligned}$$

The method of solution follows that described above. An application of the Laplace transform gives:

Transformed differential equations

$$K_1 \frac{d^2\bar{\theta}_1}{dx^2}(x, p) - p\bar{\theta}_1(x, p) + \theta_o + \frac{(1/h_o + x/k_1)}{(1/h_o + l_1/k_1 + 1/h + \overline{l_2 - l_1}/k_2 + 1/h_i)}(\theta_i - \theta_o) = 0 \quad (2.8)$$

$$K_2 \frac{d^2\bar{\theta}_2}{dx^2}(x, p) - p\bar{\theta}_2(x, p) + \theta_o + \frac{(1/h_o + l_1/k_1 + 1/h + \overline{x - l_1}/k_2)}{(1/h_o + l_1/k_1 + 1/h + \overline{l_2 - l_1}/k_2 + 1/h_i)}(\theta_i - \theta_o) = 0 \quad (2.9)$$

Transformed boundary equation

$$k_1 \frac{d\bar{\theta}_1}{dx}(x, p) = h_o\bar{\theta}_1(x, p), \qquad x = 0 \qquad (2.10)$$

$$k_1 \frac{d\bar{\theta}_1}{dx}(x, p) = h\{\bar{\theta}_2(x, p) - \bar{\theta}_1(x, p)\}, \qquad x = l_1 \qquad (2.11)$$

$$k_2 \frac{d\bar{\theta}_2}{dx}(x, p) = h\{\bar{\theta}_2(x, p) - \bar{\theta}_1(x, p)\}, \qquad x = l_1 \qquad (2.12)$$

$$k_2 \frac{d\bar{\theta}_2}{dx}(x, p) = h_i\left\{\frac{\theta_i}{p} - \bar{\theta}_2(x, p)\right\}, \qquad x = l_2 \qquad (2.13)$$

The solution of (2.8) and (2.9) subject to (2.10) through (2.13) is

$$\begin{aligned}\bar{\theta}_1(x, p) = &-\frac{a_0\theta_o}{pf(p)}\left\{\cosh\sqrt{\left(\frac{p}{K_1}\right)}(l_1 - x) + \frac{a}{\sqrt{p}}\sinh\sqrt{\left(\frac{p}{K_1}\right)}(l_1 - x)\right\} \\ &-\frac{a_0 a a_1\theta_o}{pf(p)g(p)}\left\{\cosh\sqrt{\left(\frac{p}{K_1}\right)}x + \frac{a_0}{\sqrt{p}}\sinh\sqrt{\left(\frac{p}{K_1}\right)}x\right\} \\ &\times\left\{\cosh\sqrt{\left(\frac{p}{K_2}\right)}(l_2 - l_1) + \frac{a_2}{\sqrt{p}}\sinh\sqrt{\left(\frac{p}{K_2}\right)}(l_2 - l_1)\right\} \\ &+\frac{R_1(x)}{R_{i.o}}\frac{(\theta_i - \theta_o)}{p} + \frac{\theta_o}{p}, \qquad 0 < x < l_1 \end{aligned} \qquad (2.14)$$

$$\begin{aligned}\bar{\theta}_2(x, p) = &-\frac{a_0 a_1\theta_o}{pg(p)}\left\{\cosh\sqrt{\left(\frac{p}{K_2}\right)}(l_2 - x) + \frac{a_2}{\sqrt{p}}\sinh\sqrt{\left(\frac{p}{K_2}\right)}(l_2 - x)\right\} \\ &+\frac{R_{1.2}(x)}{R_{i.o}}\frac{(\theta_i - \theta_o)}{p} + \frac{\theta_o}{p}, \qquad l_1 < x < l_2 \end{aligned} \qquad (2.15)$$

where

$$f(p)=\sqrt{p}\sinh\sqrt{\left(\frac{p}{K_1}\right)}l_1+(a_0+a)\cosh\sqrt{\left(\frac{p}{K_1}\right)}l_1+\frac{a_0a}{\sqrt{p}}\sinh\sqrt{\left(\frac{p}{K_1}\right)}l_1$$

$$\begin{aligned}g(p)=&(p+a_0a+a_1a_2)\sinh\sqrt{\left(\frac{p}{K_1}\right)}l_1\sinh\sqrt{\left(\frac{p}{K_2}\right)}(l_2-l_1)\\&+\left\{\sqrt{p}(a_1+a_2)+\frac{a_0aa_2}{\sqrt{p}}\right\}\sinh\sqrt{\left(\frac{p}{K_1}\right)}l_1\cosh\sqrt{\left(\frac{p}{K_2}\right)}(l_2-l_1)\\&+\left\{\sqrt{p}(a_0+a)+\frac{a_0a_1a_2}{\sqrt{p}}\right\}\cosh\sqrt{\left(\frac{p}{K_1}\right)}l_1\sinh\sqrt{\left(\frac{p}{K_2}\right)}(l_2-l_1)\\&+(a_0a_1+a_0a_2+aa_2)\cosh\sqrt{\left(\frac{p}{K_1}\right)}l_1\cosh\sqrt{\left(\frac{p}{K_2}\right)}(l_2-l_1)\end{aligned}$$

$$a_0=h_o\frac{\sqrt{K_1}}{k_1};\qquad a=h\frac{\sqrt{K_1}}{k_1};\qquad a_1=h\frac{\sqrt{K_2}}{k_2};\qquad a_2=h_i\frac{\sqrt{K_2}}{k_2}$$

$$R_1(x)=1/h_o+x/k_1,\qquad 0<x<l_1$$

$$R_{1.2}(x)=1/h_o+l_1/k_1+1/h+(x-l_1)/k_2,\qquad l_1<x<l_2$$

$$R_{i.o}=1/h_o+l_1/k_1+1/h+(l_2-l_1)/k_2+1/h_i$$

The solution for the temperature distribution through the cavity wall is obtained from (2.14) and (2.15) in the usual way using the inversion theorem. The integrands are single-valued functions with simple poles at $p=0$, $p=-\alpha_n^2$, and at $p=-\beta_m^2$, where α_n^2, $n=1,2,\ldots$, are the positive roots of

$$\begin{aligned}&\left\{\alpha_n(a_1+a_2)-\frac{a_0aa_2}{\alpha_n}\right\}\tan\frac{\alpha_nl_1}{\sqrt{K_1}}+\left\{\alpha_n(a_0+a)-\frac{a_0a_1a_2}{\alpha_n}\right\}\tan\frac{\alpha_n(l_2-l_1)}{\sqrt{K_2}}\\&\quad+\{(a_0a+a_1a_2)-\alpha_n^2\}\tan\frac{\alpha_nl_1}{\sqrt{K_1}}\tan\frac{\alpha_n(l_2-l_1)}{\sqrt{K_2}}=a_0a_1+a_0a_2+aa_2\end{aligned}\tag{2.16}$$

and β_m, $m=1,2,\ldots$, are the positive roots of

$$\tan\frac{\beta_ml_1}{\sqrt{K_1}}=\frac{(a_0+a)\beta_m}{\beta_m^2-a_0a}\tag{2.17}$$

For the outer leaf, $0<x<l_1$, $t>0$

$$\begin{aligned}\theta_1(x,t)=&\frac{R_1(x)}{R_{i.o}}\theta_i+2a_0\theta_o\sum_{m=1}^{\infty}\\&\times\frac{\{\cos\beta_m(l_1-x)/\sqrt{K_1}+(a/\beta_m)\sin\beta_m(l_1-x)/\sqrt{K_1}\}\exp[-\beta_m^2t]}{\beta_m^2\cdot\psi(\beta_m)}\\&+2a_0aa_1\theta_o\sum_{m=1}^{\infty}\end{aligned}$$

$$\times \frac{\{\cos \beta_m x/\sqrt{K_1} + (a_0/\beta_m) \sin \beta_m x/\sqrt{K_1}\}\{\cos \beta_m (l_2 - l_1)/\sqrt{K_2} + (a_2/\sqrt{K_2}) \sin \beta_m (l_2 - l_1)/\sqrt{K_2}\} \exp[-\beta_m^2 t]}{\beta_m^2 \{g(p)\}_{p=-\beta_m^2} \psi(\beta_m)}$$

$$+ 2a_0 a a_1 \theta_o \sum_{n=1}^{\infty}$$

$$\frac{\{\cos \alpha_n x/\sqrt{K_1} + (a_0/\alpha_n) \sin \alpha_n x/\sqrt{K_1}\}\{\cos \alpha_n (l_2 - l_1)/\sqrt{K_2} + (a_2/\sqrt{K_2}) \sin \alpha_n (l_2 - l_1)/\sqrt{K_2}\} \exp[-\alpha_n^2 t]}{\alpha_n^2 \{f(p)\}_{p=-\alpha_n^2} \phi(\alpha_n)} \tag{2.18}$$

For the inner leaf, $l_1 < x < l_2$, $t > 0$

$$\theta_2(x, t) = \frac{R_{1.2}(x)}{R_{i.o}} \theta_i + 2a_0 a_1 \theta_o \sum_{n=1}^{\infty} \times \frac{\{\cos \alpha_n (l_2 - x)/\sqrt{K_2} + (a_0/\alpha_n) \sin \alpha_n (l_2 - x)/\sqrt{K_2}\} \exp[-\alpha_n^2 t]}{\phi(\alpha_n)} \tag{2.19}$$

where

$$\psi(\beta_m) = \frac{1}{\beta_m}\left(1 + \frac{a_0 + a}{\sqrt{K_1}} + \frac{a_0 a}{\beta_m^2}\right) \sin \frac{\beta_m l_1}{\sqrt{K_1}} + \frac{l_1}{\sqrt{K_1}}\left(1 - \frac{a_0 a}{\beta_m^2}\right) \cos \frac{\beta_m l_1}{\sqrt{K_1}}$$

$$\begin{aligned}\phi(\alpha_n) = {} & [\{\alpha_n^2(a_1 + a_2) - a_0 a a_2\} l_1/\sqrt{K_1} + \{\alpha_n^2(a_0 + a) - a_0 a_1 a_2\}(l_2 - l_1)/\sqrt{K_2}] \\ & \times \cos \alpha_n l_1/\sqrt{K_1} \cos \alpha_n (l_2 - l_1)/ \\ & \sqrt{K_2} + [\alpha_n(a_0 + a) + a_0 a_1 a_2/\alpha_n + \{\alpha_n(a_0 a + a_1 a_2) - \alpha_n^3\} l_1/ \\ & \sqrt{K_1} + \alpha_n(a_0 a_1 + a_0 a_2 + a a_2)(l_2 - l_1)/\sqrt{K_2}] \cos \alpha_n l_1/ \\ & \sqrt{K_1} \sin \alpha_n (l_2 - l_1)/\sqrt{K_2} + [\alpha_n(a_1 + a_2) + a_0 a a_2/ \\ & \alpha_n + \{\alpha_n(a_0 a + a_1 a_2) - \alpha_n^3\}(l_2 - l_1)/ \\ & \sqrt{K_2} + \alpha_n(a_0 a_1 + a_0 a_2 + a a_2) l_1/\sqrt{K_1}] \sin \alpha_n l_1/ \\ & \sqrt{K_1} \cos \alpha_n (l_2 - l_1)/\sqrt{K_2} - [2\alpha_n^2 + \{\alpha_n^2(a_0 + a) - a_0 a_1 a_2\} l_1/ \\ & \sqrt{K_1} + \{\alpha_n^2(a_1 + a_2) - a_0 a a_2\}(l_2 - l_1)/\sqrt{K_2}] \\ & \times \sin \alpha_n l_1/\sqrt{K_1} \sin \alpha_n (l_2 - l_1)\sqrt{K_2}\end{aligned} \tag{2.20}$$

For practical purposes the particular solution $\theta_2(l_2, t)$ representing the temperature of the inside surface of the wall facing the constant temperature room is the most interesting. From (2.19)

$$\theta_2(l_2, t) = \frac{R_{1.2}(l_2)}{R_{i.o}} \theta_i + 2a_0 a_1 \theta_o \sum_{n=1}^{\infty} \frac{\exp[-\alpha_n^2 t]}{\phi(\alpha_n)} \tag{2.21}$$

Numerical calculation of the exact solution (2.21) is lengthy due largely to the labour involved in calculating the roots, α_n, defined by (2.16). An approximate and more easily calculable solution for $\theta_2(l_2, t)$ may be obtained by writing $g(p)$ as a polynomial with distinct zeros at $p = -\alpha_1^2$,

$-\alpha_2^2, \ldots$, and representing this function as an infinite product of linear factors. The inversion $L^{-1}\{\bar{\theta}_2(l_2, p)\}$ then takes the form

$$L^{-1}\{\bar{\theta}_2(l_2, p)\} = \frac{R_{1.2}(l_2)}{R_{\mathrm{i.o}}}\,\theta_{\mathrm{i}} - \sum_{i=1}^{\infty} \frac{a_0 a_1 \theta_{\mathrm{o}}(p+\alpha_i^2)\exp[-\alpha_i^2 t]}{pC\prod_{j=1}^{\infty}(1+p/\alpha_j^2)}, \qquad i \neq j \tag{2.22}$$

where

$$C = a_0 a_1 + a_0 a_2 + a a_2 + a_0 a a_2 l_1/\sqrt{K_1} + a_0 a_1 a_2 (l_2 - l_1)/\sqrt{K_2}$$

Expanding $g(p)$ as a polynomial in p to $\mathcal{O}(p^3)$ and inverting as indicated, the solution for the temperature of the inside surface may be written approximately as

$$\theta_2(l_2, t) \simeq \frac{R_{1.2}(l_2)}{R_{\mathrm{i.o}}}\,\theta_{\mathrm{i}} + \frac{a_0 a_1 \theta_{\mathrm{o}}}{C}\left(\frac{\alpha_2^2 \exp[-\alpha_1^2 t]}{\alpha_2^2 - \alpha_1^2} - \frac{\alpha_1^2 \exp[-\alpha_2^2 t]}{\alpha_2^2 - \alpha_1^2}\right) \tag{2.23}$$

in which $p = -\alpha_1^2, -\alpha_2^2$ are the roots of $g(p)$ expressed simply as a quadratic in p. In general, $\alpha_2^2 \geqslant \alpha_1^2$, so that for large values of t equation (2.23) reduces to

$$\theta_2(l_2, t) \simeq \frac{R_{1.2}(l_2)}{R_{\mathrm{i.o}}}\,\theta_{\mathrm{i}} + \frac{a_0 a_1 \theta_{\mathrm{o}} \exp[-\alpha_1^2 t]}{C} \tag{2.24}$$

It can be shown that $a_0 a_1/C$ simplifies to $(1/h_{\mathrm{i}})/R_{\mathrm{i.o}}$; substituting this equivalent ratio of resistances in (2.24) the approximate solution may now be re-written in the more convenient form

$$\frac{\theta_2(l_2, 0) - \theta_2(l_2, t)}{\theta_{\mathrm{o}}} \simeq \frac{1/h_{\mathrm{i}}}{R_{\mathrm{i.o}}}(1 - \exp[-\alpha_1^2 t]), \quad t \text{ large} \tag{2.25}$$

Equation (2.25) expresses the drop in the temperature of the inside surface as a fraction of the step-function change in the ambient temperature uniformly applied at the outside surface. The quantity $1/\alpha_1^2$ will be recognized as the time taken for the fractional change in the surface temperature to reach $(1-1/e)$ or 63 percent of the ultimate steady state value represented as $(1/h_{\mathrm{i}})\theta_{\mathrm{o}}/R_{\mathrm{i.o}}$; $1/a_1^2$ may be defined therefore as the time-constant of the wall for the problem considered. Writing $g(p)$ as a linear function and isolating p it can be shown that

$$\alpha_1^2 = \left(\frac{1}{2}\left\{W_1 \frac{l_1}{k_1} + W_2 \frac{(l_2 - l_1)}{k_2}\right\} + \frac{1}{R_{\mathrm{i.o}}}\left\{\frac{W_1 + W_2}{h_{\mathrm{i}} h_{\mathrm{o}}} + \frac{W_1}{h_{\mathrm{o}}}\left(\frac{1}{h} + \frac{l_2 - l_1}{k_2}\right) + \frac{W_2}{h_{\mathrm{i}}}\left(\frac{1}{h} + \frac{l_1}{k_1}\right) - \frac{1}{3}\left[W_1\left(\frac{l_1}{k_1}\right)^2 + W_2\left(\frac{l_2 - l_1}{k_2}\right)^2\right]\right\}\right)^{-1} \tag{2.26}$$

where W_1 and W_2 denote the respective heat capacities per unit area of the outer and inner leaves. The result in (2.26) is an approximate expression enabling α_1^2 to be calculated more readily than by graphical solution of the

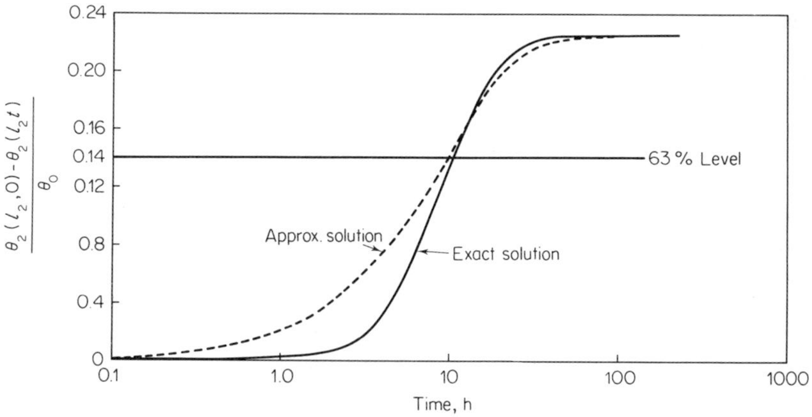

Fig. 1. Transient cooling of a 280 mm cavity brick wall

transcendental equation (2.15). The accuracy of the approximate method may be assessed by comparing numerical results given by (2.24) and the exact solution (2.19). Fig. 1 shows the result of such a calculation applied to the case of an 11 in. cavity wall of brick ($k_1 = k_2$, $K_1 = K_2$, $l_1 = l_2 - l_1$, $W_1 = W_2$) facing a constant temperature enclosure at the inside surface and exposed at the outside surface to a sudden drop through θ_o in the ambient temperature. After an initial time interval the large difference between the results of the two methods begins to decrease and at a value of t equal to the time constant for the cavity wall the approximate solution agrees with the exact solution to within about 5 percent. Thereafter the agreement is close.

The equation (2.26) offers a simple form of solution to a relatively complex calculation of dynamic heat flow in buildings that may be used for readily comparing the temperature response of different wall constructions to a sudden change in the outdoor temperature. At the inside surface the ambient temperature is assumed to be constant. In Chapter 5 a similar problem is considered but with the boundary condition at the inside surface representing heat exchange with an enclosed air-space to which heat is added at a constant rate; for this the time-constant is simply the ratio (heat stored/heat transmitted) in steady state.

It will be recognized that the above solutions for a cavity wall may readily be adjusted to give the corresponding results for homogeneous and two-layer structures. Moreover, if one of the layers is designed to have a high resistance to heat flow then because of the low density of material used for this purpose the thermal capacity of the layer may be regarded as negligibly small. If this is accepted then the solutions for the cavity wall apply without modification to a wall with cavity-fill insulation, a form of construction that is widely used for achieving a high standard of heat insulation in new and existing buildings.

The diurnal variation of the outdoor air temperature is roughly periodic and for calculations of dynamic heat flow in buildings is represented normally by a sine curve rising from a minimum night-time value to a maximum around noon. The additional thermal effects due to the simultaneous influence of radiation and other climatic factors on heat transmission through the building fabric may conveniently be included by defining an equivalent outdoor temperature (eot), as explained in Chapter 1. The variation of this quantity is generally very different from that of the air. This may be seen by comparing the dotted line with the continuous line in Fig. 2 (Pratt, 1958). The dotted line traces the change with time in the temperature of the thin metal cladding to a highly insulated roof and the full line that of the outdoor air. The thin metal sheet on a substrate of high thermal resistance approximates to the requirement for measuring the equivalent temperature as that of an adiabatic surface. Although the dotted curve is not a true measure of the eot, the approximation is sufficient to indicate the manner of its variation.

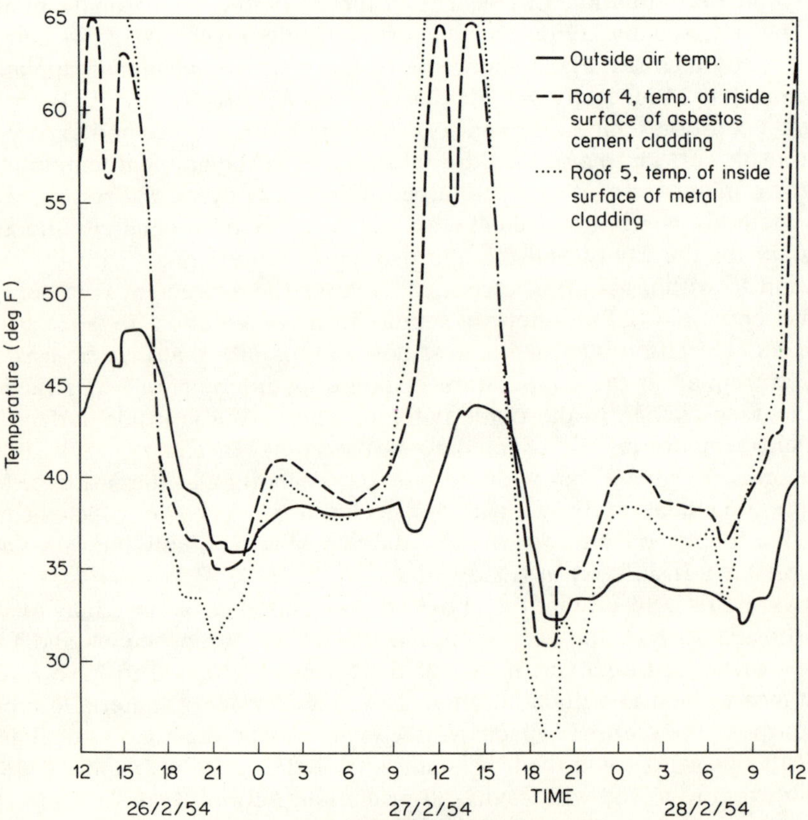

Fig. 2. Graph comparing temperature of metal cladding and of asbestos cement cladding, to insulated roof with outside air temperature

It is interesting to note that throughout the two-day period of observation recorded in Fig. 2 there was little air movement and the sky was unclouded exposing the metal cladding to comparatively strong insolation and a low ambient temperature for long wave radiation heat exchange. This weather condition accounts for the rapid drop in the temperature of the cladding from a high day time value to 3 degC below the temperature of the air after insolation has ceased.

In general the equivalent outdoor temperature varies in a more complex and irregular manner than the air temperature and has a more pronounced effect on the control of temperature in buildings determined by heat conduction through the external fabric. A Fourier series would require several terms to represent the variation with adequate precision in a conventional harmonic analysis of its passage through the structure. The calculation would be laborious and the method restricted to relatively short times. Since heat flow in buildings is considered linear a simpler alternative approach is to regard the curve as a superposition of step-function changes in the temperature, thereby dividing into rectangles the space enclosed by the curve and the time axis in a graph of the temperature variation. Each rectangle represents a pulse of temperature expressed by

$$\theta_o(t) = \theta_o\{H(t) - H(t - t_1)\} \tag{2.27}$$

where θ_o is the sudden change in the ambient temperature and $H(t)$ is the Heaviside Unit Function defined as

$$\begin{aligned} H(t) &= 1, \qquad 0 < t \leq t_1 \\ &= 0, \qquad t \leq 0, \quad t > t_1 \end{aligned}$$

Consider a change θ_o in the ambient temperature represented by (2.27) applied to the outside surface of a solid homogeneous wall, initially in steady state and facing a constant temperature at the inside surface. The resulting temperature of the wall, $\theta(x, t)$, is given by equations (2.1.26) and (2.1.27) in Chapter 1. The first of these shows a rise in temperature at a rate that diminishes exponentially with time increasing towards an ultimate steady state value. After a period t_1, measured from initial time, the outdoor temperature is suddenly reduced by θ_o and the resultant temperature of the wall, i.e. for $t > t_1$, is given by (2.1.27). This expression shows the solution for the effect of the second change in the ambient temperature superimposed on that due to the first.

A result obtained by the method is shown in Fig. 3, where the upper curve represents a sinusoidal variation in the ambient temperatue at the outside surface of a solid brick wall 225 mm thick. This input wave is shown divided into rectangular pulses of temperature as previously described, the number depending upon the degree of precision required to determine the temperature variation at the position selected, in this instance the inside surface. The passage of each pulse through the wall has been traced using the transient temperature response curve for a homogeneous solid in Fig. 4. The curve

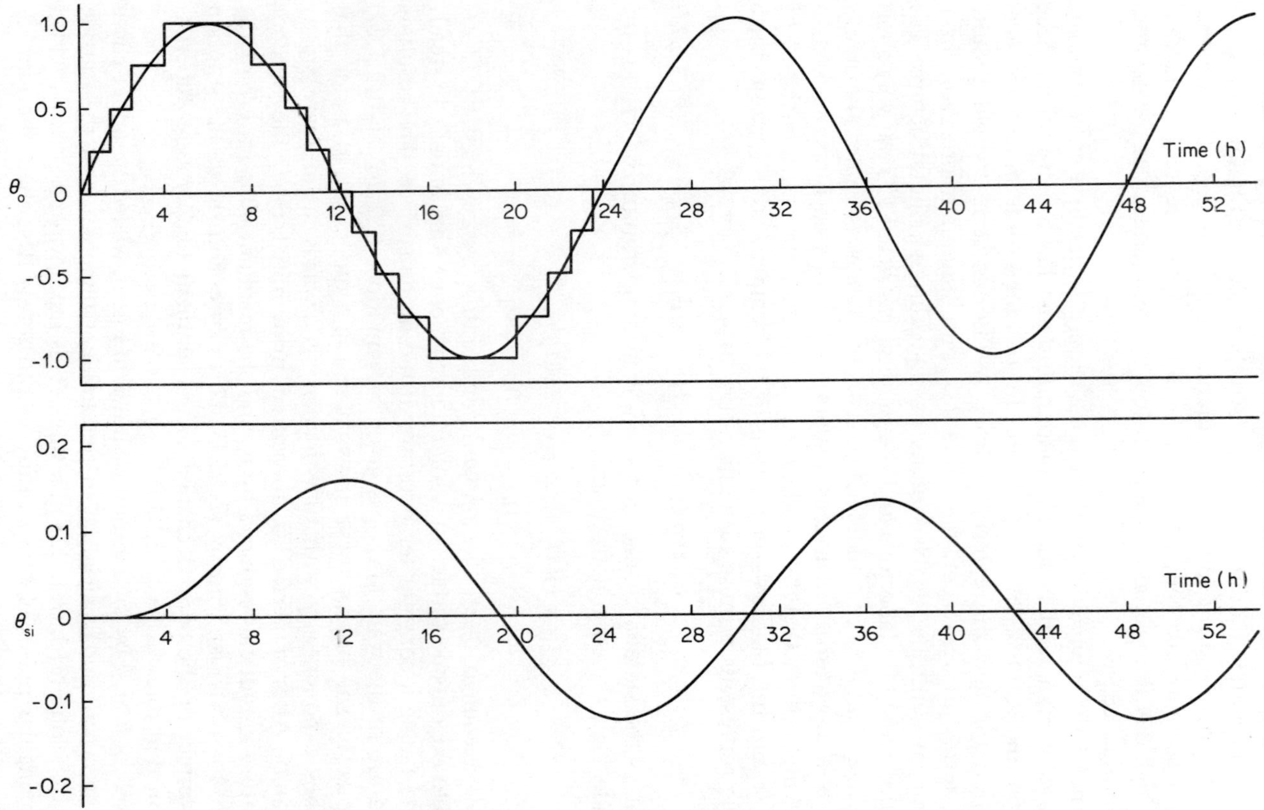

Fig. 3. 225 mm solid brick wall: response to sine-wave temperature variation

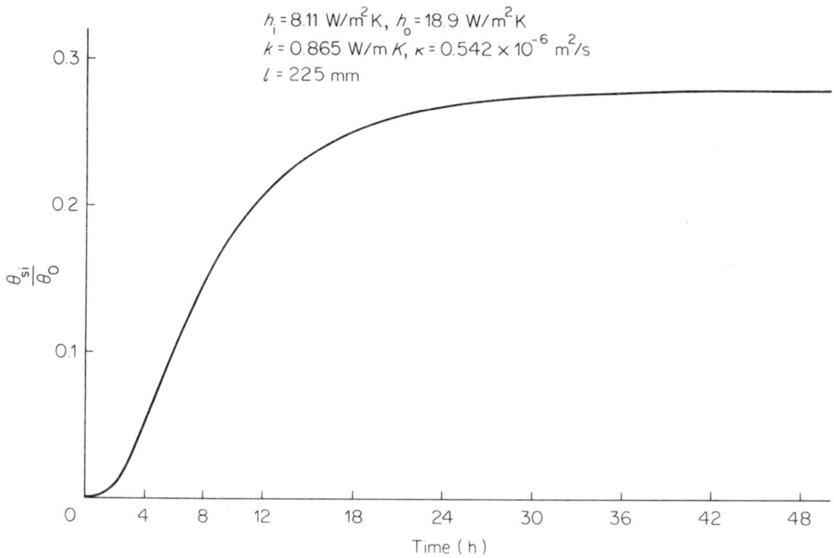

Fig. 4. Response of inside surface temperature of 220 mm solid brick wall to step-change in outside air temperature

has been calculated from equation (2.1.26) in Chapter 1 with $\xi = 1$ corresponding to the inside surface of the wall where the temperature is denoted by θ_{si}. Values of the time-lag and amplitude ratio are obtained by comparing the input and output curves in Fig. 3 after the first complete cycle (24 h) when the system has settled into a steady periodic state. The reader may check the agreement of these values with those given by Fig. 5. and Fig. 6. in Chapter 1 for the same example.

An input wave, also of regular form that may be likened to the daily high peaks in the equivalent outdoor temperature due to insolation, is the 'rectified' sine curve in Fig. 5. The effect of this on the temperature of the inside surface of a 225 mm thick brick wall as determined by the above method is shown by the lower curve in the diagram.

Further examples are given in Figs. 6 and 7 which show the result of a 'saw-tooth' wave form, periodicity 12 h, applied to a solid 225 mm brick wall and a 280 mm cavity brick wall, respectively. After about 36 h when the initial rise in the temperature of the wall has ceased, the steady oscillation in θ_{si} is virtually sinusoidal, a result that has been used with advantage in experiments to determine the thermal diffusivity of walls (see Chapter 9). The transient response curve for the cavity wall shown in Fig. 8 and used in preparing Fig. 7 was calculated with equation (2.20).

The smaller amplitude ratio and longer time-lag will be noted in θ_{si} for the cavity wall of heavier construction and lower thermal conductance.

The only analytical result required for plotting $\theta(x, t)$ by the superimposition principle outlined above is the solution for the variation of temperature

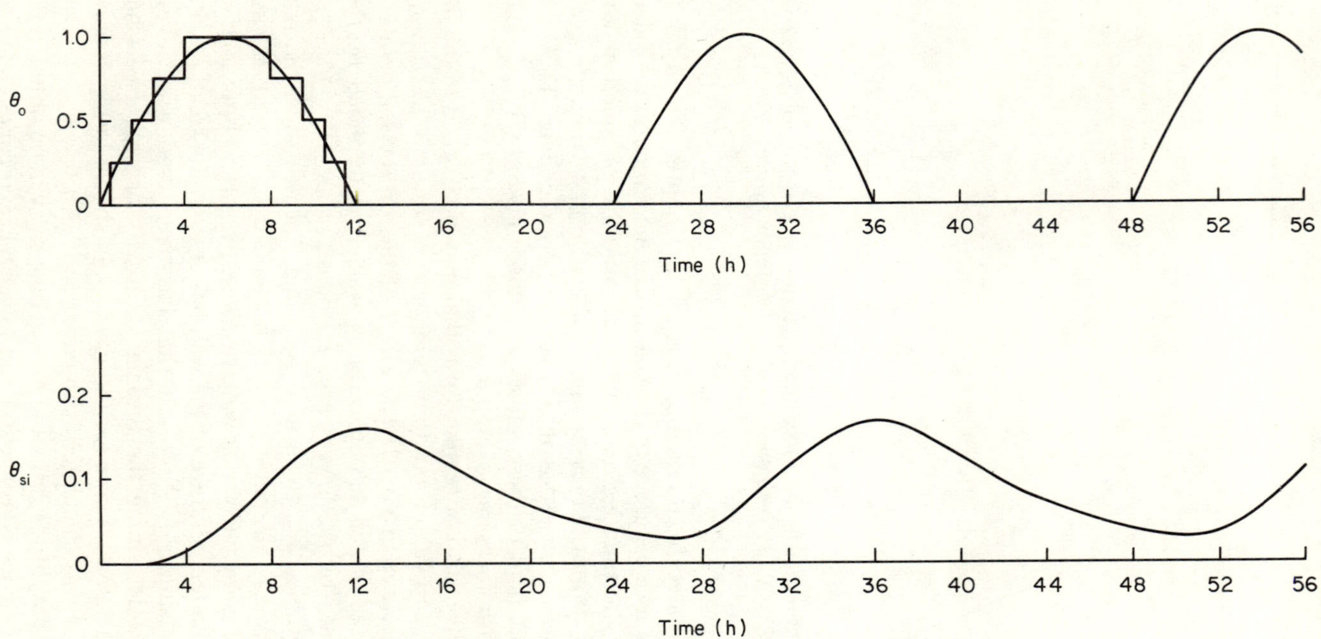

Fig. 5. 225 mm solid brick wall: response to rectified sine wave

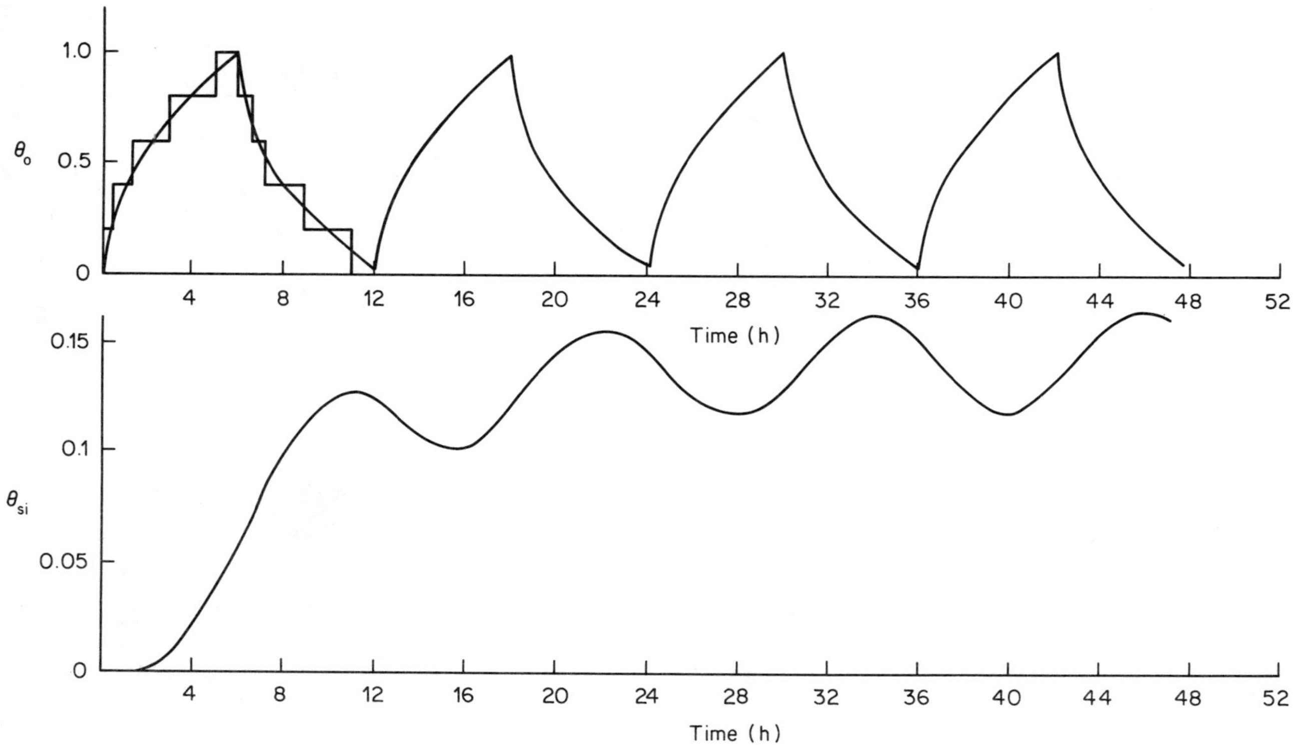

Fig. 6. 225 mm solid brick wall: response to 'saw tooth' wave

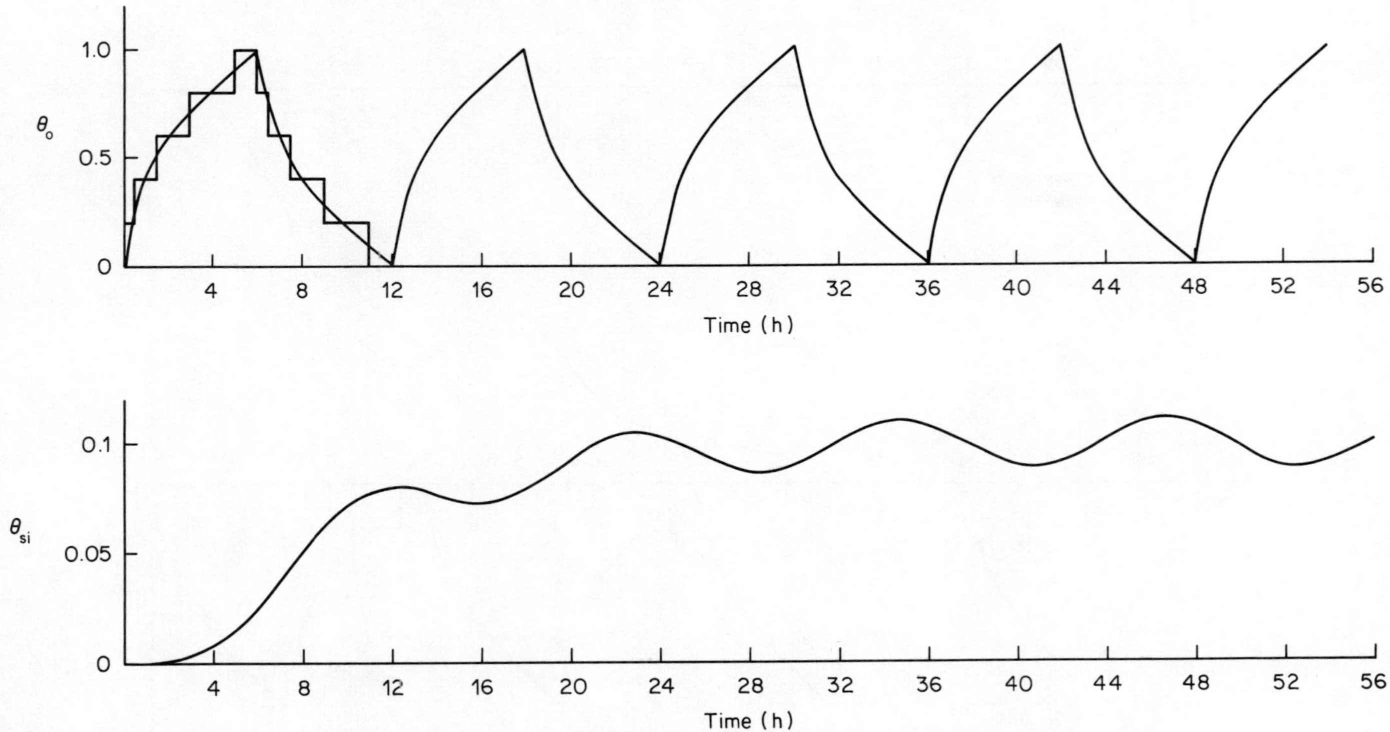

Fig. 7. 280 mm cavity brick wall: response to 'saw tooth' wave

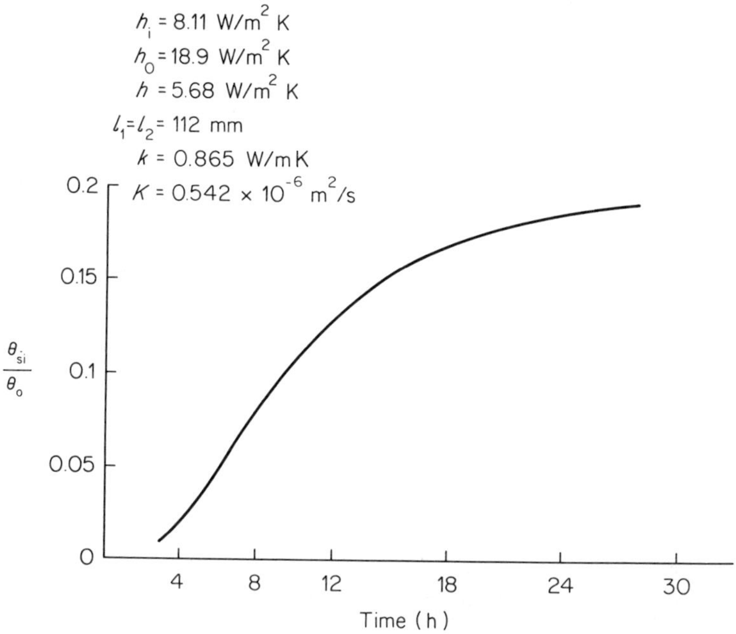

Fig. 8. Response of inside surface temperature of 280 mm cavity brick wall to step-change in outside air temperature

with time in the structure due to a step-function change in the ambient temperature applied to one surface. The step changes may be positive or negative representing an increase or decrease, respectively, in the applied ambient temperature and the results combined accordingly to give the resultant temperature pattern in the structure. The technique is reducible to a simple drill, the amount of labour involved depending upon the form of the input function and increasing slightly, though without added complication, with the accuracy required. A relatively crude approximation to the applied boundary function, regarded as a superimposition of rectangular pulses, gives a measure of agreement with the exact thermal response that may be considered acceptable for many problems of variable heat flow in buildings. Moreover, the input temperature variation—or heat flux if required (see equations (2.1.48) and (2.1.49) in Chapter 1)—may be regular or random and applied singly to either surface, or to both simultaneously, provided the appropriate transient response curves are available. These curves and the U-value together afford a comprehensive evaluation of heat flow in structures.

REFERENCES

Bruckmayer, F. (1944) *Ges. Ing.*, **67,** 147.

Carslaw, H. S. and Jaeger, J. C. (1959) *Conduction of Heat in Solids,* Clarendon Press, Oxford.

Crank, J. (1956) *The Mathematics of Diffusion*, Clarendon Press, Oxford.
Esser, W. and Krischer, O. (1930) *The Computation of the Warming-up and Cooling of Plane and Cylindrical Walls (Houses and Piping)*, Springer, Berlin.
Mackey, A. M. and Wright, L. T. (1946) *Heat. Pip. Air Cond.*, **18,** 6, 107.
Pratt, A. W. (1958) *Condensation in Sheeted Roofs*, National Building Studies Research Paper no. 23, HMSO, London.
Shklover, A. M. (1945) *Method of Calculating Heat Transmission in Buildings*, Academy of Architecture of the USSR, Inst. of Building Technology, Moscow.

CHAPTER 7

Solar Heat Gain to Buildings through Glass

Glass has the distinctive property of being transparent to short-wave radiation. The spectral transmission curve of 6 mm Float glass for radiation at normal incidence, shown in Fig. 1, is typical of the clear glass commonly used in windows. The transmitted radiation lies within the range of wavelengths 0.3–3 μm approximately. Visible radiation lies within the range 0.4–0.8 μm and, as Fig. 2 shows, the peak intensity of the solar energy received at the earth's surface lies within this narrow band of the solar spectrum. To fix orders of magnitude the maximum flux of direct solar energy likely to be received on a clear June day in southern England on a surface normal to the sun's rays is about 630 W/m^2 (Stagg, 1950). Assuming an average figure of 85 percent for the transmission factor for solar radiation passing through clear glass this value corresponds to an instantaneous flux of about 270 W/m^2 transmitted through a vertical window facing south. The corresponding December maximum is also about 270 W/m^2; the values are similar since the reduction in solar radiation intensity is accompanied by a more nearly normal incidence upon a vertical plane.

In addition to this direct radiation an exposed surface receives diffuse radiation more or less evenly distributed over the whole sky vault. Stagg (1950) reports measurements which show that the diffuse solar radiation received on a horizontal surface can attain a maximum intensity of about 570 W/m^2, giving a thermal flux through a vertical window of about 235 W/m^2. All the figures quoted above are peak values; mean values on the whole will be much lower.

A proportion of the solar radiation transmitted through glass into an enclosure is absorbed at the opaque interior surfaces, the remainder being reflected back through the glass. Inside a building, the heat gain to the air resulting from the absorption of solar radiation by the internal opaque surfaces adds to the cooling load in summer and reduces the heating load in winter. The relationship between the thermal load and the solar heat entering an enclosure through glass is examined theoretically below. Two situations of practical interest are considered. In one the temperature of the

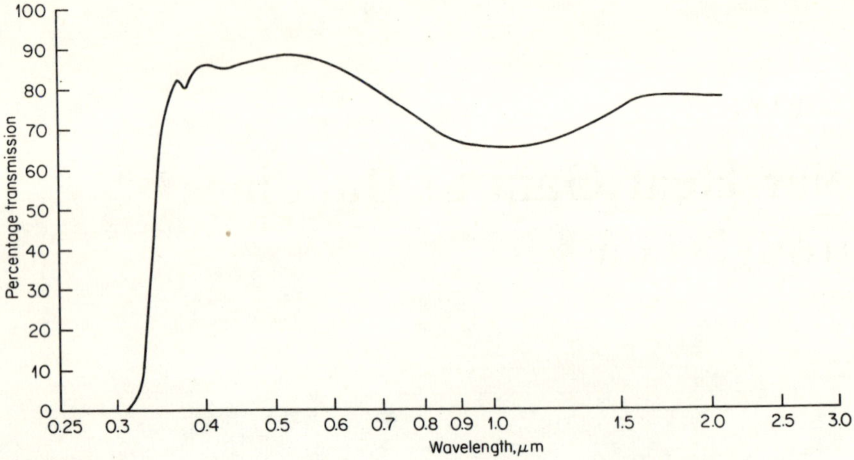

Fig. 1. Spectral transmission curve of 6 mm Pilkington glass. (Reproduced by kind permission of Pilkington Flat Glass Limited, Merseyside

air within the enclosure is allowed to vary freely; in the other the air temperature is thermostatically controlled at a fixed value.

The analysis assumes that the incoming solar radiation is uniformly distributed over the opaque interior surfaces thereby restricting the results to radiation that is wholly diffuse. The assumption clearly cannot apply to direct radiation although a proportion of this may be regarded as diffuse after repeated internal reflection.

A theoretical examination of the cooling load on a sunlit, thermostatically controlled enclosure is reported by Mackey and Gay (1952). The present

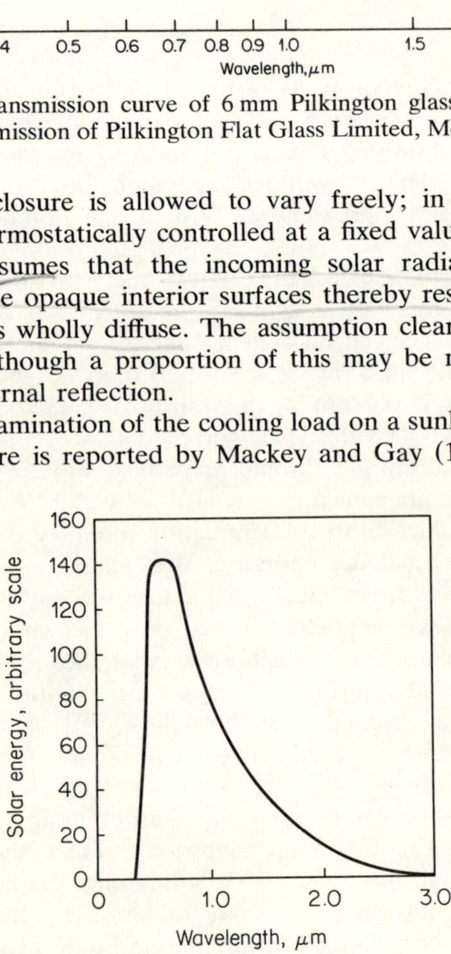

Fig. 2. Spectral distribution of solar energy at sea-level. Typical curve ignoring atmospheric absorption bands.

analysis examines the problem in greater detail by introducing boundary conditions which define transient and periodic heat flow in a form that takes account of ventilation of the enclosure and conduction heat transfer through the glazing as affected by the solar heat absorbing properties of the glass. In addition the calculation is extended to include the period after insolation has ceased when the solar energy stored in the enclosure wall continues to be effective though at a diminishing rate.

1. AIR TEMPERATURE CHANGE IN A VENTILATED ENCLOSURE DUE TO SOLAR HEAT TRANSMISSION THROUGH GLASS

The analysis examines the temperature change with time in an enclosure (Fig. 3) contained by a wall composed partly of glass and partly of opaque material. The area of the glass per unit area of the opaque wall surface is denoted by A_g. The opaque wall is represented as a large, homogeneous, finite slab, thickness l of isotropic material of thermal conductivity k and thermal diffusivity K. Initially the air inside and outside the enclosure is at the same temperaure which, for convenience, is put at zero level. For all values of time t the ambient temperature at the outside surface is uniformly distributed and maintained at zero level. Initially the system is in steady state. Subsequently, during the interval $0<t\leq t_1$ the glass is exposed to a time-variation of solar energy and a flux $I(t)$ of short-wave radiation enters the outside surface of the glass. A fraction of this amount, ϕ for single glazing, $\phi_1+\phi_2$ for double glazing, is absorbed by the glass. The remainder of the radiation enters the enclosure through the inside surface of the glass and is assumed to be uniformly distributed over the opaque wall surface which has an absorptivity α for short-wave radiation. The interval $t=0$ to $t=t_1$ will be described as the period of insolation. For $t>t_1$, $I(t)=0$. The surface of the opaque wall, temperature $\theta(0, t)$, transfers heat by convection

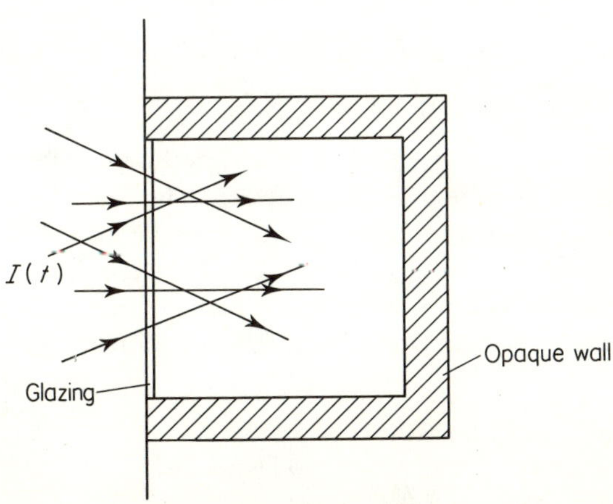

Fig. 3. Insolated enclosure

to the inside air, temperature $\theta_i(t)$, at a rate h_{ci} times the temperature difference per unit area of this wall surface. Heat is also exchanged by radiation between the opaque surface and the inside surface of the glass which is at a temperature $\theta_g(t)$. Natural convection is assumed and the temperature dependence of the convection transfer coefficient, h_{ci}, is neglected; the temperature dependence of the radiation coefficients is also neglected so that the system of defining equations is linear throughout. At the surface $x=l$ the opaque wall loses heat at a rate h_{so} times the temperature of the surface $\theta(l, t)$ per unit area of the wall. Ventilation to the outside air removes heat from the enclosure at a rate C per degree temperature difference and per unit area of the opaque surface for all $t>0$. The analysis ignores the thermal capacity of the air inside the enclosure since its effect on the temperature change will be negligible after a few minutes measured from initial time.

Neglecting edge losses and representing the heat as flowing in the x-direction only through the wall it is required to calculate a solution for $\theta_i(t)$ that satisfies the following formulation

$$\frac{\partial\theta(x,t)}{\partial t}=K\frac{\partial^2\theta(x,t)}{\partial x^2}, \qquad 0<x<l, \quad t>0 \tag{1.1}$$

$$-k\frac{\partial\theta(x,t)}{\partial x}=\Gamma I(t)-h_{ci}\{\theta(x,t)-\theta_i(t)\}$$

$$-\sigma(F_aF_e)_{x.g}\{[T(x,t)]^4-[T_g(t)]^4\}, \qquad x=0, \quad 0<t\leq t_1 \tag{1.2}$$

$$-k\frac{\partial\theta(x,t)}{\partial x}=h_{so}\theta(x,t), \qquad x=l, \quad \text{all } t \tag{1.3}$$

$$\theta(x,t)=0, \qquad 0<x<l, \quad t\leq 0$$

where

$$\Gamma=\begin{cases}\alpha(1+A_g)(1-\phi)A_g/(\alpha+A_g), & \text{for single glazing}\\ \alpha(1+A_g)(1-\phi_1-\phi_2)A_g/(\alpha+A_g), & \text{for double glazing}\end{cases}$$

The symbol Γ denotes the fraction of the radiation incident at the outside surface of the glazing which is absorbed at the interior surfaces of the enclosure. The expressions for this quantity are derived below.

Referring to Fig. 3, the total radiation entering the enclosure is $A_gA_w\Phi I$. Radiation absorbed at interior surfaces on *first* incidence $=\alpha A_gA_w\Phi I$. Radiation reflected from interior surfaces at first incidence $=(1-\alpha)A_gA_w\Phi I$, of which an amount

$$\frac{A_g}{(1+A_g)}(1-\alpha)A_gA_w\Phi I$$

is lost from the enclosure through the glazing, leaving

$$\frac{(1-\alpha)A_g A_w \Phi I}{(1+A_g)}$$

for subsequent reflection or absorption.

Radiation absorbed at interior surfaces on *second* incidence

$$=\frac{\alpha(1-\alpha)A_g A_w \Phi I}{(1+A_g)}$$

Continuing as above the total radiation absorbed in the enclosure is

$$A_g A_w \Phi I\alpha\left[1+\frac{(1-\alpha)}{(1+A_g)}+\frac{(1-\alpha)^2}{(1+A_g)^2}+\cdots\right]$$

This is a convergent geometric series and may be represented therefore as

$$\frac{A_g A_w(1+A_g)\Phi I\alpha}{(\alpha+A_g)}$$

Thus, the fraction of the incident radiation absorbed per unit area of the opaque wall surface is

$$\frac{\Phi A_g(1+A_g)\alpha}{(\alpha+A_g)}=\Gamma$$

Putting

$$\Phi=\begin{cases}1-\phi, & \text{for single glazing}\\ 1-\phi_1-\phi_2, & \text{for double glazing}\end{cases}$$

the expressions for Γ following equation (1.3) are obtained.

For convenience the quantities defining radiation heat exchange in the boundary equation (1.2) will be abbreviated to the expression $h_{ri}[\theta(x, t)-\theta_g]$, where h_{ri} denotes a coefficient of radiation transfer per unit area of the opaque surface.

Before proceeding to solve (1.1) through (1.3) it is necessary to make a suitable substitution for $\theta_i(t)$. This is obtained by considering heat flow through glass and assuming that it takes place in steady state for all values of time. The implication thereby that the thermal capacity of the glazing may be regarded as negligibly small is examined below. A form of glazing with the widest practical interest is considered. It consists of two plane sheets of glass, mounted vertically and spaced a small distance apart to contain an airspace of uniform thickness that may be sealed or ventilated. The cavity thickness is small compared with the edge dimensions and, where ventilated, the air change is to the outside air. The arrangement is shown in Fig. 4: from this diagram it will be seen that, in steady heat flow, the following system of

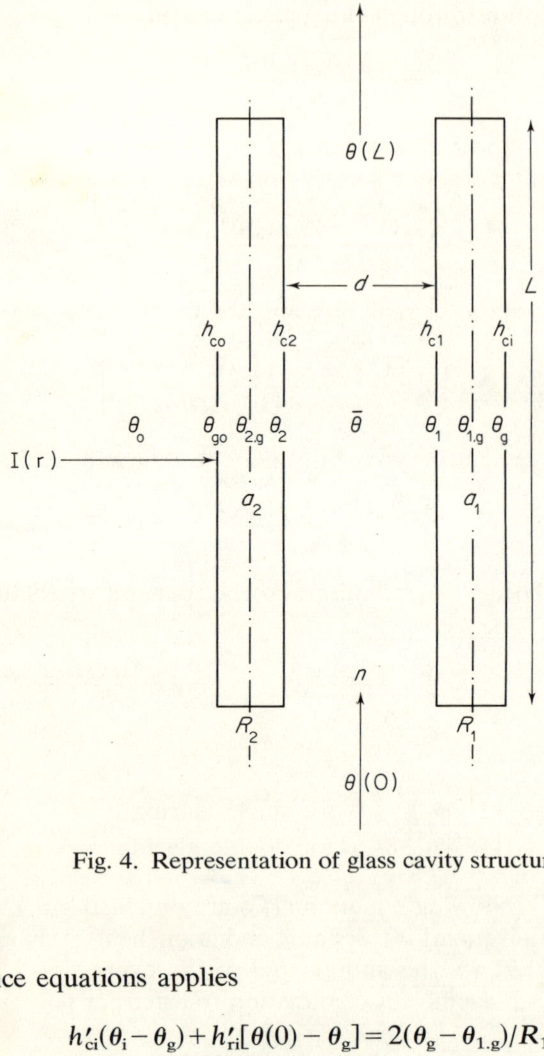

Fig. 4. Representation of glass cavity structure

heat balance equations applies

$$
\begin{aligned}
&h'_{ci}(\theta_i-\theta_g)+h'_{ri}[\theta(0)-\theta_g]=2(\theta_g-\theta_{1.g})/R_1\\
&2(\theta_g-\theta_{1.g})/R_1+\phi_1 I(t)=2(\theta_{1.g}-\theta_1)/R_1\\
&2(\theta_{1.g}-\theta_1)/R_1=h_{c1}(\theta_1-\theta_c)+h_{ra}(\theta_1-\theta_2)\\
&h_{c1}(\theta_1-\theta_c)=nws[\theta(L)-\theta(0)]+h_{c2}(\theta_c-\theta_2)\\
&h_{c2}(\theta_c-\theta_2)+h_{ra}(\theta_1-\theta_2)=2(\theta_2-\theta_{2.g})/R_2\\
&2(\theta_2-\theta_{2.g})/R_2+\phi_2 I(t)=2(\theta_{2.g}-\theta_{go})/R_2\\
&2(\theta_{2.g}-\theta_{go})/R_2=h'_{co}(\theta_{go}-\theta_o)+h'_{ro}(\theta_{go}-\theta_{s.o})
\end{aligned}
\tag{1.4}
$$

The quantities $\phi_1 I(t)$ and $\phi_2 I(t)$ denote the amounts of solar radiation absorbed by each sheet of glass and are represented as equivalent heat sources located at the mid-planes of the inner and outer sheets, respectively.

The quantities h'_{ci}, h'_{co}, h'_{ri}, and h'_{ro}, are heat transfer coefficients corresponding to h_{ci}, h_{co}, h_{ri}, and h_{ro}, but referring to the surfaces of the glazing and based upon the unit area of glazing.

For those cases where the airspace is ventilated the mean temperature of the air in the cavity, θ_c, may be calculated by considering the heat balance for an air stream bounded by two plane isothermal surfaces of length L in the direction of airflow, forming a channel of uniform width w that is small compared with the edge dimensions. It can then be shown (see Chapter 3) that

$$\theta_c = \frac{1}{L}\int_0^L \{\psi + [\theta(0) - \psi]\exp[-x(h_{c1} + h_{c2})/vsw]\}\,dx$$

where $\psi = (h_{c1}\theta_1 + h_{c2}\theta_2)/(h_{c1} + h_{c2})$, v is the air velocity, and s is the volumetric specific heat of air. This expression for θ_c assumes that the convection coefficients h_{c1} and h_{c2} are constant.

Defining

$$U(0) = \{h'_{ci}(\theta_i - \theta_g) + h'_{ri}[\theta(0) - \theta_g]\}/(\theta_i - \theta_o)$$

as the steady state thermal transmittance (air-to-air heat transfer coefficient) of the glazing when shaded from solar radiation the equations (1.4) yield the following results for the rate $H_g(t)$ at which heat is conducted through unit area of the glass:

for single glazing, surface-to-surface resistance R,

$$H_g(t) = U(0)(\theta_i - \theta_o) - U(0)\phi I(t)(R_{so} + R/2);$$

for double glazing, cavity sealed,

$$H_g(t) = U(0)(\theta_i - \theta_o) - U(0)I(t)[\phi_1(R_{so} + R_1/2) + \phi_2(R_{so} + R_1 + R_c + R_2/2)]$$

where R_c, the thermal resistance of the cavity is

$$R_c = (\theta_1 - \theta_2)/[h_{c1}(\theta_1 - \theta_c) + h_{ra}(\theta_1 - \theta_2)],$$

the thermal resistance at the outside surface is

$$R_{so} = (\theta_{go} - \theta_o)/[h_{co}(\theta_{go} - \theta_o) + h_{ro}(\theta_{go} - \theta_{s.o})],$$

and R_1, R_2 denote the thermal resistance of the inner and outer sheets of glass, respectively.

Now, at time t, the instantaneous heat balance of the enclosure is

$$h_{ci}[\theta(0, t) - \theta_i(t)] + h_{ri}[\theta(0, t) - \theta_g(t)]$$
$$= [C + U(0)A_g]\theta_i(t) - U(0)A_g\beta I(t), \qquad 0 < t \leq t_1 \qquad (1.5)$$

where β is a characteristic of the glass defined as follows:

for single glazing

$$\beta = \phi(R_{so} + R/2)$$

for double glazing

$$\beta = \phi_1(R_{so}+R_1/2)+\phi_2(R_{so}+R_1+R_c+R_2/2)$$

Consider the interval $0<t\leq t_1$; this is the period of insolation when the temperature of the inside surface of the glass is

$$\theta_g(t)=[1-U(0)R_{si}]\theta_i(t)+U(0)R_{si}\beta I(t) \tag{1.6}$$

where R_{si} is the thermal resistance at this inside surface. Eliminating $\theta_g(t)$ between (1.5) and (1.6) gives the following expression for the indoor air temperature:

$$\theta_i(t)=\frac{h_{si}\theta(0, t)-U(0)\beta I(t)(h_{ri}R_{si}-A_g)}{h_{si}+C-U(0)(h_{ri}R_{si}-A_g)}, \qquad 0<t\leq t_1 \tag{1.7}$$

where $h_{si}=h_{ri}+h_{ci}$, denoting the surface transfer coefficient at $x=0$.

With the aid of (1.6) and (1.7) the formulation (1.1) through (1.3) is now written more conveniently as

$$\frac{\partial\theta(x, t)}{\partial t}=K\frac{\partial^2\theta(x, t)}{\partial x^2}, \qquad 0<x<l, \quad t>0 \tag{1.8}$$

$$-k\frac{\partial\theta(x, t)}{\partial x}=-y\theta(x, t)+\eta I(t), \qquad x=0, \quad t>0 \tag{1.9}$$

$$k\frac{\partial\theta(x, t)}{\partial x}=h_{so}\theta(x, t), \qquad x=l, \quad \text{all } t \tag{1.10}$$

where

$$y=h_{si}[C+U(0)A_g]/[h_{si}+C+U(0)A_g-h_{ri}U(0)R_{si}]$$

and

$$\eta=\Gamma+U(0)\beta(A_g h_{si}+h_{ri}R_{si}C)/[h_{si}+C+U(0)A_g-h_{ri}U(0)R_{si}]$$

The quantity y has the dimensions of a heat transfer coefficient, and η is a constant; both quantities represent mean conditions over the intervals indicated.

An application of the Laplace transform to the above formulation gives the following subsidiary equations

$$K\frac{d^2\bar{\theta}(x, p)}{dx^2}-p\bar{\theta}(x, p)=0, \qquad 0<x<l \tag{1.11}$$

$$-k\frac{d\bar{\theta}(x, p)}{dx}=-y\bar{\theta}(x, p)+\eta L[I(t)], \qquad x=0 \tag{1.12}$$

$$-k\frac{d\bar{\theta}(x, p)}{dx}=h_{so}\bar{\theta}(x, p), \qquad x=l \tag{1.13}$$

where

$$\bar{\theta}(x, p)=L[\theta(x, t)]$$

the operator L implying the transformation

$$L[\theta(x, t)] = \int_0^\infty e^{-pt}\theta(x, t)\, dt$$

with the inverse relationship

$$\theta(x, t) = \frac{1}{2\pi i}\int_{\gamma - i\infty}^{\gamma + i\infty} L[\theta(x, t)]e^{pt}\, dp \qquad (\gamma > 0)$$

The function $I(t)$ may assume a variety of different forms: two forms will be considered, a rectangular pulse (1.i) and a sinusoidal pulse (1.ii), each with two representations (a) and (b) denoting a single pulse and a regular series of pulses respectively.

[*1.i*(*a*)]—A single rectangular pulse of height proportional to a constant value I_0 which is 'on' for the duration $t = 0$ to $t = t_1$, and 'off' for $t > t_1$.

The functional representation is

$$I(t) = I_0[H(t) - H(t - t_1)] \tag{1.14}$$

where $H(t)$ is the Heaviside unit function which is unity for $t \geqslant 0$ and zero for $t < 0$. Then

$$L[I(t)] = \frac{I_0}{p}(1 - e^{-pt_1}) \tag{1.15}$$

Consider the interval $0 < t \leqslant t_1$; the general solution of (1.11) satisfying the boundary conditions (1.12) and (1.13) is of the form

$$\bar{\theta}(x, p) = \eta \frac{I_0}{p}\frac{f(p)}{g(p)} \tag{1.16}$$

where

$$\begin{aligned} f(p) &= k\sqrt{(p/K)} \cosh \sqrt{(p/K)}(l - x) + h_{so} \sinh \sqrt{(p/K)}(l - x) \\ g(p) &= k\sqrt{(p/K)}[k\sqrt{(p/K)} \sinh \sqrt{(p/K)}l + h_{so} \cosh \sqrt{(p/K)}l] \\ &\quad + y[k\sqrt{(p/K)} \cosh \sqrt{(p/K)}l + h_{so} \sinh \sqrt{(p/K)}l] \end{aligned} \tag{1.17}$$

The transform solution (1.16) is inverted without difficulty to give

$$\theta(x, t) = \frac{\eta I_0(1/h_{so} + (l - x)/k)}{1 + y(1/h_{so} + l/k)} + \eta I_0 \sum_{n=1}^{\infty} \frac{F(\alpha_n)}{G(\alpha_n)} \frac{e^{-\alpha_n^2 t}}{(-\alpha_n^2)} \tag{1.18}$$

where

$$\begin{aligned} F(\alpha_n) &= h_{so}\sqrt{(K)} \sin \alpha_n(l - x)/\sqrt{K} + k\alpha_n \cos \alpha_n(l - x)/\sqrt{K} \\ G(\alpha_n) &= \frac{(2k^2 + lkh_{so} + lky)}{2\sqrt{K}} \sin \alpha_n l/\sqrt{K} \\ &\quad - \frac{[K(lh_{so}y + kh_{so} + ky) - lk^2\alpha_n^2]}{2\alpha_n K} \cos \alpha_n l/\sqrt{K} \end{aligned} \tag{1.19}$$

and the zeros α_n, $n = 1, 2, \ldots$, are defined by

$$\tan \frac{\alpha_n l}{\sqrt{K}} = \frac{(h_{so} + y)k\alpha_n \sqrt{K}}{k^2\alpha_n^2 - h_{so}Ky} \tag{1.20}$$

Writing $\theta(0, t)$ from (1.18) and substituting in (1.7) there follows the required solution (1.21) for the rise in the inside air temperature during the period of insolation

$$\theta_i(t) = \frac{I_0 y}{C + U(0)A_g} \left\{ \frac{\eta(1/h_{so} + l/k)}{1 + y(1/h_{so} + l/k)} - \frac{U(0)\beta}{h_{si}} (h_{ri}R_{si} - A_g) \right.$$
$$+ \eta \sum_{n=1}^{\infty} [k\alpha_n \cos \alpha_n l/\sqrt{K} + h_{so}\sqrt{K}$$
$$\left. \times \sin \alpha_n l/\sqrt{K} e^{-\alpha_n^2 t}/[-\alpha_n^2 G(\alpha_n)] \right\}, \qquad 0 < t \leq t_1 \tag{1.21}$$

For the period $t > t_1$, i.e. when insolation has ceased, the result (1.18) leads to the solution for the temperature of the opaque surface at $x = 0$ which, from (1.15), may be written down immediately as

$$\theta(0, t) = \eta I_0 \sum_{n=1}^{\infty} \frac{k\alpha_n \cos \alpha_n l/\sqrt{K} + h_{so}\sqrt{K} \sin \alpha_n l/\sqrt{K}}{(-\alpha_n^2)G(\alpha_n)} [e^{-\alpha_n^2 t} - e^{-\alpha_n^2 (t - t_1)}],$$
$$t > t_1 \tag{1.22}$$

When $t > t_1$, $I(t) = 0$ and from (1.7) it follows that the temperature of the air in the enclosure is then

$$\theta_i(t) = h_{si}\theta(0, t)/[h_{si} + C - U(0)(h_{ri}R_{si} - A_g)], \qquad t > t_1 \tag{1.23}$$

It will be noted that the temperature of the inside surface of the glass may be obtained from substituting the solution for $\theta_i(t)$ into (1.6). This temperature may be important in considerations of indoor climate especially if the glass is of the solar heat absorbing variety.

[1.*i*(*b*)]—The result obtained above for the single rectangular pulse will now be used to extend the analysis by calculating the solution for the temperature variation in the enclosure when $I(t)$ describes a periodic rectangular wave form represented by the function

$$\begin{aligned} I(t) &= 0, & t &< 0 \\ &= I_0, & nT &< t < nT + t_1, \quad n = 0, 1, 2, \ldots \\ &= 0, & nT + t_1 &< t < (n+1)T \end{aligned} \tag{1.24}$$

where the period $T = t_1 + t_2$, with t_2 the time interval between successive periods, duration t_1, of insolation. Taking the transform

$$L[I(t)] = \frac{I_0}{p} \left(\frac{1 - e^{-pt_1}}{1 - e^{-pT}} \right) \tag{1.25}$$

and substituting (1.25) for I_0/p in (1.16) gives the general solution for the temperature of the enclosure wall as

$$\theta(x, t) = \frac{1}{2\pi i} \int_{\gamma - i\infty}^{\gamma + i\infty} \frac{\eta I_0}{p g(p)} \left(\frac{1 - e^{-pt_1}}{1 - e^{-pT}}\right) \{k\sqrt{(p/K)} \cosh \sqrt{(p/K)}(l - x)$$
$$+ h_{so} \sinh \sqrt{(p/K)}(l - x)\} e^{pt} \, dp \qquad (1.26)$$

Taking the inversion (1.26) and following Carslaw and Jaeger (1959, p. 401), the result is written

$$\theta(x, t) = \theta_s(x, t) + \eta I \sum_{n=1}^{\infty} \frac{F(\alpha_n)}{(-\alpha_n^2) G(\alpha_n)} \left[\frac{e^{-\alpha_n^2 t} - e^{-\alpha_n^2 (t - t_1)}}{1 - e^{\alpha_n^2 (t_1 + t_2)}}\right] \qquad (1.27)$$

consisting of a steady periodic part $\theta_s(x, t)$ and a transient part.

By equating (1.18) and (1.27) it follows that the temperature of the enclosure wall during insolation when the system is in a steady periodic condition is given by

$$\theta_s(x, t) = \frac{\eta I_0 [1/h_{so} + (l - x)/k]}{1 + y(1/h_{so} + l/k)} - \eta I_0 \sum_{n=1}^{\infty} \frac{F(\alpha_n)}{(-\alpha_n^2) G(\alpha_n)} \left[\frac{e^{\alpha_n^2 (t_1 - t)} - e^{\alpha_n^2 (T - t)}}{1 - e^{\alpha_n^2 T}}\right],$$
$$0 < t \leqslant t_1 \qquad (1.28)$$

Similarly, from (1.22) and (1.27) the temperature during the regular intervals without insolation is

$$\theta_s(x, t) = \eta I_0 \sum_{n=1}^{\infty} \frac{F(\alpha_n)}{(-\alpha_n^2) G(\alpha_n)} \left[\frac{e^{\alpha_n^2 (T + t_1 - t)} - e^{\alpha_n^2 (T - t)}}{1 - e^{\alpha_n^2 T}}\right], \qquad t_1 < t \leqslant T$$
$$(1.29)$$

The corresponding values of the steady periodic variation in the air temperature and that of the inside surface of the glass follow from the appropriate formulae as described above in the treatment of the single rectangular pulse.

[*1.ii*(*a*)]—The diurnal variation in the intensity of solar radiation suggests that, alternatively, $I(t)$ may be considered as the positive half of a sine wave with nodes at $t = 0$ and $t = t_1$ so that

$$I(t) = \text{Im}\{\eta I_0 e^{i\omega t}[H(t) - H(t - t_1)]\} \qquad (1.30)$$

where ηI_0 denotes the peak value of the solar energy flux absorbed at the wall surface and $\omega = \pi/t_1$. From (1.30) taking the transform

$$L[I(t)] = \eta I_0 (1 + e^{-\pi p/\omega})/(p - i\omega),$$

substituting in (1.12) and solving for $\bar{\theta}(x, p)$, then by the inversion theorem,

$$\theta(x, t) = \frac{1}{2\pi i} \int_{\gamma - i\infty}^{\gamma + i\infty} \eta I_0 \left(\frac{1 - e^{-\pi p/\omega}}{p - i\omega}\right) \frac{f(p)}{g(p)} e^{pt} \, dp$$

which yields the result

$$\theta(x, t) = \begin{cases} \eta I_0 \sqrt{\left(\dfrac{a^2+b^2}{c^2+d^2}\right)} \sin(\omega t - \lambda) \\ \qquad\qquad + \eta I_0 \displaystyle\sum_{n=1}^{\infty} \frac{\omega}{(\alpha_n^4+\omega^2)} \frac{F(\alpha_n)}{G(\alpha_n)} e^{-\alpha_n^2 t}, & 0 < t \leq t_1 \quad (1.31) \\ \eta I_0 \displaystyle\sum_{n=1}^{\infty} \frac{\omega}{(\alpha_n^4+\omega^2)} \frac{F(\alpha_n)}{G(\alpha_n)} [e^{-\alpha_n^2 t} + e^{-\alpha_n^2 (t-t_1)}], & t > t_1 \quad (1.32) \end{cases}$$

where $F(\alpha_n)$ and $G(\alpha_n)$ are defined by (1.19),

$$\lambda = \tan^{-1}\left(\frac{ad-bc}{ac+bd}\right),$$

and

$$\begin{aligned} a = {} & k\sqrt{(\omega/2K)}[\cosh(l-x)\sqrt{(\omega/2K)} \cos(l-x) \\ & \times \sqrt{(\omega/2K)} - \sinh(l-x)\sqrt{(\omega/2K)} \sin(l-x) \\ & \times \sqrt{(\omega/2K)}] + h_{so} \sinh(l-x)\sqrt{(\omega/2K)} \cos(l-x)\sqrt{(\omega/2K)} \\ b = {} & k\sqrt{(\omega/2K)}[\cosh(l-x)\sqrt{(\omega/2K)} \cos(l-x) \\ & \times \sqrt{(\omega/2K)} + \sinh(l-x)\sqrt{(\omega/2K)} \sin(l-x) \\ & \times \sqrt{(\omega/2K)}] + h_{so} \cosh(l-x)\sqrt{(\omega/2K)} \sin(l-x)\sqrt{(\omega/2K)} \\ c = {} & k\sqrt{(\omega/2K)}(h_{so}+y)[\cosh l\sqrt{(\omega/2K)} \cos l \\ & \times \sqrt{(\omega/2K)} - \sinh l\sqrt{(\omega/2K)} \sin l\sqrt{(\omega/2K)}] \\ & + h_{so} y \sinh l\sqrt{(\omega/2K)} \cos l\sqrt{(\omega/2K)} \\ & - (k^2\omega/K) \cosh l\sqrt{(\omega/2K)} \sin l\sqrt{(\omega/2K)} \\ d = {} & k\sqrt{(\omega/2K)}(h_{so}+y)[\cosh l\sqrt{(\omega/2K)} \cos l \\ & \times \sqrt{(\omega/2K)} - \sinh l\sqrt{(\omega/2K)} \sin l\sqrt{(\omega/2K)}] \\ & + (k^2\omega/K) \sinh l\sqrt{(\omega/2K)} \cos l\sqrt{(\omega/2K)} \end{aligned}$$

[*1.ii*(*b*)]—Corresponding solutions for $\theta(x, t)$ when the function (1.30) is established as a steady periodic variation may be calculated by following the procedure detailed in [1.i(b)].

2. SOLAR HEAT GAIN THROUGH GLASS TO A THERMOSTATICALLY CONTROLLED ENCLOSURE

This case considers the contribution made by the quantity $\Gamma I(t)$ to the heating, or cooling, load of an enclosure. The boundary conditions specify heat input to the inside air, regulated by thermostatic control at the instantaneous rate $q(t)$ per unit area of the opaque wall, necessary to maintain the air temperature constant at θ_i for all values of time. The ambient temperature at the outside surface of the opaque wall ($x=l$) is assumed to be uniform and constant at θ_i for all t. A similar condition applies at the outside surface of the glass where the ambient temperature is $\theta_o(<\theta_i)$. In all other respects the situation is similar to that described under 1 and has the following formulation,

$$\frac{\partial\theta(x,t)}{\partial t}=K\frac{\partial^2\theta}{\partial x^2}(x,t), \qquad 0<x<l, \quad t>0 \tag{2.1}$$

$$-k\frac{\partial\theta(x,t)}{\partial x}=\Gamma I(t)-h_{ci}[\theta(x,t)-\theta_i]-\sigma(F_aF_e)_{x.g}\{[T(x,t)]^4-[T_g(t)]^4\},$$

$$x=0, \quad 0<t\leq t_1 \tag{2.2}$$

$$-k\frac{\partial\theta(x,t)}{\partial x}=h_{so}[\theta(x,t)-\theta_i], \qquad x=l, \quad \text{all } t \tag{2.3}$$

Initially the system is in steady state so that

$$\frac{\theta(x,0)-\theta(0,0)}{x/k}=\frac{\theta_i-\theta(0,0)}{1/h_{so}+l/k} \tag{2.4}$$

giving the initial temperature distribution

$$\theta(x,0)=\theta(0,0)+\frac{x/k}{1/h_{so}+l/k}[\theta_i-\theta(0,0)]$$

in which, by writing

$$\frac{\theta_i-\theta(0,0)}{l/h_{so}+l/k}=h_{ci}[\theta(0,0)-\theta_i]+h_{ri}[\theta(0,0)-\theta_g(0)]$$

it follows that

$$\theta(0,0)=\theta_i-\frac{h_{ri}}{h_{si}}\left(\frac{1/h_{so}+l/k}{R_{a.a}}\right)(\theta_i-\theta_g(0)) \tag{2.5}$$

where

$$h_{si}=h_{ci}+h_{ri}$$

$$R_{a.a}=1/h_{so}+l/k+1/h_{si}$$

and

$$\theta_g(0)=\theta_i-R_{si}U(0)(\theta_i-\theta_o)$$

At this point it may be noted that

$$\theta_g(t)=[1-U(0)R_{si}]\theta_i+R_{si}U(0)[\theta_o+\beta I(t)] \tag{2.6}$$

Since the temperature of the air in the enclosure is constant for all values of time the boundary condition (2.2) may be expressed more conveniently by introducing an equivalent temperature. Accordingly, the heat balance at the surface $x=0$ during $0<t\leqslant t_1$ is written

$$h_{e.1}[\theta_{e.1}(t)-\theta(0,t)]=\Gamma I(t)-h_{ci}[\theta(0,t)-\theta_i]-h_{ri}[\theta(0,t)-\theta_g(t)] \quad (2.7)$$

defining an equivalent temperature $\theta_{e.1}$ of the interior medium for heat transfer at the inside surface of the opaque wall, with $h_{e.1}$ the corresponding surface transfer coefficient. Neglecting the temperature dependence of the surface coefficient by writing $h_{e.1}=h_{ci}+h_{ri}$, the surface temperature $\theta(0,t)$ is thereby eliminated from (2.7) to give

$$\theta_{e.1}(t)=\theta_i-\frac{h_{ri}}{h_{e.1}}[\theta_i-\theta_g(t)]+\frac{\Gamma I(t)}{h_{e.1}}, \qquad 0<t\leqslant t_1 \quad (2.8)$$

Similarly,

$$\theta_{e.2}(t)=\theta_i-\frac{h_{ri}}{h_{e.2}}[\theta_i-\theta_g(t)], \qquad t>t_1 \quad (2.9)$$

Writing

$$\theta_e(t)=\theta_{e.1}(t)-\theta_{e.2}; \qquad \theta_{i.e}=\theta_i-\theta_{e.2}$$

and for convenience putting $\theta_{e.2}=0$, the formulation (2.1) *et seq.* now reads

$$\frac{\partial\theta(x,t)}{\partial t}=K\frac{\partial^2\theta(x,t)}{\partial x^2}, \qquad 0<x<l, \quad t>0 \quad (2.10)$$

$$-k\frac{\partial\theta(x,t)}{\partial x}=h_{si}[\theta_e(t)-\theta(x,t)], \qquad x=0, \quad 0<t\leqslant t_1 \quad (2.11)$$

$$-k\frac{\partial\theta(x,t)}{\partial x}=h_{so}[\theta(x,t)-\theta_{i.e}], \qquad x=l, \quad \text{all } t \quad (2.12)$$

$$\theta_e=0; \qquad t\leqslant 0; \qquad t>t_1$$

in which $h_{si}=h_{e.1}=h_{e.2}$ to conform with the assumption made throughout that the temperature dependence of the surface transfer coefficients may be neglected.

The expressions (2.10) through (2.12) are abbreviated further by introducing the dimensionless moduli

$$\xi=x/l; \qquad B_i=h_{si}l/k; \qquad B_0=h_{so}l/k; \qquad \tau=Kt/l^2$$

Applying the Laplace transform to the resulting differential equation and boundary conditions gives the subsidiary equations

$$\frac{d^2\bar{\theta}(\xi,p)}{d\xi^2}-p\bar{\theta}(\xi,p)+\theta(\xi,0)=0, \qquad 0<\xi<1 \quad (2.13)$$

$$-\frac{d\bar{\theta}(\xi,p)}{d\xi}=B_i\{L[\theta_e(\tau)]-\bar{\theta}(\xi,p)\}, \qquad \xi=0 \quad (2.14)$$

$$-\frac{d\bar{\theta}(\xi,p)}{d\xi}=B_o\bar{\theta}(\xi,p), \qquad \xi=1 \quad (2.15)$$

where

$$\theta(\xi, 0)=\theta(0, 0)+\frac{\xi}{1+1/B_o}[\theta_i-\theta(0, 0)]$$

with

$$\theta(0, 0)=\theta_i=\frac{h_{ri}}{h_{si}}\left(\frac{1+1/B_o}{1+1/B_i+1/B_o}\right)[\theta_i-\theta_g(0)]$$

It will be clear that $\theta_e(\tau)$ takes the form of the function $I(t)$ as considered under [1.i(a) and (b)] and [1.ii(a) and (b)]. Hence, for the single pulse of rectangular form

$$L[\theta_e(\tau)]=\frac{\theta_e}{p}(1-e^{-p\tau_1}), \tag{2.16}$$

and for the positive half of a single sine-wave pulse

$$L[\theta_e(\tau)]=\text{Im}\left[\frac{\theta_e(1+e^{-\pi p/\omega})}{(p-i\omega)}\right] \tag{2.17}$$

2.i.(a) Single rectangular pulse—Making the substitution of (2.16) in (2.14) the formal solution for the temperature rise $\theta(\xi, \tau)$ satisfying the conditions (2.14) and (2.15) is

$$\theta(\xi, \tau)=\frac{1}{2\pi i}\int_{\gamma-i\infty}^{\gamma+i\infty} B_i\frac{\theta_e}{p}(1-e^{-p\tau_1}) \times\left[\frac{B_o \sinh\sqrt{p}(1-\xi)+\sqrt{p}\cosh\sqrt{p}(1-\xi)}{(p+B_iB_o)\sinh\sqrt{p}+(B_i+B_o)\sqrt{p}\cosh\sqrt{p}}\right]e^{p\tau}\,dp \tag{2.18}$$

Taking the inversion (2.18) yields the following result for the temperature rise of the opaque wall

$$\theta(\xi, \tau)=\begin{cases}\dfrac{(1+1/B_o)\theta_e}{1+1/B_i+1/B_o}-2B_i\theta_e\displaystyle\sum_{n=1}^{\infty}\alpha_n^2\frac{u(\alpha_n)}{v(\alpha_n)}e^{-\alpha_n^2\tau}, & 0<\tau\leqslant\tau_1 \quad (2.19)\\ -2B_i\theta_e\displaystyle\sum_{n=1}^{\infty}\alpha_n^2\frac{u(\alpha_n)}{v(\alpha_n)}(e^{-\alpha_n^2\tau}-e^{-\alpha_n^2(\tau-\tau_1)}), & \tau>\tau_1 \quad (2.20)\end{cases}$$

where

$$u(\alpha_n)=\left(\frac{1-B_iB_o}{\alpha_n^2}\right)\left[\cos\alpha_n(1-\xi)+\frac{B_o}{\alpha_n}\sin\alpha_n(1-\xi)\right]$$

$$v(\alpha_n)=\alpha_n^4+\alpha_n^2[B_i(1+B_i)+B_o(1+B_o)]+B_iB_o(B_i+B_o+B_iB_o)$$

and the summation extends over the positive roots α_n, n integral, of

$$\frac{\alpha^2(B_i+B_o)}{\alpha^2-B_iB_o}=\alpha\tan\alpha \tag{2.21}$$

Multiplying the expressions (2.19) and (2.20) by the convection transfer coefficient, written

$$B_{ci}(=h_{ci}l/k)$$

in dimensionless notation, gives the instantaneous rate at which heat due to solar radiation transmitted by the glass and absorbed by the wall is transferred locally to the air in the enclosure. The total amount of solar heat $Q(t_1)$, written $Q(\tau_1)$ in dimensionless notation, absorbed by the opaque wall and transferred to the air per unit area of the wall surface over the period $\tau=0$, to $\tau=\tau_1$ may be obtained by calculating the inversion

$$Q(\tau_1)=\frac{1}{2\pi i}\int_{\gamma-i\infty}^{\gamma+i\infty} B_{ci}\left[\frac{\bar{\theta}(\xi, p)}{p}\right]_{\xi=0} e^{p\tau}\, dp$$

where $Q(\tau_1)$ represents the dimensionless group $Q(t_1)K/lk$ and $\bar{\theta}(\xi, p)$ is the integrand in (2.18). The result is

$$Q(\tau_1)=\frac{B_{ci}(1+1/B_o)\theta_e\tau_1}{1+1/B_i+1/B_o}-\frac{B_{ci}(1/3+1/B_o+1/B_0^2)\theta_e}{B_i(1+1/B_o+1/B_i)^2}$$
$$+2B_iB_{ci}\theta_e\sum_{n=1}^{\infty}\frac{(1+B_o^2/\alpha_n^2)}{v(\alpha_n)}e^{-\alpha_n^2\tau_1} \qquad (2.22)$$

It will be seen that $Q(\tau_1)$ approaches a linear asymptote that may be useful for making approximate calculations of heat flow especially when τ_1 is large. For $\tau>\tau_1$

$$Q(\tau_1)=\frac{B_{ci}(1+1/B_o)\theta_e\tau_1}{1+1/B_i+1/B_o}+2B_iB_{ci}\theta_e\sum_{n=1}^{\infty}\frac{(1+B_0^2/\alpha_n^2)}{v(\alpha_n)}$$
$$\times[e^{-\alpha_n^2\tau}-e^{-\alpha_n^2(\tau-\tau_1)}] \qquad (2.23)$$

2.i.(*b*) *Periodic rectangular wave form*—The solution for the steady periodic temperature $\theta_s(\xi, \tau)$ in the opaque portion of the enclosure wall follows from dividing the integrand in (2.18) by the periodicity factor $[1-e^{-p(\tau_1+\tau_2)}]$, calculating the inversion and proceeding as detailed under [1.i.(b)], equations (1.27) through (1.29). The calculation yields

$$\theta_s(\xi,\tau)=\begin{cases}\dfrac{(\xi+1/B_o)\theta_e}{1+1/B_i+1/B_o}-2B_i\theta_e\displaystyle\sum_{n=1}^{\infty}\alpha_n^2\frac{u(\alpha_n)}{v(\alpha_n)}\left[\frac{e^{\alpha_n^2(\tau_1-\tau)}-e^{\alpha_n^2(\tau_1+\tau_2-\tau)}}{1-e^{\alpha_n^2(\tau_1+\tau_2)}}\right], & 0<\tau\leq\tau_1 \quad (2.24)\\[2ex] -2B_i\theta_e\displaystyle\sum_{n=1}^{\infty}\alpha_n^2\frac{u(\alpha_n)}{v(\alpha_n)}\left[\frac{e^{\alpha_n^2(2\tau_1+\tau_2-\tau)}-e^{\alpha_n^2(\tau_1+\tau_2-\tau)}}{1-e^{\alpha_n^2(\tau_1+\tau_2)}}\right], & \tau_1<\tau\leq\tau_2\end{cases}$$
$$(2.25)$$

2.ii.(*a*) *Single pulse, positive half of a sine wave*—The method of calculation will be clear from the corresponding analysis [under 1.ii.(a)] and gives the result

$$\theta(\xi,\tau)=\begin{cases} B_i\theta_e\sqrt{\left(\dfrac{a^2+b^2}{c^2+d^2}\right)}\sin(\omega\tau-\lambda)+2B_i\theta_e\sum\limits_{n=1}^{\infty}\dfrac{\alpha_n^2\omega}{(\alpha_n^4+\omega^2)}\dfrac{u(\alpha_n)}{v(\alpha_n)}e^{-\alpha_n^2\tau}, \\ \qquad\qquad\qquad\qquad\qquad\qquad\qquad\qquad\qquad 0<\tau\leqslant\pi/\omega \\ 2B_i\theta_e\sum\limits_{n=1}^{\infty}\dfrac{\alpha_n^3\omega}{(\alpha_n^4+\omega^2)}\dfrac{u(\alpha_n)}{v(\alpha_n)}[e^{-\alpha_n^2\tau}+e^{-\alpha_n^2(\tau-\pi/\omega)}], \qquad \tau>\pi/\omega \end{cases}$$

Corresponding expressions for a regular series of the pulse under [2.ii.(a)] follow similarly from the treatment described under case 1.

For the present case 2 the indoor air temperature is assumed to be constant at θ_i for all t. The instantaneous heat input $q(t)$ (or extraction if $q(t)$ is a cooling load) necessary to maintain this condition is obtained by substituting the result for $\theta(0, t)$ in the following heat balance for the enclosure

$$q(t)+h_{ci}[\theta(0,t)-\theta_i]+h_{ri}[\theta(0,t)-\theta_g(t)] \\ =(C+U(0)A_g)(\theta_i-\theta_o)-U(0)A_g\beta I(t)$$

The above theoretical examination of the thermal load on an enclosure due to solar energy transmitted through glass depends essentially on calculating the temperature in the insolated opaque portion of the enclosure wall. By assuming that the solar energy admitted to the enclosure through the glass is distributed as a uniform flux of heat over the opaque wall surface it is shown that the expression for the temperature in the wall due to insolation has a basic form which may be written

$$\theta(x,t)=\text{steady-state term}+I(t)\sum_{n=1}^{\infty}Y(\alpha_n)e^{-\alpha_n^2 t} \tag{3.1}$$

where $I(t)$ represents the boundary function expressing the variation of short-wave radiation intensity with time at the outside surface of the glass. The quantity $Y(\alpha_n)$ is the ratio of functions of α_n determined by the boundary conditions and the geometry and thermal properties of the system. The form of $Y(\alpha_n)$ is characteristic of the problem considered; it may be evaluated and the results tabulated for rapid computation. Numerical results have not been considered, interest is confined to a method of analysis for obtaining solutions to a number of idealized cases selected to represent situations of practical interest. They include a rectangular pulse and a pulse of sinusoidal form, each considered separately both as a single pulse and as a series of repeated pulses representing an idealized variation in the ambient conditions. For arbitrary functions representing actual variations in $I(t)$ a solution for $\theta(x, t)$ may be obtained as a superimposition of step functions using the results given above for a rectangular pulse. By this method (see Chapter 6) the solution for an arbitrary function, under the conditions of case 1 for example, is written

$$\theta(x, t) = \eta\Delta I_r(1/h_{so} + \overline{l - x}/k)/[1 + y(1/h_{so} + l/k)] + \eta\sum_{r=1}^{m}\Delta I_r\left[\sum_{n=1}^{\infty} Y(\alpha_n)e^{-\alpha_n^2 t}\right] \tag{3.2}$$

in which m is the number of rectangular pulses into which the amplitude of the temperature change is divided, ΔI_r is the height of the pulse, and t_r is the time interval measured from the commencement of the rth pulse and within the duration of that pulse.

The analysis assumes the thermal capacity of the glass to be negligibly small. This may be justified by considering a step-function drop in the ambient temperature at the outside surface of a cavity structure and comparing the subsequent transient change in temperaure at the inside surface (facing a constant temperature enclosure) for glazing, and for a masonry wall to correspond with the opaque wall of the enclosure. The relative cooling rates are compared on the basis of a thermal time-constant defining the time taken for the change in the surface temperature, measured from initial time, to reach $(1-1/e)$, or 63 percent, of its ultimate steady state value. Fig. 5 shows the cooling curves and Table 1 gives the calculated thermal time-constants, for single and double glazing using $\frac{1}{4}$ in. polished plate glass and a brick wall 9 in. thick.

Though primarily the object of the analysis is to determine the indoor temperature change due to solar heat gain through glass it may be of interest to include the effect of an accompanying diurnal change in the outdoor air temperature. This applies to the examples considered under case 1, in which

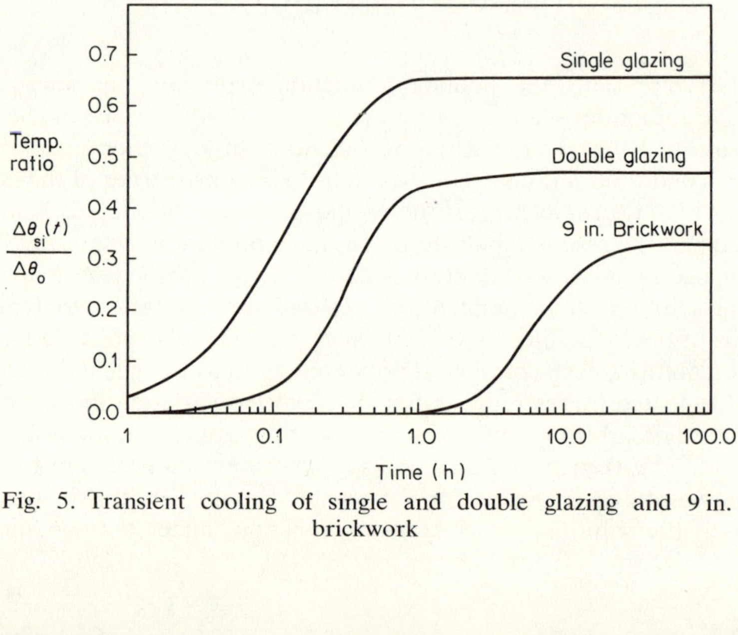

Fig. 5. Transient cooling of single and double glazing and 9 in. brickwork

Table 1 Thermal time-constants for single and double glazing, and 9 in. brickwork facing a constant temperature enclosure

Single glazing	8 min
Double glazing	23 min
9 in. brickwork	8 h

the boundary statement (1.2) is modified accordingly to read

$$-k\frac{\partial\theta(x, t)}{\partial x}=\Gamma I(t)-h_{ci}[\theta(x, t)-\theta_i(t)-\theta_o(t)]-h_{ri}[\theta(x, t)-\theta_g(t)],$$

$$x=0, \quad t>0 \quad (3.3)$$

where $\theta_o(t)$ represents the time variation in the temperature of the outdoor or ventilating air. The heat balance of the enclosure is

$$h_{ci}[\theta(x, t)-\theta_i(t)-\theta_o(t)]+h_{ri}[\theta(x, t)-\theta_g(t)]$$
$$=[C+U(0)A_g]\theta_i(t)-U(0)A_g\beta I(t) \quad (3.4)$$

in which, neglecting the thermal capacity of the glass,

$$\theta_g(t)=[1-U(0)R_{si}]\theta_i(t)+U(0)R_{si}\beta I(t)+\theta_o(t) \quad (3.5)$$

By substitution from (3.4) and (3.5), (3.3) becomes

$$-k\frac{\partial\theta(x, t)}{\partial x}=-y\theta(x, t)+\lambda(t), \qquad x=0, \quad t>0 \quad (3.6)$$

where

$$\lambda(t)=y\theta_o(t)+\eta I(t)$$

The equation (3.6) is of similar form to (1.9) so that the solution for the air and surface temperatures in the enclosure when the outdoor air is a function of time may be written down from previous results.

SYMBOLS

A_g	area of glazing per unit area of opaque wall
A_w	area of opaque wall
B_i	Biot number for heat transfer at inside ($x=0$) surface of opaque wall ($=h_{si}l/k$)
B_o	Biot number for heat transfer at outside ($x=l$) surface of opaque wall ($=h_{so}l/k$)
C	rate of heat loss by ventilation per unit temperature difference and per unit area of wall
F_a	configuration factor for radiation heat exchange

F_e	emissivity factor for radiation heat exchange
h_{ci}	inside wall surface ($x=0$) convection heat transfer coefficient
h_{co}	outside wall surface ($x=l$) convection heat transfer coefficient
h_{ri}	inside wall surface ($x=0$) radiation heat transfer coefficient
h_{ro}	outside wall surface ($x=l$) radiation heat transfer coefficient
h'_{ci}	inside glazing surface convection heat transfer coefficient
h'_{co}	outside glazing surface convection heat transfer coefficient
h'_{ri}	inside glazing surface radiation heat transfer coefficient
h_{ro}	outside glazing surface radiation heat transfer coefficient
h_{so}	outside wall surface ($x=l$) overall heat transfer coefficient
$h_{e.1}$, $h_{e.2}$	coefficient of heat transfer between medium at temperature $\theta_{e.1}$, $\theta_{e.2}$ and inside wall surface ($0<t\leqslant t_1$, $t>t_1$, respectively)
h_{si}	inside wall surface ($x=0$) overall heat transfer coefficient (neglecting dependence upon temperature $=h_{e.1}=h_{e.2}$)
$h_{c.1}$	convection heat transfer coefficient at cavity-facing surface of inner pane of glass in double glazing
$h_{c.2}$	convection heat transfer coefficient at cavity facing surface of outer pane of glass in double glazing
h_{ra}	coefficient of radiation heat exchange between the cavity surfaces in double glazing
$H_g(t)$	heat flux through glazing
$H(t)$	the Heaviside unit function
I	flux of solar radiation per unit area of wall
k	thermal conductivity of wall material
K	thermal diffusivity of wall material
l	thickness of wall
L	length of cavity in double glazing in direction of airflow
n	ventilation rate of airspace in double glazing, air changes/h
$q(t)$	instantaneous rate of heat input per unit area of wall
$Q(t_1)$	total heat flow in interval $t=0$ to $t=t_1$
$Q(\tau_1)$	dimensionless group $=Q(t_1)K/lk$
R	thermal resistance of glass in single glazing
R_1, R_2	thermal resistance of inner and outer panes of glass respectively in double glazing
R_c	thermal resistance of cavity in double glazing
R_{so}	thermal resistance at outside surface of glazing
R_{si}	thermal resistance at inside surface glazing
R_{aa}	air-to-air resistance of wall ($=1/h_{so}+l/k+1/h_{si}$)
s	volumetric specific heat of air
t	time
$0<t\leqslant t_1$	period of insolation
$t>t_1$	period after insolation
T	repetition period of rectangular pulse waveform
$T(x, t)$	absolute temperature at time t and position x in wall
T_g	absolute temperature of inside surface of glazing

$U(0)$	steady state thermal transmittance of shaded glazing
v	velocity of air flow in airspace in double glazing
w	width of airspace in double glazing
x	position in wall
α	absorptivity of inside surface $(x=0)$ of wall for short-wave radiation
α_n	$n=1, 2, 3, \ldots$, successive roots of relevant function
β	$\begin{cases} =\phi(R_{so}+R/2) \text{ single glazing,} \\ =\phi_1(R_{so}+R_1/2)+\phi_2(R_{so}+R_1+R_c+R_2/2) \text{ double glazing} \end{cases}$
Γ	fraction of short-wave radiation incident upon outside surface of glazing which is absorbed at opaque wall surfaces
$\theta(x, t)$	temperature at time t and position x in wall
θ_i	inside air temperature
θ_o	outside air temperature
$\theta_{s.o}$	mean radiation (long-wave) temperature of medium facing outside surface of glass
θ_g	inside surface temperature of glass
θ_{go}	outside surface temperature of glass
θ_1	temperature of cavity-facing surface of inner pane of glass in double glazing
θ_2	temperature of cavity-facing surface of outer pane of glass in double glazing
$\theta_{1.g}$	temperature at mid-plane of inner pane of glass in double glazing
$\theta_{2.g}$	temperature at mid-plane of outer pane of glass in double glazing
$\theta(0), \theta(L)$	temperaure of ventilating air respectively entering and leaving the airspace in double glazing
θ_c	mean temperature of air in airspace in double glazing
$\theta_s(x, t)$	steady periodic component of temperature in wall under cyclic insolation
$\theta_{e.1}, \theta_{e.2}$	equivalent temperature of medium for heat exchange at inside surface of wall $(0<t\leqslant t_1,\ t>t_1$ respectively)
θ_e	temperature difference $\theta_{e.1}-\theta_{e.2}$
$\theta_{i.e}$	temperature difference $\theta_i-\theta_{e.2}$
λ	angular phase displacement
$\lambda(t)$	function defined in text
ξ	dimensionless group $=x/l$
ρ	density of air
σ	Stefan's constant
τ	dimensionless group $=Kt/l^2$
ϕ	fraction of incident solar radiation absorbed by glass in single glazing
ϕ_1, ϕ_2	fraction of incident solar radiation absorbed by inner and outer panes of glass respectively in double glazing

Φ	fraction of incident solar radiation transmitted through glazing
ψ	$=(h_{c1}\theta_1+h_{c2}\theta_2)/(h_{c1}+h_{c2})$
ω	angular frequency of sinusoidal insolation function

REFERENCES

Carslaw, H. S. and Jaeger, J. C. (1959) *Conduction of Heat in Solids*, Second edition, Clarendon Press, Oxford.

Mackey, C. O. and Gay, N. R. (1952) *ASHVE Transactions*, **58,** 321.

Stagg, J. M. (1950) *Solar Radiation at Kew Observatory*, Geophysical Memoirs no. 86, HMSO.

CHAPTER 8

Thermal Transmittance: Measurement and Results

1. Direct measurements of thermal transmittance are made in a number of ways which may be broadly grouped into (a) methods using a controlled 'cold' exposure, and (b) methods using a natural exposure to the external weather.

The controlled cold exposure methods are typified by those described in ASTM No. C 236-54T (1954). The section of test wall is sandwiched between two boxes, one of which is controlled at the desired high temperature, usually around 21 degC, and the other at a low temperature which may be about −18 degC. The air flow over the surfaces is controlled by fans and baffles to give a constant rate of air movement. The results are obtained in the form of a surface-to-surface conductance and converted to a U-value using standard values of the surface-to-air conductances.

A variant of this test is one in which the specimen is exposed on one side to a heated plate and on the other to the air in a cool room, without any artificial chilling, or is bounded by a cold plate in contact with the wall surface (Griffiths, 1942).

A later development is a large hot box–cold box apparatus, accommodating walls up to 2.4 m square, constructed by the Division of Building Research, Ottawa (Solvason, 1959). In the design of this equipment provision is made for variable heat flow studies, and to eliminate errors inherent in some other methods.

A variety of methods has been developed for measuring U-values of walls which have one face exposed to natural weather conditions. They fall into the following groups:

(a) small houses incorporating the test walls on all four sides;

(b) free-standing walls with guarded hot box or hot plate on one face;

(c) wall laboratories with test walls heated on one face; and

(d) field measurements using guarded hot box or hot plate in actual buildings.

In several countries sets of small research houses have been built (Haller, 1957; Reiher, 1954; Holmquist, 1948). In the German and Swiss experiments the buildings were about 3.5 to 4 m square inside. The walls were constructed of the test materials and the floors and roofs well insulated. The houses were spaced well apart so that there was little interference to the surrounding wind, rain, and solar radiation. Measurements were made over extended periods, three complete winters November to February in the Swiss experiments, and mean U-values computed from the data.

In the German experiments there was considerable emphasis on humidity measurements as affecting condensation.

Other experiments of a similar nature but using much smaller structures have been made in Norway, Sweden, and Canada.

Some work has been done in Britain on free-standing test walls with a guarded hot-box clamped on one face. The object of this was to examine the effect of orientation on heat transmission through exposed walls but the method proved to be of limited value.

Wall laboratories take many forms. Two different types have been used in Norway. In one of these a heated enclosure is clamped to the test wall and the heat flowing through the wall is measured by a thermo-electric heat meter plate on the heated face of the wall (Tveit, 1953). Measurements are restricted to periods when the outside weather conditions are steady. A similar arrangement is used in Finland. The other one is a long building with a series of test panels on the two sides facing east and west. On each test panel is clamped a hot box. The laboratory as a whole is controlled in temperature and acts as a guard chamber to the hot boxes. A similar laboratory has been used in Sweden. All the heat supplied to the hot box is assumed to flow through the test area of the wall, as in the usual guarded hot box experiments.

Another kind of laboratory built in Britain contains two series of walls, one set facing north and the other south. Heat is supplied to the test walls by guarded hot plates with a thermo-electric heat meter between the heater and the wall to measure the heat flow directly (Dinnie, Beard, and Richards, 1959).

Finally there was the laboratory at the Building Research Station (BRS) at Garston in Britain, formerly of the Department of Scientific and Industrial Research (DSIR). Test walls, 2.4 m square, exposed on the north face of the building, had a heated room behind them, the individual rooms acting as thermal guards to those either side. Each room was essentially a guarded hot box, with relatively small heat losses through floor and ceiling which were separately metered.

Work in the last group (field measurements) has been reported by Ryder and McIlraith (1946) and Bastings and Benseman (1950) in New Zealand. They used a guarded hot box which could be transported readily to the building where the wall was to be measured.

Of the various methods, only those in which the warm side of the wall is exposed to warm air for approximately room height, that is about 2.4 m, give the thermal transmittance directly. Experience has shown that for test specimens of smaller type the inside surface transfer coefficient differs from the generally accepted values due to abnormal warm air movement over the surface. Only in the Garston wall laboratory and in the larger of the test houses did the whole face of the wall form the test area, fully exposed to the warm air of a room of normal height.

In all the other experimental arrangements the measurements made give the conductance from the warm surface of the wall to the outside air. To obtain the overall air-to-air conductance, called the thermal transmittance or *U*-value, a nominal value for the inside surface conductance is added.

The Heat Transmission Wall Laboratory at the Building Research Station

Investigations of heat transmission through walls have been carried out over a period of many years in Britain. Experimental studies began in 1936 when a laboratory specially designed for the purpose was constructed at the Building Research Station, now the Building Research Establishment (BRE).

A number of considerations led to the design of the laboratory. It was decided to have a number of test walls exposed to the weather because a criticism of current thermal transmittance values was that all were calculated from conductivities which had been determined in laboratory apparatus using specimens whose moisture content, if known at all, was usually different from that of actual walls. This had caused Dufton and Sheard (1937) in the first report on the new laboratory to write as follows:

> ... The information at present available on the heat-transmission coefficients of walls is incomplete and in many cases not very accurate. Many of the existing figures have been determined under laboratory conditions with specimens artifically dried and may be quite different from the values for the same materials under normal conditions of exposure and moisture content. A new transmission laboratory has therefore been erected at the Station in order that the data may be revised and extended . . . .

To give a uniform exposure to the test walls, six in number, they were arranged in a row, side-by-side, facing slightly west of north. This could be expected to give the most severe exposure available while avoiding the complication of having direct solar radiation falling on the walls during the winter test period.

Because building materials are so variable, while variations in building technique cause further lack of uniformity in the final assemblage, it was desirable that the walls should be as large as possible. It was decided that the test specimens should be 2.4 m square, thus corresponding to storey height, and each was to form the outer wall of a heated room, as in a normal installation. This had the advanage of exposing the specimens to air currents similar to those in a normal room.

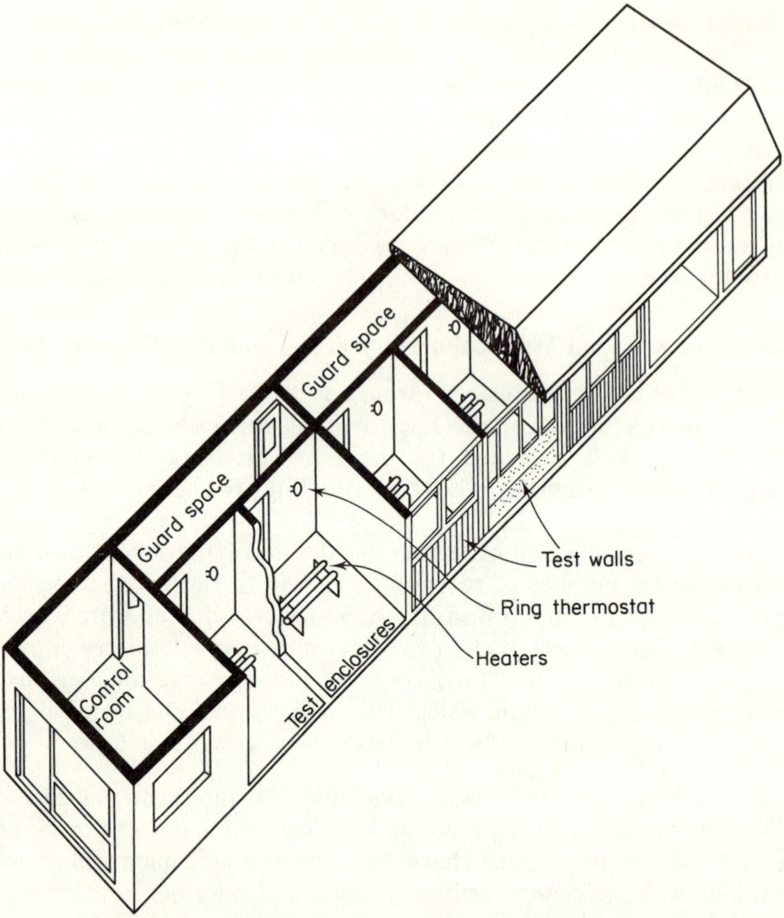

Fig. 1. Isometric sketch of BRS Wall Laboratory in its final form.

It was not possible to measure directly the heat flow through the test walls without interfering with their relevant physical properties, and it was decided therefore to determine the heat flow indirectly. This was done by measuring the electrical heat input to the room, and metering the losses through the floor and ceiling. Loss of heat through the side walls was to be prevented by heating all rooms in the laboratory to the same constant temperature. Thus, the general layout of the laboratory (Fig. 1) was closely determined by the experimental requirements.

The detailed construction of the laboratory is shown in Fig. 2. The outer walls were of brick, with a concrete slab floor, all insulated with 50 mm thick cork slabs. Internal partitions were of cork and fibreboard on wood framing. Originally the roof was almost flat, as shown in Fig. 2, but was replaced in 1954 by a double-pitched roof. The 2.4 m square test walls rested on 50 mm thick cork over the foundations and could be readily removed and replaced by new ones. The openings could accommodate walls up to about 300 mm thick.

The felt-lined fascia boards were fixed after the walls were erected and served to reduce air flow past the edges of the test panels into the test rooms. As a further precaution against air leakage the doors at the ends of the corridor, as well as the two outside doors, were weather stripped. It was important to keep the rate of air flow through the laboratory small, for one air change per hour corresponded to a heat loss from one room of about 60 W with an air-to-air temperature difference of 11 degC, which is nearly as great as the heat flow through a test wall with a thermal transmittance of 1.1 W/m^2K.

Fig. 2 shows a layer of 'B.P. boarding' on the floor of the test rooms; this is a type of laminated fibreboard with several layers of bituminous material separating these layers of fibre. This material was chosen for use as a heat meter in the laboratory because it is little affected by moisture, so that its thermal conductivity remains substantially constant. Thermocouples fixed to the two faces of this board measured the temperature drop across it and hence the amount of heat flowing from the room into the floor. A similar panel above the ceiling measured the heat flow upwards from the room into the roof. The sum of these two quantities was deducted from the electrical heat input to the room, so giving the amount of heat flowing through the test wall. There were small errors in the heat loss measurements due to leakage through corners, and these are discussed in Chapter 4. Each test room was electrically heated with two 2.1 m tubular heaters, rated at about 500 W each, placed parallel to the test wall and about 1.5 m from it. This arrangement was chosen to give as uniform an air temperature as possible and Figs. 3(a) and (b), show the air temperature distribution in the room with heaters on and off with a 225 mm solid brick test wall. A more uniform distribution of temperature would be expected with a better insulated wall. The heaters were switched by a mercury-tube magnetic relay under the control of a bimetal ring thermostat fixed on the south wall and the amount of power delivered to the room measured by an ordinary domestic-type kilowatt-hour meter.

No attempt was made to control the humidity in the test rooms. Measurements showed that the relative humidity in the rooms settled down to about 45 percent during the heating season. This represents a moisture content slightly lower than in domestic living rooms and is about the same as that in offices.

The temperature of the room and outside surfaces of the test walls were also measured by thermocouples connected to them. There were five junctions on each face with reference junctions in the laboratory, all connected in series. From the readings surface-to-surface and air surface temperature differences were determined, and hence the corresponding thermal resistances.

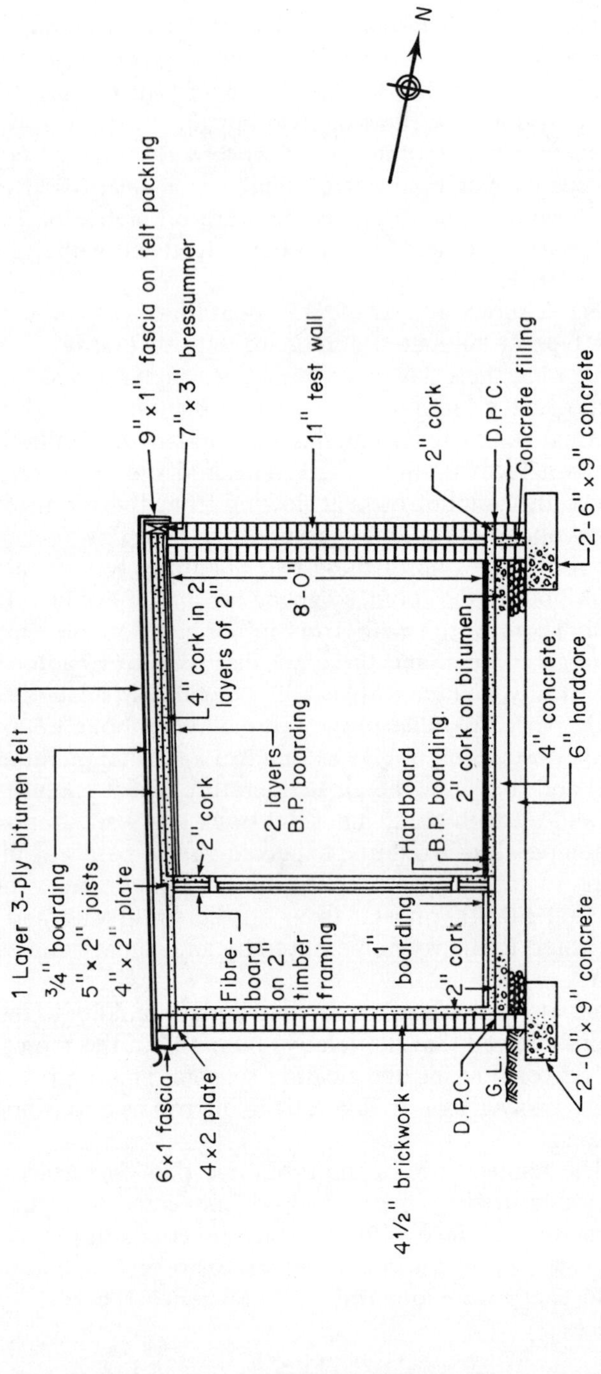
N
1 Layer 3-ply bitumen felt
3/4" boarding
5" x 2" joists
4" x 2" plate
9" x 1" fascia on felt packing
7" x 3" bressummer
11" test wall
2" cork
D. P. C.
Concrete filling
2'-6" x 9" concrete
8'-0"
4" cork in 2
layers of 2"
2" cork
2 layers
B. P. boarding
Hardboard
B. P. boarding.
2" cork on bitumen
4" concrete.
6" hardcore
Fibre-
board
on 2"
timber
framing
1"
boarding
2" cork
2'-0" x 9" concrete
6 x 1 fascia
4 x 2 plate
4 1/2" brickwork
D. P. C.
G. L.

SECTION AA

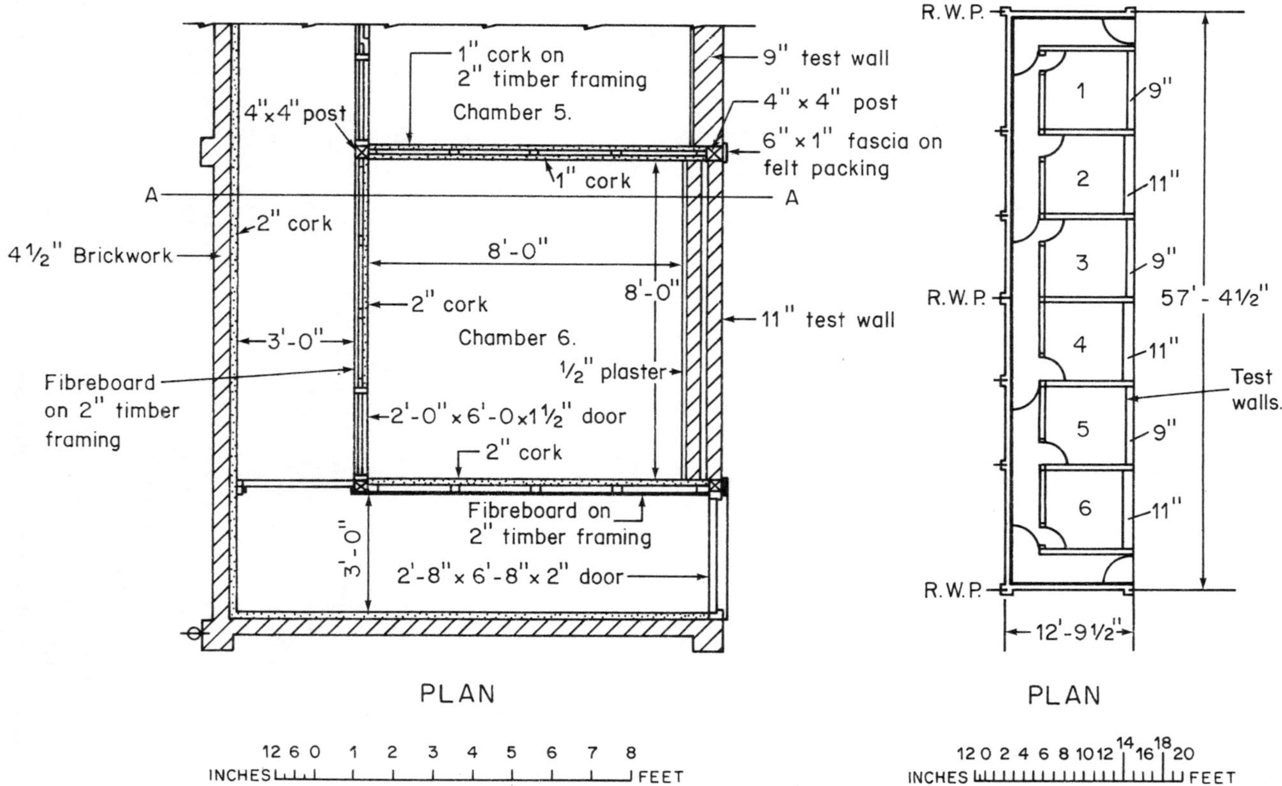

Fig. 2. Wall Laboratory

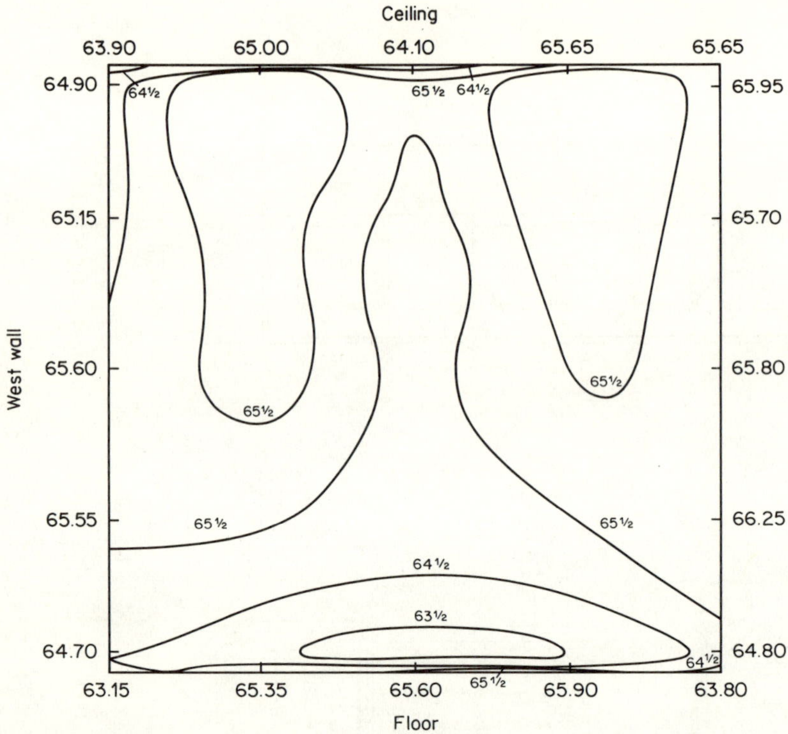

Fig. 3a. East–west section through centre of room, showing air temperature (degF) distribution with heaters on. (Survey made on 18 December 1939)

Associated with the laboratory was a meteorological site for the measurement of the external climate. The plan (Fig. 4) shows the layout, and indicates, in addition, the situation of the various nearby buildings that affected the exposure of the test walls. It will be seen that the walls were in a relatively sheltered situation, corresponding perhaps to a fairly open suburban density of buildings. From an experimental viewpoint a less sheltered exposure would have been preferable, but shortage of space determined the building layout.

The principal measurements made at the meteorological site were of air temperature, rainfall, wind, sunshine, and solar radiation. Most of the readings were made on standard instruments of Meteorological Office pattern (Meteorological Office, 1956a,b). As far as possible the climatic observations were made in accordance with established procedure, in particular with these specified by the British Meteorological Office. This was essential in order that the experimental results from the laboratory may be related directly to climatic conditions at another place whose climate is specified.

The basic method of operation of the Wall Laboratory remained the same

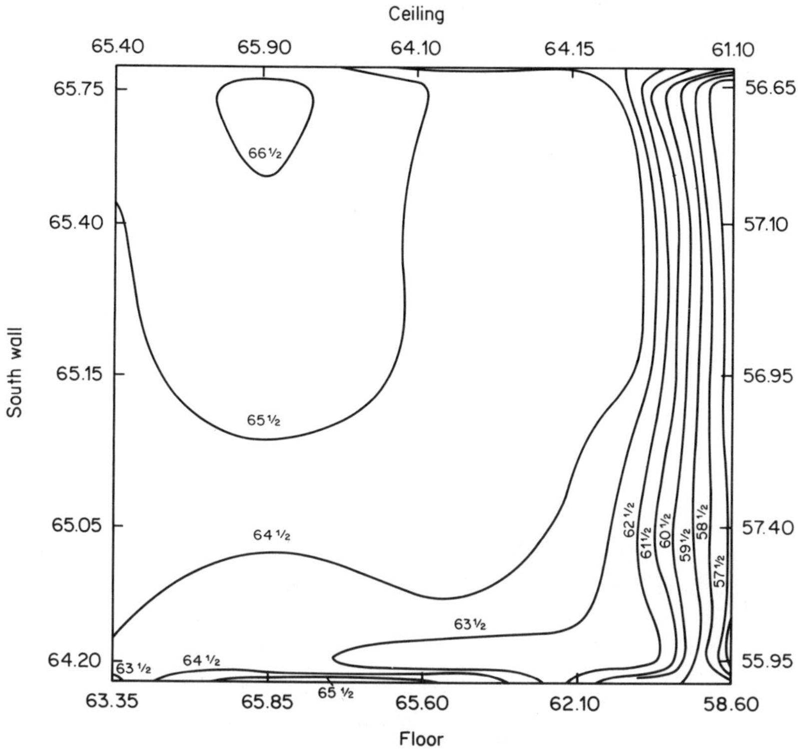

Fig. 3b. North–south section through centre of room showing air temperature (degF) distribution with heaters off. (Survey made on 18 December 1939)

throughout the whole period of operation, although there were many detailed alterations. All the rooms were controlled at a constant air temperature of 18.3 degC for the complete heating season—nominally from the beginning of October to the end of March. The air temperature in each room was measured with maximum and minimum thermometers which were read daily at 10.00 h at the same time as the power meters, the mean of maximum and minimum temperatures being taken as the 24 h mean temperature of the room. Outside-air temperature was obtained by averaging the trace of a bimetallic thermograph exposed in the ventilated screen of the nearby meteorological enclosure.

Daily mean values of the thermal transmittance were computed from the observations.

Tests were carried out in the laboratory for seventeen winters between 1936 and 1961 when soon afterwards it was taken out of commission. More than seventy different wall constructions were invetigated. Many of these were complex structures whose U-value could not be computed with sufficient accuracy for practical calculations of heat transfer.

Other measurements included surface coefficients, cavity resistances and,

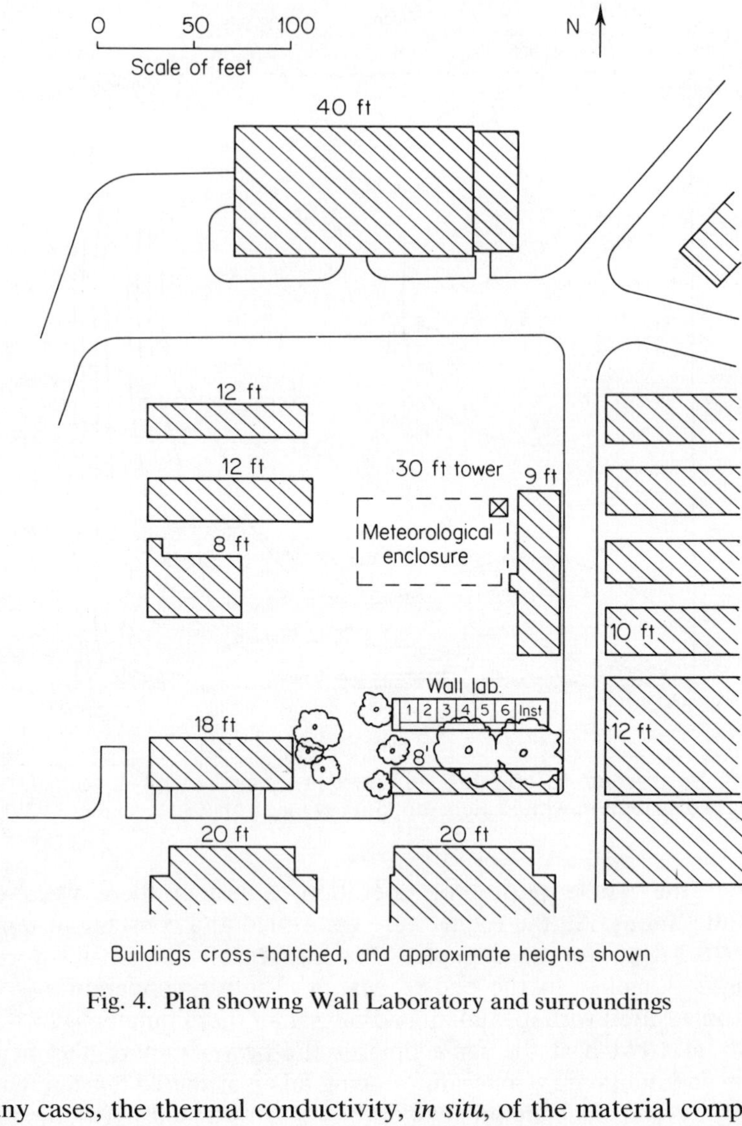

Buildings cross-hatched, and approximate heights shown

Fig. 4. Plan showing Wall Laboratory and surroundings

in many cases, the thermal conductivity, *in situ*, of the material composing a homogeneous layer. Some measurements were also made of the moisture content and distribution in walls, and of the periodic heat-flow characteristics of some constructions.

2. The main results of the experiments carried out in the Wall Laboratory are summarized in Table 1 and in more detail on data sheets. They include thermal transmittance, computed thermal conductivities or conductances, periodic heat flow, and other relevant properties. The performance of the walls in periodic heat flow is determined by the fundamental time-lag and amplitude decrement. The fundamental time-lag is given as the delay in the

temperature variation at the inside surface following a sinusoidal temperature variation of 24 h period applied at the outside surface. The amplitude decrement is the ratio of the peak values of temperature at the two surfaces. A detailed account of the measurement of periodic heat flow through certain of the walls is given in Chapter 9.

Most of the information refers to a complete heating season, or such part of it as was available, and is presented as experimental results uncorrected for effects such as those caused by moisture in the materials. Where possible allowance has been made for the effect of moisture content. The moisture dependence of thermal conductivity observed in these measurements is discussed in Chapter 9 and is the basis upon which the adjustments have been made. The result is a series of 'normalized' U-values. These, it is considered, are the ultimate values that the walls would reach after they had been in position long enough for the materials to have lost their initial water of construction. It has been assumed, for example, that most of the lightweight concretes would finally have a moisture content of 5 percent by volume. These are the values listed in Table 1 where the walls are classified according to the form of construction.

As an introduction to the summary of results it is useful to consider briefly the chief types of materials used in the construction of walls, and their most significant physical properties. The materials may be conveniently divided into seven classes, as shown in Table 2.

(a) Fired clay bricks differ widely in their properties, the most dense engineering-type bricks having a porosity (volume fraction of air) of only about 1 percent, while dry-pressed Fletton bricks may have over 40 percent porosity. About two-thirds of the bricks used in Britain are of the Fletton type with a density of 1600–1760 kg/m^3.

(b) Dense concretes are those with heavy aggregate such as gravel, brick, or whinstone, and the group includes 'no-fines' concretes.

(c) Lightweight concretes are much more variable in composition; foamed-slag, clinker, expanded clay and similar aggregates give concretes with densities ranging from 1040 to 1280 kg/m^3, while foamed or aerated concretes have densities in the range 480–800 kg/m^3.

(d) Most of the timber used in wall construction is low density material, the denser hardwoods being used mainly for window framing.

(e) Steel and aluminium are the main metals used, high thermal conductivity being their most notable characteristic. They are used so much because their great strength permits the use of thin sections, but the heat loss through thin sections is still high.

(f) Glass is remarkable for the fact that it is almost completely transparent to solar (short-wave) radiation, while being almost opaque to low-temperature (long-wave) radiation. In this respect it differs from all other materials and the heat transfer problems become correspondingly more complex.

(g) Foamed and fibrous materials comprise both natural materials and a

Table 1 Brief description of test walls and 'normalized' thermal transmittances, with sheltered north exposure

Wall no.	Year of test	Description	'Normalized' (Btu/ft^2h deg F)	*U*-value (W/m^2K)
1	1936–7 to 1948–9	9 in. Fletton	0.40	2.27
2	1937	9 in. Fletton, plastered	0.39	2.21
3	1937, 1937–8	9 in. Fletton, rendered	0.39	2.21
4	1938–9, 1947–8	9 in. Fletton, rendered, strapped and plastered	0.28	1.59
Cavity brick walls				
5	1936–7 & 1937–8	11 in. Fletton + Fletton	0.29	1.65
6	1937, 1937–8	11 in. Fletton + Fletton ventilated cavity (iron grills)	0.32	1.82
7	1959–60, 60–1	11 in. Fletton + Fletton plastered	0.29	1.65
8	1954–5	11 in. dense bricks, plastered (133 lb/ft^3 bricks)	0.34	1.93
9	1954–5	11 in. dense bricks, plastered (133 lb/ft^3 bricks) cavity filled with resin-bonded glass fibre	0.15	0.85
10	1954–5	11 in. dense bricks perforated (48 holes, 17% void) plastered	0.29	1.65
11	1954–5	11 in. Flettons, frog down + 6-hole clay blocks, plastered	0.23	1.31
12	1951–2	11 in. Eecon 4-hole clay blocks rendered and plastered	0.23	1.31
Single-leaf walls of perforated bricks				
13	1948–9	6 in. L.B.C. horizontally-perforated bricks, rendered	0.37	2.10
		the same, with lining of 1 in. wood wool-cement slabs on mortar	0.24	1.36
14	1956–7, 1957–8	8 in. V.2. vertically perforated bricks, plastered	0.26	1.48
15	1957–8	8 in. V.4 vertically perforated bricks, plastered	0.26	1.48
16	1958–9 & 1959–60	9 in. V.5 vertically perforated bricks, plastered	0.26	1.48
17	1959–60	9 in. V.6 vertically perforated bricks, plastered	0.30	1.70
18	1960–1	9 in. V.6/2 vertically perforated bricks, plastered	0.37	2.10
19	1957–8 & 1958–9	10 in B.25 vertically perforated bricks, rendered and plastered	0.21	1.19
20	1958–9	9 in. H.2 horizontally perforated bricks, plastered	0.35	1.99

Double-leaf cavity walls, brick outer leaf				
21	1948–9	11 in. Fletton brick + 3 in. foamed-slag concrete blocks, plastered	0.26	1.48
22	1951–2	11 in. Fletton brick + $4\frac{1}{4}$ in. clinker concrete blocks, plastered	0.27	1.53
23	1957–60	11 in. Fletton brick + 4 in. aerated concrete (Thermalite) blocks, plastered	0.23	1.31
24	1959–61	11 in. Fletton brick + 4 in. Aglite concrete blocks, plastered	0.26	1.48
25	1950–1	11 in. Fletton brick + 4 in. cored clinker concrete blocks, plastered	0.26	1.48
26	1950–1	As above but with different pattern of blocks	0.27	1.53
27	1957–8	As No. 23, but with insulated cavity and battening + plasterboard	0.10	0.57
28	1953–4	11 in. Fletton brick + 4 in. Bellrock honeycomb gypsum slabs	0.22	1.25
Single-leaf walls of concrete cast in situ, *8 in. thick*				
29	1947–8	Gravel no-fines aggregate, rendered and plastered	0.42	2.39
30	1938–9	Whinstone aggregate no-fines	0.41	2.33
31	1947–8	Ham River gravel aggregate 1:2:4	0.59	3.35
32	1948–9	Ham River gravel aggregate 1:2:4	0.63	3.58
33	1948–9	Accrington brick aggregate 1:2:4	0.47	2.67
34	1947–8	Foamed-slag aggregate $1:2\frac{1}{2}:7\frac{1}{2}$ rendered and plastered	0.21	1.19
35	1951–3	Foamed-slag aggregate $1:2\frac{1}{2}:7\frac{1}{2}$ rendered and plastered	0.21	1.19
36	1950–1	Foamed-slag aggregate 1:3:6 rendered and plastered	0.23	1.31
37	1952–3	Foamed-slag aggregate 1:3:6 rendered and plastered	0.23	1.31
38	1952–3	Foamed-slag aggregate 1:2:8 rendered and plastered	0.19	1.08
Single-leaf walls of concrete blocks or slabs				
39	1951–2	6 in. thick Thermalite' aerated concrete blocks, rendered and plastered	0.26	1.48
40	1947–8	9 in. foamed-slag concrete cored blocks with battens and lining rendered	0.26	1.48
41	1952–3	9 in. foamed-slag concrete Besser cored blocks	0.27	1.53
42	1953–4	9 in. clinker concrete Besser cored blocks	0.29	1.65
43	1960–1	8 in. Siporex aerated concrete storey-height slabs, rendered and plastered	0.15	0.85

Table 1 (*Continued*)

Wall no.	Year of test	Description	'Normalized' (Btu/ft^{2}h deg F)	U-value (W/m^{2}K)
Double-leaf walls of concrete blocks				
44	1948–9	18×9×3 in. foamed-slag concrete blocks, 3 in. cavity, rendered and plastered	0.25	1.42
45	1952–3, 53–4, 54–5	18×9×4 in, foamed-slag concrete blocks—plastered, rendered in Aug. 1953	0.20	1.13
46	1953–4	18×9×4 in. LECA concrete blocks; rendered and plastered	0.23	1.31
47	1958–9, 59–60	18×9×4 in. pulverised fuel-ash concrete blocks, rendered and plastered	0.22	1.25
Prefabricated walls				
48	1949–50	Woolaway: panels of aerated concrete bolted to concrete posts to form cavity, rendered outside	0.43	2.44
49	1949–50	Woolaway: the same without posts	0.37	2.10
50	1949–50	Airey: concrete slabs wired to concrete posts, plasterboard lining, with aluminium foil forming double cavity	0.22	1.25
51	1949–50	Reema: double-leaf dense concrete with integral spacer webs $\frac{1}{2}$ in. fibreboard on inner face	0.27	1.53
52	1950–1	Orlit: dense concrete slabs on posts, $5\frac{1}{2}$ in. cavity, $2\frac{1}{2}$ in. foamed-slag concrete blocks, plastered	0.35	1.99
53	1951–2	Wilson: double-leaf blocks of dense concrete with inner lining plaster-board on integral battens	0.30	1.70
		The same with aluminium-faced plasterboard	0.23	1.31
54	1949–50	Weir: steel outer sheet, cavity plasterboard, cavity, aluminium faced plasterboard	0.31	1.76
55	1949–50	Hawkesley: aluminium sheet on wood frame, plasterboard inner lining, glass fibre fill	0.17	0.97
56	1953–4	Bristol: aluminium sheet on aluminium frame, hardboard inner covering, glass fibre fill	0.38	2.16
57	1958–9	Thain QSA: corrugated aluminium sheet glass fibre mat, steel inner sheet	0.23	1.31

58	1957–8	Jane Modolite: plywood with aluminium foil cavity, 2 in. glass fibre, cavity, aluminium faced plasterboard, wood frames	0.12	0.68
59	1955–6	T.D.A.: 1 in. wood panel + 1 in. sawdust cement, cavity, plasterboard on wood frames	0.26	1.48
Heavy glass walls				
60	1937–8	4 in. thick, hollow glass blocks	0.44	2.50
61	1937–8	Window, 3 ft 3 in. × 4 ft, in steel frames, set in 9 in. solid Fletton brick wall	0.48	2.72
		The same, double-glazed	0.41	2.33
Curtain walls				
62	1955–6 & 1956–7	Quicktho aluminium frame, Seco wood-wool cement panel, single-glazed	0.60	3.41
63	1955–6	Hills patent-glazed: steel frame and subframe, gypsum inner spandel panel, single-glazed	0.62	3.52
64	1956–7	The same: double-glazed	0.44	2.50
65	1956–7	Tomo-Stramit: wood, cavity, 2 in. strawboard on wood frame. Wood frame double window	0.27	1.53
66	1956–7	Holoplast: aluminium subframe, hollow plastic spandel panels, double-glazed	0.37	2.10
67	1955–6	LCC Picton Street: well-insulated spandrel panel on wood frame, single-glazed	0.52	2.90
68	1955–6	Herts C.C. 8 ft 3 in. school: concrete slabs and insulation for spandrel on steel frame, aluminium-framed single windows	0.61	3.46
69	1956–7	The same, with added inner glazing on wood frame	0.51	1.76
70	1955–6	B.R.S. wood aluminium subframe, woodwool cement + aluminium faced plasterboard spandrel. Single-glazed	0.57	3.24
71	1938–9	Wood-framed single-glazed window	0.58	3.29
72	1938–9	The same, double-glazed	0.30	1.70
73	1960–1	Wood-framed double-glazed window	0.45	2.55

Table 2 The properties of materials used in buildings

Material	Thermal conductivity (k) (W/m K)	Specific heat (c) (J/kg K)	Bulk density (ρ) (kg/m^3)	Porosity (per cent by volume)
Bricks (fired clay)	0.36–1.2	837	1440–2320	1–50
Concrete (dense)	0.58–1.7	837	1760–2400	2–5
Concrete (lightweight)	0.14–0.58	837	480–1280	10–50
Timber	0.10–0.17	1675	465–530	63–67
Metal	43–230	420	2700–8000	0
Glass	1.0–1.2	628	2240–2560	0
Foamed and fibrous materials	0.03–0.07	837	48–290	40–98
Air	0.025	1005	1.28	—
Water	0.58	4190	1001	—

wide range of synthetic ones, including glass or mineral fibres and foamed plastics. All enclose small cells or pockets of still air and hence have a high degree of thermal insulation.

Finally Table 2 shows the corresponding physical properties of air and water.

In comparison with the other properties given in Table 2 the range of specific heat is quite small. Variations in the thermal diffusivity $k/\rho c$, which determines the temperature changes during conditions of varying heat flow, are therefore largely dependent upon the values of the conductivity and the density.

Walls may also be divided into categories according to their mode of construction. Masonry walls may be solid (single leaf) or cavity structures with a further sub-division according to the placing of the cavity.

Non-traditional walls may be defined as pre-fabricated constructions, either load-bearing or non-load-bearing, the latter being usually the so-called 'curtain walls'. Most of the load-bearing pre-fabricated walls tested in the Wall Laboratory had been developed as contributions towards post-war housing reconstruction.

3. The summary of U-values would be incomplete without a survey of the various factors that affect the exchange of heat between the test walls and the ambient climate. Table 3 lists therefore, for each of the seventeen winters during which tests were made in the Wall Laboratory, monthly mean values of the outside air temperature, the daily range of air temperature, the wind speed, total rainfall, the driving rain received by a north-facing vertical rain gauge, and the water vapour pressure. Although the Wall Laboratory was not in operation during the winter of 1946/47, data for January to March at this time have been included because the spell of cold wather in this period was more severe than any other in the seventeen winters

Table 3 Summary of climatic data for Garston in winters 1936–7 to 1960–1

Month	Mean daily Temp (degF)	Mean daily range of Temp (degF)	Mean wind speed* (mile/h)	Total rain-fall (mm)	Driving rain from north (mm)	Vapour pressure at 9 GMT (mbar)
1936						
October	48.1	15.8	4.3	47.8	8.3	
November	41.7	12.8	4.7	96.5	3.0	
December	40.6	11.8	5.2	48.3	2.0	
1937						
January	41.1	10.3	5.8	110.0	2.2	
February	42.6	10.8	5.4	113.0	2.2	
March	39.3	13.0	4.7	78.2	9.1	
October	51.0	14.4	3.4	62.0	3.8	
November	41.3	12.5	3.4	32.8	4.2	
December	36.5	9.9	4.6	83.1	6.7	
1938						
January	41.7	10.6	6.2	84.8	7.2	
February	40.5	11.6	6.6	12.4	2.0	
March	48.1	21.9	4.2	8.6	0.6	
October	49.2	15.9	4.5	86.4	2.7	
November	48.3	12.1	5.0	76.7	0.7	
December	38.9	11.7	5.4	69.6	3.8	
1939						
January	40.1	9.4	6.5	118.6	15.3	
February	42.3	13.2	5.2	16.3	0.4	
March	42.3	12.8	5.5	37.1	5.9	
April (15–30)	46.3	18.2		39.6	(5)	
1947						
January	34.8	12.5	5.3	45	0.9	
February	28.4	8.8	5.8	46	2.7	
March	39.7	13.4	5.5	148	5.1	
October	50.5	20.5	3.1	2	0.1	
November	44.6	12.8	5.0	31	0.5	
December	41.9	11.2	4.8	67	2.9	
1948						
January	42.2	12.2	6.2	122	1.8	
February	40.5	12.3	6.2	29	0.4	
March	47.3	20.6	4.1	20	0.1	
October	49.0	16.5	3.2	54	4.9	
November	44.0	13.1	3.3	45	2.1	
December	42.9	11.3	5.9	66	0.8	
1949						
January	41.6	11.5	4.6	27	0.9	7.5
February	41.3	17.5	4.7	36	2.0	7.2
March	41.0	15.6	4.9	31	1.0	6.8

Table 3 (*Contd.*)

Month	Mean daily Temp (degF)	Mean daily range of Temp (degF)	Mean wind speed* (mile/h)	Total rain-fall (mm)	Driving rain from north (mm)	Vapour pressure at 9 GMT (mbar)
October	53.4	16.8	3.1	156	9.1	12.0
November	43.3	13.0	3.5	63	3.2	8.1
December	42.3	11.1	4.6	44	2.1	7.7
1950						
January	39.3	10.3	3.9	21	1.5	7.5
February	42.7	13.4	5.8	115	2.4	7.7
March	45.5	17.3	4.0	20	1.1	7.8
October	49.4	15.0	3.2	8	0.5	10.2
November	42.9	10.9	4.8	124	8.3	8.2
December	33.5	8.9	4.0	38	10.6	5.9
1951						
January	39.3	10.5	4.5	90	0.5	7.3
February	38.5	10.9	4.7	122	9.9	7.2
March	40.2	12.8	4.8	91	3.3	6.9
October	48.9	16.3	2.7	24	4.4	10.4
November	47.1	11.7	4.7	131	1.7	10.0
December	41.7	11.1	4.5	44	2.1	7.9
1952						
January	37.5	11.1	4.8	43	0.8	6.3
February	37.6	13.4	3.1	15	1.6	6.2
March	44.0	13.3	4.3	70	8.4	8.1
October	48.1	13.2	4.0	76	2.8	10.0
November	39.1	10.9	4.7	90	11.4	6.8
December	36.5	11.2	3.6	72	3.8	6.9
1953						
January	37.3	8.5	3.1	21	2.5	6.6
February	38.9	12.2	4.6	45	9.7	6.9
March	42.5	20.0	3.4	11	0.3	7.1
October	50.5	14.3	2.4	70	1.5	11.1
November	46.5	10.7	3.5	37	1.0	9.6
December	44.7	8.7	3.0	12	2.3	9.2
1954						
January	36.7	9.8	5.3	21	1.0	6.4
February	37.1	11.3	4.6	54	1.3	6.6
March	43.1	14.3	4.3	63	2.5	7.8
October	54.1	12.7	4.9	51	0.3	12.5
November	44.3	13.7	4.7	123	7.4	8.9
December	43.6	10.8	5.8	48	0.3	8.3
1955						
January	35.9	10.1	5.2	63	8.1	6.6
February	34.8	10.2	5.7	46	5.1	6.0
March	38.0	15.1	5.6	31	2.8	6.1

Table 3 (*Contd.*)

Month	Mean daily Temp (degF)	Mean daily range of Temp (degF)	Mean wind speed* (mile/h)	Total rain-fall (mm)	Driving rain from north (mm)	Vapour pressure at 9 GMT (mbar)
October	48.6	15.2	3.2	91	6.6	10.0
November	44.7	11.9	3.9	61	2.0	8.9
December	42.2	12.8	4.8	53	3.6	8.3
1956						
January	38.3	11.1	4.8	104	14.2	7.1
February	30.6	12.2	5.0	7	1.5	5.0
March	43.5	16.5	4.7	25	1.0	7.1
October	49.2	13.0	4.1	52	4.2	10.1
November	42.3	11.1	4.3	12	3.0	8.0
December	42.3	8.2	4.7	98	1.8	8.5
1957						
January	41.9	10.4	6.4	34	2.0	8.0
February	42.3	11.6	4.2	84	1.6	8.0
March	49.3	16.3	4.0	33	1.0	9.7
October	51.9	14.9	3.7	45	0	11.4
November	43.5	8.9	4.4	53	0.6	8.0
December	38.9	11.5	5.2	59	26.6	7.1
1958						
January	38.4	10.4	4.3	72	0	6.9
February	41.3	12.3	5.4	71	2.0	7.6
March	39.3	13.0	3.8	35	1.0	6.4
October	52.0	11.9	3.8	63	1.0	11.7
November	43.3	9.1	2.7	58	3.4	8.9
December	41.1	9.9	3.7	81	11.4	8.0
1959						
January	34.9	11.2	5.1	72	4.6	5.8
February	39.0	11.4	3.7	2	0	7.1
March	45.4	14.3	5.0	59	0.8	8.6
October	54.3	18.0	4.0	57	1.8	11.8
November	43.9	13.2	3.9	54	5.0	8.7
December	42.1	11.1	6.6	115	0.8	8.2
1960						
January	38.8	9.3	5.2	60	5.0	7.5
February	39.7	11.3	5.5	58	7.8	7.1
March	43.7	11.3	4.9	44	9.0	8.0
October	51.5	11.1	3.7	149	17.2	11.6
November	45.7	11.0	5.2	107	2.2	9.5
December	39.4	8.6	4.5	83	5.8	7.6
1961						
January	38.7	8.5	5.6	73	TR	6.9
February	45.1	10.0	6.1	65	0.8	8.7
March	46.6	20.0	5.1	2	0	8.2

* October 1936 to March 1956 wind speed computed from Kew data.

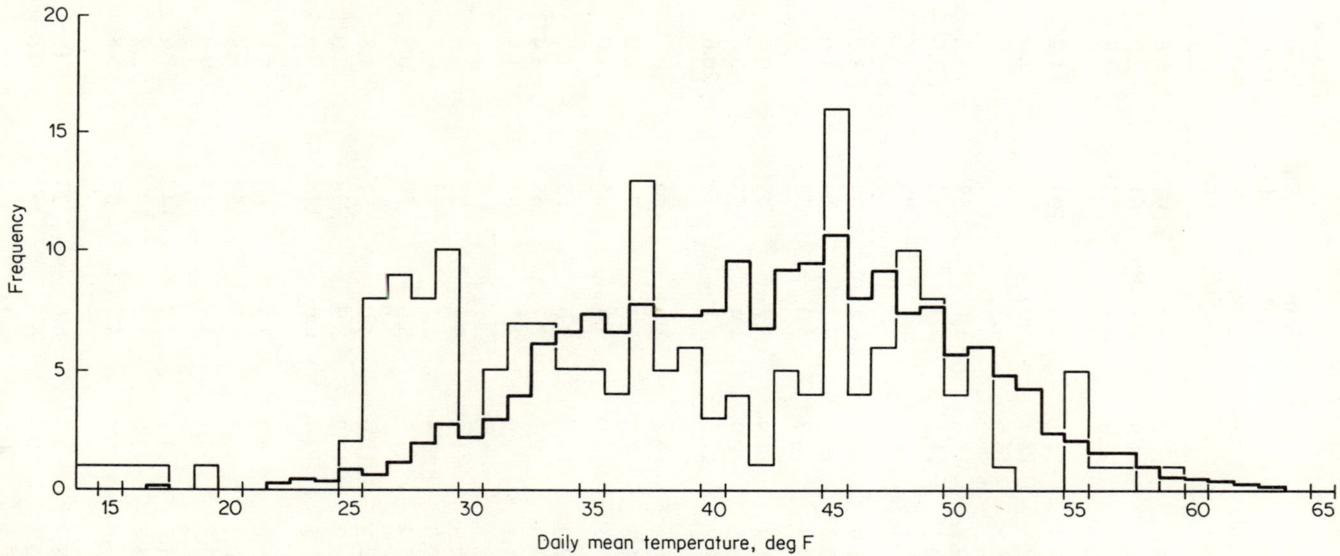

Fig. 5. October to March frequency diagram of daily mean temperatures. Mean for eleven winters 1948–9 to 1958–9, with, for comparison, a very cold winter—1946–7 (thin lines)

considered; indeed, probably more severe than any other for a hundred years or more. It thus provides a useful indication of the worst weather likely to be experienced in south-east England.

The distribution of daily mean air temperatures in winter is shown in Fig. 5. This is based on eleven winters, 1948–9 to 1958–9, the months October to March being used. For comparison the frequency distribution for the very cold winter 1946–7 is shown on the same figure.

Solar Radiation on a North-facing Wall

Values of solar radiation falling on a north-facing wall are not listed in Table 3 because direct measurements are available for one winter only, that of 1960–1. During the period of the observations only diffuse radiation reached the test walls, and analysis of the 1960–1 observations showed that on average the solar radiation falling each day on the north wall at Garston was 36 percent of the diffuse solar radiation falling on a horizontal surface at Kew Observatory. The comparison with climatic measurements at Kew is made because these are customarily adopted as being representative of south-east England. The correlation coefficient between ninety-three pairs of daily readings was 0.89 indicating that it would be adequate for present purposes to assume that the radiation falling on a north wall at Garston in winter was a fixed proportion of the diffuse on the horizontal at Kew.

Fig. 6 shows five-year mean diffuse solar radiation on the horizontal for the period September to April at Kew Observatory (after Blackwell, 1954), and Table 4 lists the computed mean solar radiation falling on the north wall

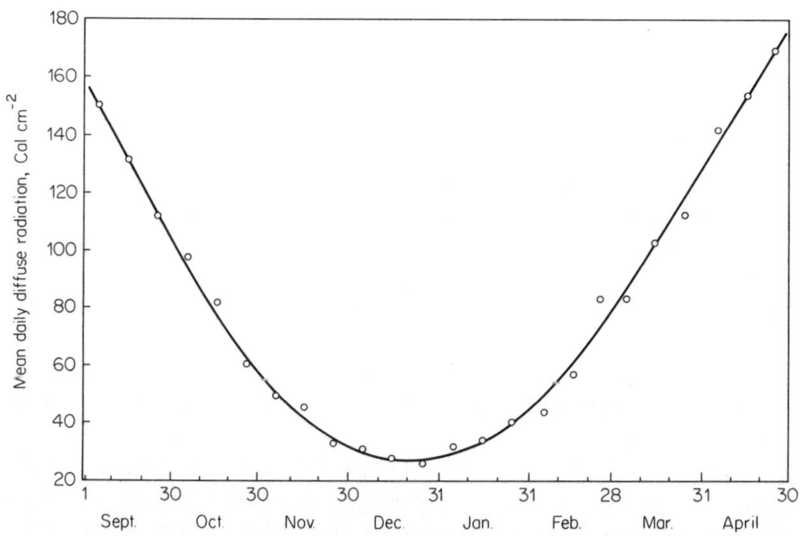

Fig. 6. Daily totals of diffuse solar radiation on horizontal; five-year (1947–51) means at Kew (Blackwell, 1954)

Table 4 Monthly mean solar radiation on north wall of Wall Laboratory in winter, including radiation reflected from ground. From data for Kew 1947–51 (Blackwell, 1954), expressed as hourly mean over whole 24 h.

Month	Solar radiation on north wall ($cal\ cm^{-2}\ h^{-1}$)
October	1.20
November	0.64
December	0.42
January	0.53
February	0.92
March	1.50

of the laboratory each month of the heating season. These values, representing 36 percent of the mean diffuse radiation on the horizontal at Kew averaged over the years 1947–51, include radiation reflected from the ground.

Long-wave Radiation Exchange

No direct measurements are available of the average long-wave (10–12 μm) radiation exchange with a naturally exposed surface. Continuous measurement of this proportion of the heat exchange at a wall surface is difficult, but it is possible to make reasonable estimates of the average value. Certain assumptions must be made:

(a) the mean temperature of the ground surface, θ_g, is the same as that of the air;

(b) the wall is exposed to half a hemisphere of sky temperature, θ_{sk}, and half a hemisphere of ground;

(c) the emissivity of ground and wall to long-wave radiation is unity;

(d) the mean cloudiness of the sky is 0.75;

(e) the overall average radiation loss is directly related to the relative proportion of clear and overcast sky;

(f) the mean net radiation to a clear sky from the ground is 188 cal cm^{-2} day^{-1}; and

(g) the mean net radiation to a completely overcast sky from the ground is 28.5 cal cm^{-2} day^{-1}.

The mean cloudiness of the sky quoted is the value for Greenwich in December and is likely to be about the same as the mean winter value for Garston.

The radiation losses to clear and overcast skies are based on the results of Dines and Dines (1927) from their extensive series of measurements at Benson, Oxfordshire. The measurements showed that with clear skies the ground surface was on average about 2 degF colder than the air, and 0.3 degF under an overcast sky. Roach (1955) confirmed that on average the monthly mean ground surface temperature at Kew was little different from air temperature. The figures of Dines and Dines are probably sufficiently accurate when the cloud cover is dense and complete, but less so where there is broken or thin cloud. No data for these occasions are available. It is reasonable to make the further assumption that the overall average radiation loss will be related directly to the relative proportion of clear and overcast sky. Hence with an average cloudiness of 0.75 (i.e. 75 percent) the mean radiation loss from ground to sky in winter

$$= 0.255 \times 188 + 0.75 \times 28.5$$

$$= \text{approx. } 70 \text{ cal cm}^{-2} \text{ day}^{-1}$$

$$= \text{approx. } 0.048 \text{ cal cm}^{-2} \text{ min}^{-1}$$

With the above assumptions and data the net radiation heat exchange $q_{s.r}$, between the wall surface at temperature θ_s and surroundings at mean temperature θ_r is

$$q_{s.r} = \sigma(T_s^4 - T_r^4) \qquad T = \theta + 273$$

$$= \frac{\sigma}{2}(T_s^4 - T_g^4) + \frac{\sigma}{2}(T_s^4 - T_{sk}^4)$$

$$= \frac{\sigma}{2}(T_s^4 - T_g^4) + \frac{\sigma}{2}(T_s^4 - T_g^4) + \frac{\sigma}{2}(T_g^4 - T_{sk}^4)$$

$$= \sigma(T_s^4 - T_o^4) + 0.024 \text{ cal cm}^{-2} \text{ min}^{-1} \tag{3.1}$$

As an example, the mean outside surface temperature of a wall was 45.7 degF (280.6 K), when the mean air temperature was 41.8 degF (278.4 K). Substituting in (3.1) it follows that $T_r = 275.0$ K, or 35.6 degF, which is 6.2 degF below mean air temperature. From the regression equation relating heat loss through the wall to the prevailing climatic parameters, the mean radiant temperature had been estimated to be about 35 degF, so that the two estimates were in good agreement.

Loss of Water from a North-facing Wall

Water is gradually lost from a wall by evaporation from the surfaces, the rate depending in part upon the humidity and speed of movement of the air in contact with the surfaces, and in part upon the nature of the materials of the wall and its moisture content. Table 5 lists, in column 3, the mean rate of decrease of the water content of some of the walls tested in the Wall Laboratory. The rate is expressed as an hourly mean value over the period shown in column 2, being obtained from measurements of moisture content at different times (see Chapter 9). It is assumed that all moisture was lost from the outside surface. The ventilation rate in the laboratory was low and evaporation to the air in the test rooms was assumed to be negligible.

Panels 35–38 were solid concrete walls, rendered outside and no. 41 was

Table 5 Water losses from walls in the Wall Laboratory

		Decrease in water content			Driving rain on wall			Solar radiation on wall	
Panel no. 1	Dates 2	$g\ cm^{-2}\ h^{-1}$ 3	heat equiv. $cal\ cm^{-2}\ h^{-1}$ 4	% of total heat-flux 5	$g\ cm^{-2}\ h^{-1}$ 6	heat equiv. $cal\ cm^{-2}\ h^{-1}$ 7	% of total heat-flux 8	$cal\ cm^{-2}\ h^{-1}$ 9	% of total heat-flux 10
7	17–23 Dec. 1959	8.3×10^{-4}	0.49	25	0	0	0	0.42	26
23 inner leaf	29 Jan. to 21 Mar. 1958	6.3×10^{-5}	0.037	3	—	—	—	1.21	89
35	July 1951 to May 1953	1.09×10^{-4}	0.065	4.6	5.73×10^{-4}	0.34	24	—	—
36	Oct 1950 to May 1951	1.63×10^{-4}	0.097	6	7.99×10^{-4}	0.47	29	1.28	78
37	July 1952 to May 1953	2.22×10^{-4}	0.131	9	8.73×10^{-4}	0.52	35	—	—
38	Sept 1952 to May 1953	1.99×10^{-4}	0.118	9	8.50×10^{-4}	0.50	39	1.36	105
41	Oct 1952	6.2×10^{-4}	0.367	19	3.8×10^{-4}	0.23	12	1.20	62
41	Oct 1952 to March 1953	3.0×10^{-4}	0.178	9	7.0×10^{-4}	0.41	22	0.87	53

built of cored concrete blocks, also rendered. All received driving rain during the period of test, the amount listed in column 6 being that caught in the north-facing aperture of the free-standing driving-rain gauge. Probably the walls received no more than about half this because of deflection of the wind by the building. Even if the amounts in column 6 were halved, the walls gained in most cases about twice as much water as the net loss, so that the total rate of evaporation was probably the net loss, or even more. No driving rain reached the face of panel 7 during December 1959 while the inner leaf of panel 23 was of course completely sheltered from driving rain so that the water losses for these two walls in column 3 were the actual evaporation water.

The heat required to evaporate the water is given in columns 4 and 7 of Table 5 and is also listed as a percentage of the mean heat flux through the wall in columns 5 and 8. The nett loss of water represents a heat flux of from 3 to 25 percent of the total heat flux through the wall.

It has been shown by Cooling (1929) for bricks, and Penman (1940) for soils that the rate of evaporation of water from a saturated surface is the same as that from free water, and is given by the formula

$$\Phi = \beta(p_o - p_s) \qquad (3.2)$$

Here Φ is the evaporation rate, β is the water vapour transfer coefficient at the surface, p_o is the water vapour pressure in the ambient air, and p_s is the saturation vapour pressure at the temperature of the surface. Values of β are given by Fig. 7 from data quoted by Illig (1952) and Penman (1940).

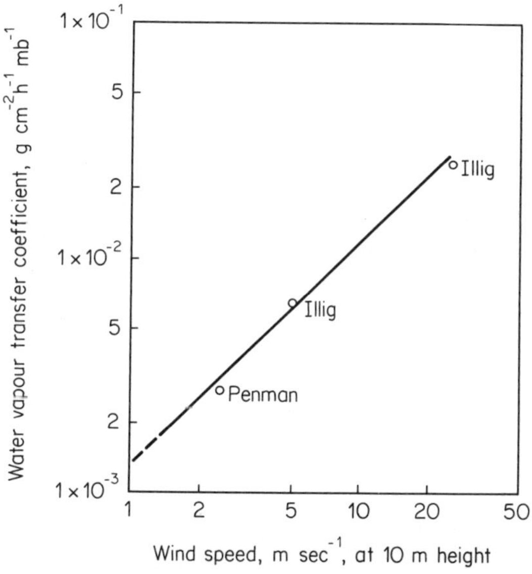

Fig. 7. Relation between surface water vapour transfer coefficient and wind speed, from data of Illig (Cammerer, 1952, and Penman, 1940)

Table 6 Data for panel 41, October 1952 to March 1953

Mean windspeed at 10 m above ground	$1.74\ m\ s^{-1}$
Surface water vapour transfer coefficient	$2.5\ g\ cm^{-2}\ h^{-1}\ mbar^{-1}$
Mean atmospheric vapour pressure	7.4 mbar
Mean air temperature	4.7 degC
Estimated mean wall surface temperature	5.9 degC
Saturation vapour presure at 5.9 degC	9.3 mbar
'Wet surface' evaporation rate	$4.75 \times 10^{-3}\ g\ cm^{-2}\ h^{-1}$

The relationship (3.2) holds only as long as the surface is wet and may be used to estimate the maximum possible evaporation rate for the walls in the laboratory. For example, in Table 6 are given the relevant data and result calculated for panel no. 41, a single-leaf wall of cored concrete blocks rendered on the outer face. The actual net loss of water from this wall in the same period was $3.0 \times 10^{-4}\ g\ cm^{-2}\ h^{-1}$ (see Table 5), to which must be added the amount of driving rain falling on the wall, estimated to be $7.0 \times 10^{-4}\ g\ cm^{-2}\ h^{-1}$, a total of $1.0 \times 10^{-3}\ g\ cm^{-2}\ h^{-1}$. This is only about 20 percent of the 'wet surface' rate estimated above, indicating that the wall surface was dry for most of the time.

It may be noted that on average the diffuse solar radiation provides much more than sufficient heat to evaporate water at these rates (Table 5, columns 9 and 10). Only in mid-winter might the solar radiation be insufficient to apply all the heat needed—estimated to be $0.42\ cal\ cm^{-2}\ h^{-1}$ which is about enough to evaporate water at the rates noted in Table 5. On a very dull day the mean solar radiation on the north wall may be as low as $0.07\ cal\ cm^{-2}\ h^{-1}$ which is insufficient to maintain these evaporation rates. Heat will then be drawn from the wall.

DATA SHEETS

The construction of each wall illustrated on the data sheets is described as originally published (Pratt, 1969). The results of the measurements have been converted to SI Units and are expressed as follows: density in kg/m^3; thermal conductivity in W/mK; thermal transmittance in W/m^2K; and thermal resistance in m^2K/W.

Panel No. 1

Single-leaf wall of solid brickwork, 9 in. thick, constructed in 1936 *when the laboratory was built. Fletton bricks set in cement mortar.*

Three similar walls were built at this time and the first set of results are the mean for all three. One wall (in room 5) *was retained unchanged until* 1949.

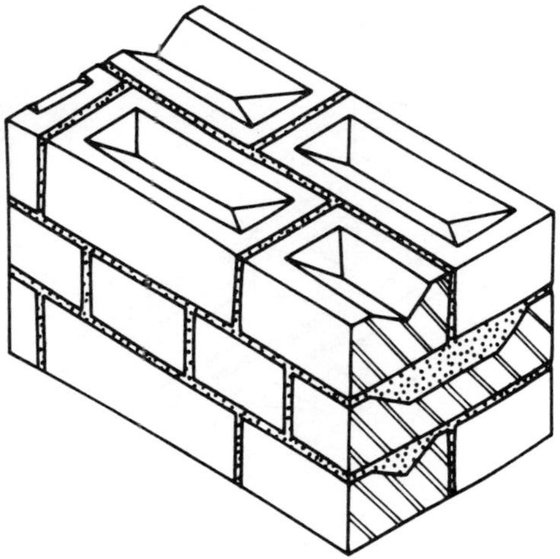

Density of bricks = 1640

(a) Measurements in November–December 1936.

Mean $U = 2.6$
Mean k of wall = 1.15

(b) Measurements in January–March 1937

Mean $U = 2.6$
Mean k of wall = 1.15

(c) Measurements in 1937–8

Mean moisture content = 4.9 percent by volume in April 1938
Mean $U = 2.5$
Mean k of wall = 1.1

(d) Measurements in 1947–8

Mean $U = 2.3$
Mean k of wall = 0.99

(e) Measurements in 1948–9

Mean $U = 2.3$
Mean k of wall = 0.98

Fundamental time-lag for 24 h period temperature wave = 5.8 h.

Panel No. 2

Single-leaf wall of solid brickwork, 9 *in. thick, constructed in* 1936. *Fletton bricks set in cement mortar. Plastered inside in early February* 1937.

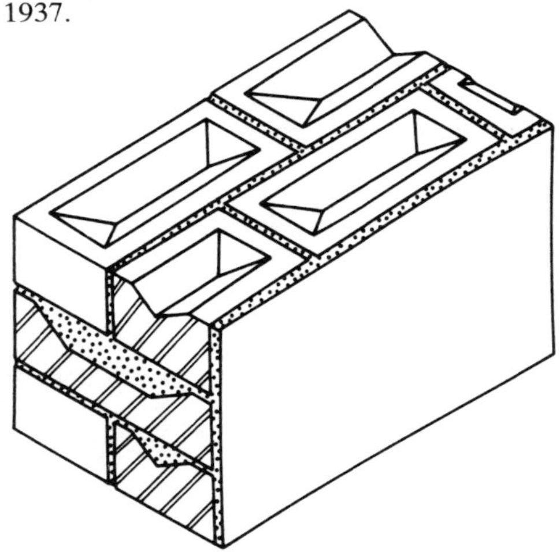

Panel No. 2 (*continued*)

Tested February–March 1937

Mean moisture content in April 1938 = 6.6 percent by vol.

Mean $U = 2.4$

Mean k of brickwork = 1.3

Panel No. 3

Single-leaf wall of Fletton bricks, 9 in. thick, built in 1936. *Rendered on outside early February* 1937.

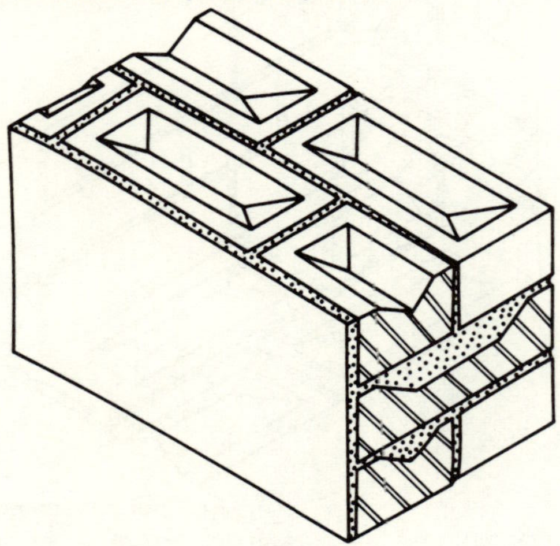

(*a*) Measurements February–March 1937

$U = 2.3$

k of brickwork = 1.3

(*b*) Measurements in 1937–8

$U = 2.4$

k of brickwork = 1.3

Moisture content in April 1938 = 6.5 percent by vol.

Panel No. 4

Single-leaf wall of Fletton bricks, 9 in. thick, built in 1946. *Rendered on outside, strapped and plastered on laths inside. Total thickness* 11 *in.*

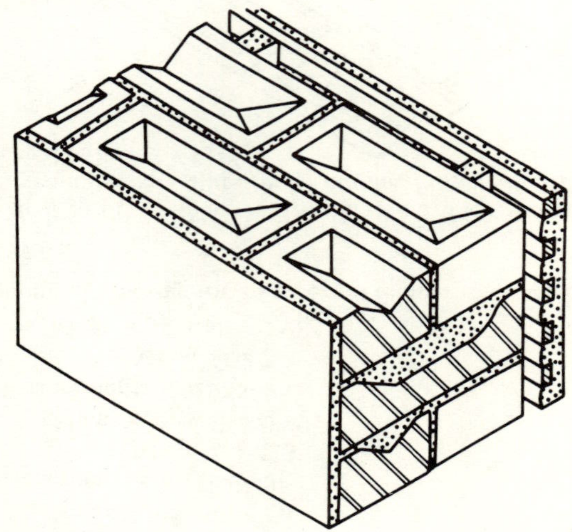

Tested in 1947–8

$U = 1.65$

Equivalent k of wall = 0.65

Panel No. 5

Double-leaf wall of Fletton bricks, total thickness 11 *in., built in* 1936. *Cavity unventilated throughout.*

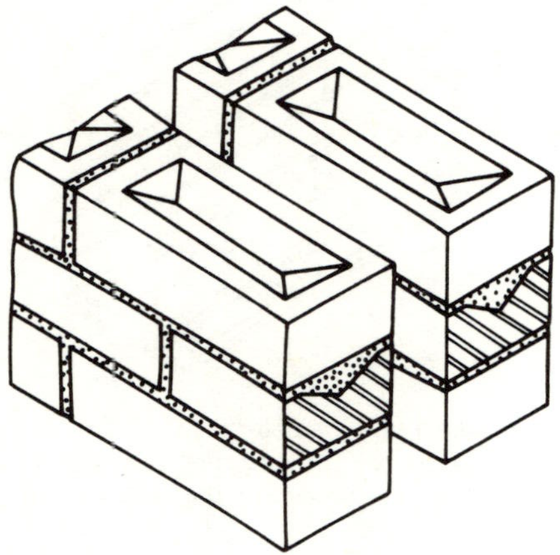

(a) Measurements in November–December 1936 (mean of 3 walls)

$$U = 1.9$$

Equivalent k of wall = 0.84

(b) Measurements in January–March 1937 (mean of 2 walls)

$$U = 1.8$$

Equivalent k of wall = 0.76

(c) Measurements in 1937–8

$$U = 1.8$$

Equivalent k of wall = 0.75

Moisture content of outer leaf in March 1938 = 8.7 percent by vol.

Moisture content of inner leaf in March 1938 = 1.0 percent by vol.

Panel No. 6

Double-leaf wall of Fletton bricks, total thickness 11 *in., built in* 1936. *Cavity ventilated as described below.*

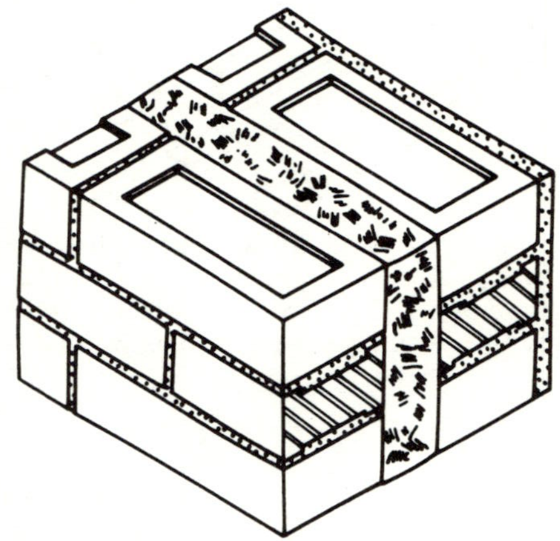

(a) Measurements in January–March 1937

Cavity ventilated with ordinary ceramic air bricks, two at both top and bottom, each with 39 holes $\frac{3}{8}$ in. square.

$$U = 1.8$$

Equivalent k of wall = 0.84

(b) Measurements in 1937–8, natural ventilation of cavity

Panel No. 6 (*continued*)

with iron grills in place of ceramic air bricks.

$U = 2.0$

(*c*) Measurements in 1937–8, forced ventilation of cavity with an air flow of about 200 ft^3/min.

$U = 2.6$

(*d*) Measurements in February–March 1938, ventilators blocked.

$U = 1.8$

Moisture content at end of March 1938

inner leaf = 1.4 percent by vol.
outer leaf = 3.4 percent by vol.

Panel No. 7

11-*in. cavity wall of Fletton bricks. Outer leaf dark sand-faced bricks, density* 109 *lb/ft*^{3} (1.75 g cm^{-3}), *inner leaf of common bricks, density* 101 *lb/ft*^{3} (1.61 g cm^{-3}). *Plastered inside. Built* 25 *August* 1959.

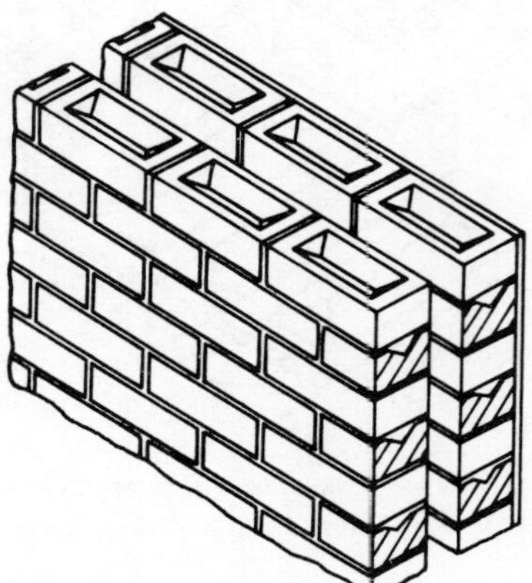

Tested in 1959–60 and 1960–61.

	1959–60	1960–1
24 h mean U-value	1.7	1.65
Day time mean U-value	2.0	1.9
Night-time mean U-value	1.6	1.5
Mean thermal conductivity		
of outer leaf	1.0	0.97
of inner leaf	0.84	0.78

Moisture content, percent by vol.
in February 1960, outer leaf = 4.5
in February 1960, inner leaf = 0.8
in March 1961, inner leaf = 1.4

Panel No. 8

Double-leaf wall of dense pressed bricks, with $2\frac{1}{2}$ *in. cavity. Plastered on inside. Built in August* 1954. *These denser bricks were chosen to be typical of many used in northern England. A* 1:2:9 *mortar mix was used to assist drying of the wall.*

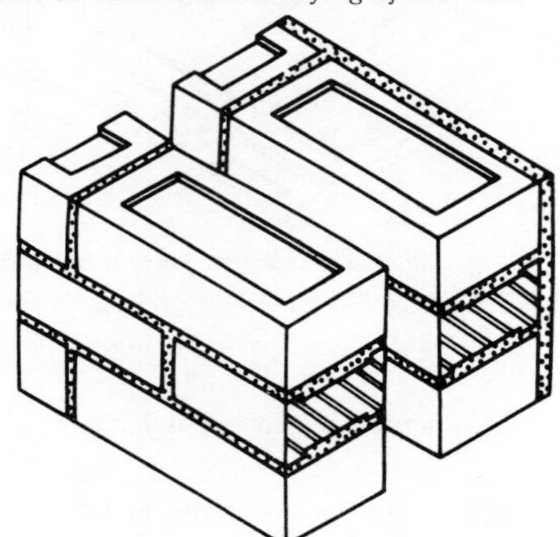

Panel No. 8 (*continued*)

Tested in 1954–5.

Density of bricks = 2130
Mean $U = 1.9$
Mean k of outer leaf = 1.6
Mean k of inner leaf = 1.25
(without plaster)

When dismantled in July 1955 the bricks were dry, there being no loss of weight on drying.

Panel No. 9

Double-leaf wall of dense pressed bricks, exactly as No. 8, *but with cavity filled with shredded resin-bonded fibreglass. The fill was put in after each* 18-*in. lift of brickwork was completed, to a mean density of* 2.94 *lb/ft*^{3}. *Built in August* 1954.

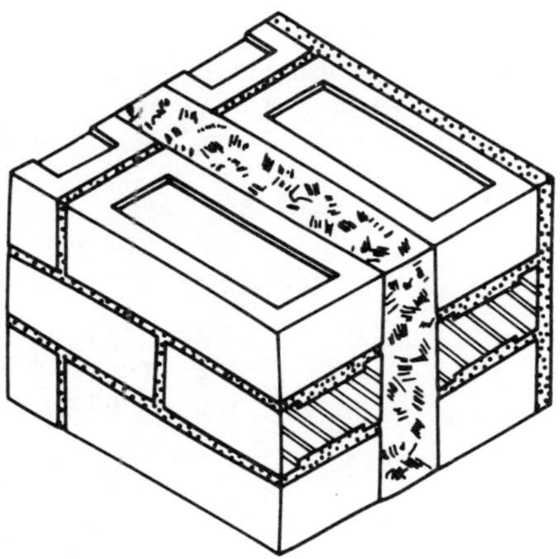

Tested in 1954–5

Density of bricks = 2131
Density of fill = 47.1
Mean $U = 0.85$
Mean k of outer leaf = 1.8
Mean k of inner leaf (without plaster) = 1.3
Mean k of cavity fill = 0.07

Mean moisture content of outer leaf, 5 April 1955, was about 4 percent by vol.

Mean moisture contents, October 1955

Outer leaf bricks = 6.1 percent by vol.
Inner leaf bricks = 0.3 percent by vol.
Cavity fill = 0.2 percent by vol.

Panel No. 10

Double-leaf wall of dense wire-cut bricks, from the same clay as panels 8 *and* 9. *Each brick perforated with* 48 *square holes in* 4 *rows giving* 17 *percent void. Plastered inside. Built August* 1954.

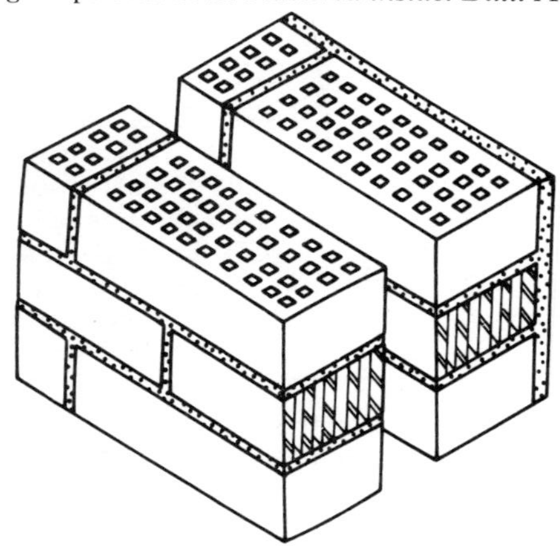

Panel No. 10 (*continued*)

Tested in 1954–5.

Density of bricks = 2083
(Effective density = 1730)

Mean $U = 1.6$
Mean k of outer leaf = 1.2
Mean k of inner leaf = 0.82

Moisture concent 25 July 1955: nil.

Panel No. 11

11-in. cavity wall: outer leaf of sand-faced Fletton bricks laid frog-down; $2\frac{1}{2}$ *in. cavity; inner leaf of 6-hole clay blocks* $9 \times 12 \times 4\frac{1}{4}$ *in. with horizontal cavities. Plastered internally. Built August* 1954.

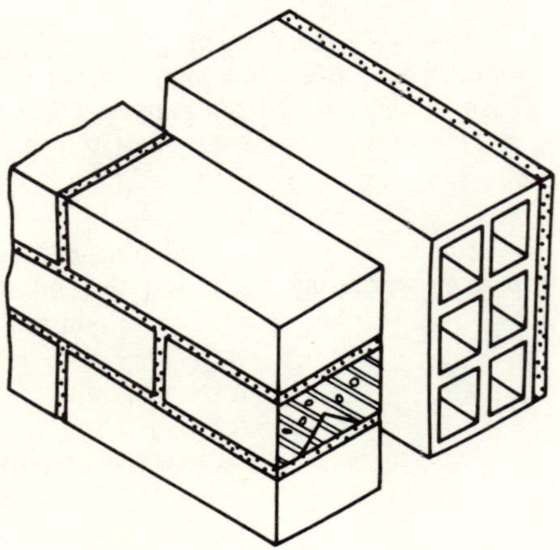

Tested in 1954–5
Mean thermal transmittance = 1.4
Resistance of clay blocks = 0.26
Conductivity of Fletton bricks = 0.92

Panel No. 12

Double-leaf wall of 4-hole clay blocks $9 \times 9 \times 4$ *in., each block delivered with factory-applied,* $\frac{1}{16}$ *in. thick coat of rendering on one face. After erection in July* 1951, *outside rendered with thin coat of* 1:1:6 *mix, inside plastered with* $\frac{1}{16}$ *in. thick undercoat of* 1:3 *Thistle Browning and Leighton sand, followed by* $\frac{1}{16}$ *in. thick Thistle plaster finish.*

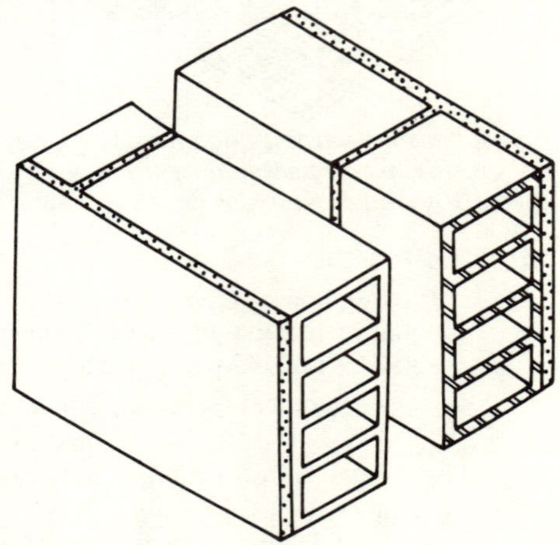

Tested in 1951–2.

Mean $U = 1.4$
Resistance of outer leaf = 0.19
Resistance of inner leaf = 0.18

Panel No. 12 (*continued*)

Fundamental time-lag to 24 h temperature wave = 4.5 h
Amplitude decrement to 24 h temperature wave = 0.053

Panel No. 13

Single-leaf wall of 6 *in. thick* × $8\frac{1}{2}$ *in. long* × $5\frac{1}{2}$ *in. high horizontally perforated bricks, rendered. Built* 1948.

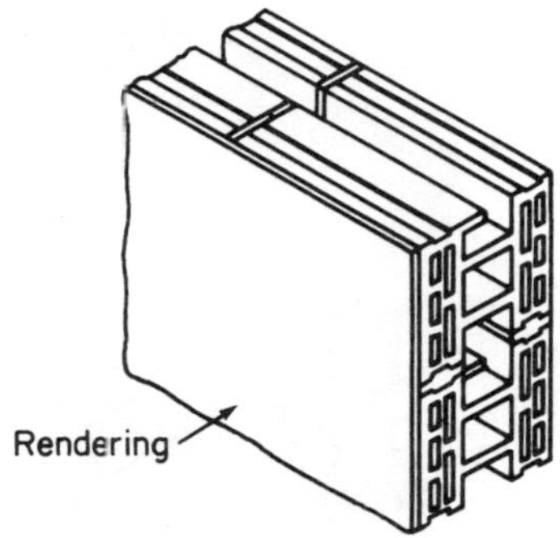

Tested in 1948–9.

Mean thermal transmittance = 2.1
Mean surface-to-surface resistance = 0.32

On 7th February 1949 the wall was rendered inside, and 1 in. thick wood-wool cement slabs were bedded on the rendering, the whole being then finished with hard plaster. Testing was resumed almost immediately with the following results:

Mean thermal transmittance (March only) = 1.6
Mean surface-to-surface resistance (March and April) = 0.46

Panel No. 14

Single-leaf wall of perforated clay bricks, B.R.S. type V.2, 8 × 8 × 4 *in. vertical perforations giveing* 41 *percent void. Clay:* 85 *percent Keuper Marl,* 15 *percent London Clay. Laid with two strips of mortar in horizontal and vertical joints. Plastered inside.*

Wall built 30 *October–*5 *November* 1956.

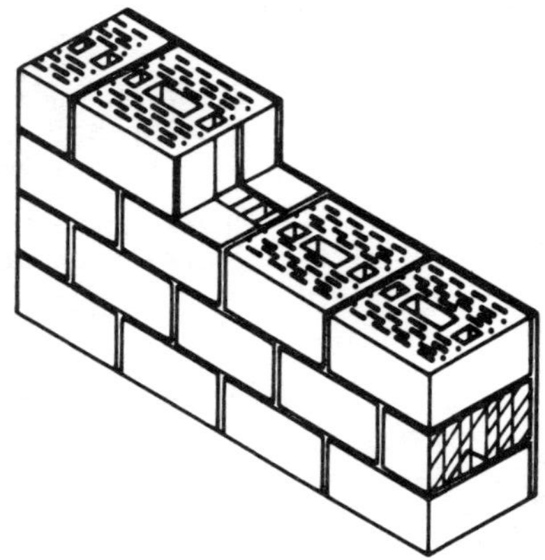

Tested in 1956–7 and 1957–8.

	1956–7	1957–8
24-h mean U-value	1.5	1.4
Day-time mean U-value	1.9	1.8
Night-time mean U-value	1.5	1.3

Thermal resistance of blocks (1957–8) = 0.48 (8.2 in. thick).

Panel No. 15

Single-leaf wall of perforated clay bricks B.R.S. V.4, type. Laid with two strips of mortar in horizontal and vertical joints. Plastered

Panel No. 15 (*continued*)

inside. Bricks made from 15 *percent London Clay,* 85 *percent Keuper Marl. Perforations give* 44 *percent void. Built August* 1957.

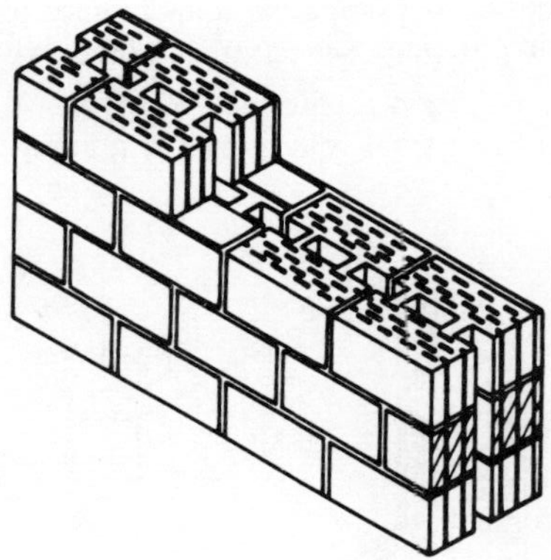

Tested in 1957–8.

24 h mean U-value	1.5
Day-time mean U-value	1.9
Night-time mean U-value	1.4

Thermal resistance of blocks = 0.49 (8.2 in. thick).

Panel No. 16

Single-leaf wall of perforated clay bricks, B.R.S. *type* V.5, *each brick* $8\frac{3}{4}\times8\frac{3}{4}\times2\frac{3}{4}$ *in., clay mix one part by volume London Clay,* 6 *parts Keuper Marl. Vertical perforation giving* 57 *percent void. Laid with two strips of mortar in horizontal and vertical joints. Plastered inside. Light self-coloured bricks outside. Built* 30 *June* 1958.

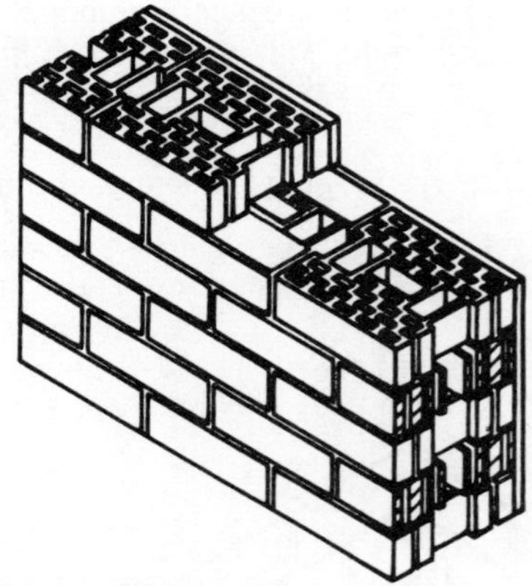

Tested in 1958–9 and 1959–60.

	1958–9	1959–60
24 h mean U-value	1.4	1.4
Day-time mean U-value	1.7	1.6
Night-time mean U-value	1.4	1.4

Moisture content in June 1960
Outer half = 2.2 percent by vol.
Inner half = 1.3 percent by vol.

Panel No. 17

Single-leaf wall of perforated clay bricks, B.R.S. *type* V.6. *Each brick* $8\frac{3}{4}\times8\frac{3}{4}\times2\frac{3}{4}$ *in., clay mix* 85 *percent Keuper Marl,* 15 *percent London Clay. Vertical perforations giving* 50 *percent void. Laid with two strips of mortar in horizontal and vertical joints. Plastered inside. Light self-coloured bricks outside. Built* 9–12 *October* 1959.

Panel No. 17 (*continued*)

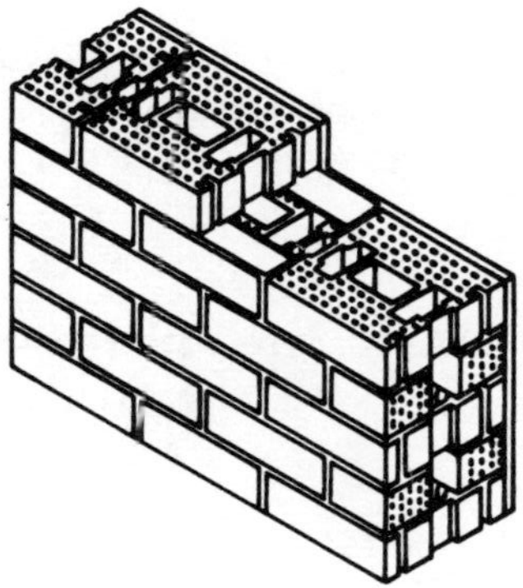

Tested in 1959–60.

24 h mean U-value = 1.8
Day-time mean U-value = 2.0
Night-time mean U-value = 1.6

Moisture content in May 1960

Outer half = 2.9 percent by vol.
Inner half = 1.5 percent by vol.

Panel No. 18

Single-leaf wall of vertically perforated bricks, each 8.6 × 8.6 × 2.6 *in., type* V.6/2. *Designed and made by Sussex & Dorking Co., of Weald clay.* 40 *percent void. Wall plastered inside. Built June* 1960.

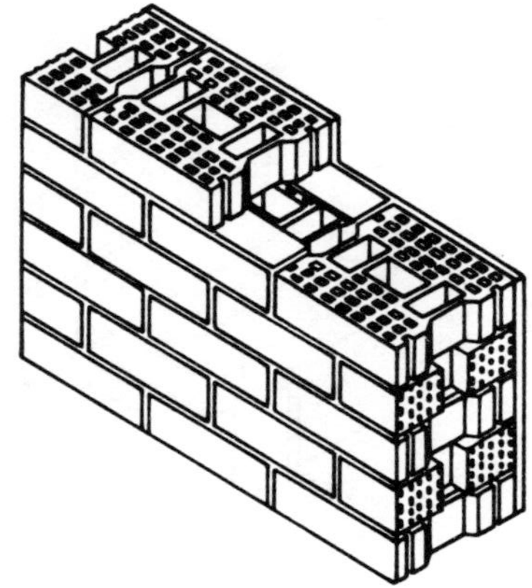

Tested in 1960–1.

Mean thermal transmittances,

	24 h	day	night
Whole season	2.0	2.2	1.9
Dec.–Jan.–Feb.	1.9	2.1	1.9

Panel No. 19

Single-leaf wall of Swiss-made perforated clay bricks type B.25. Each brick 30 *cm wide and* 13.5 *cm high, perforations running full height of brick giving* 37 *percent void. Laid with two horizontal strips of mortar, none in vertical joints, but bituminous felt inserts in vertical slots. Rendered and plastered. Built July* 1957.

Panel No. 19 (*continued*)

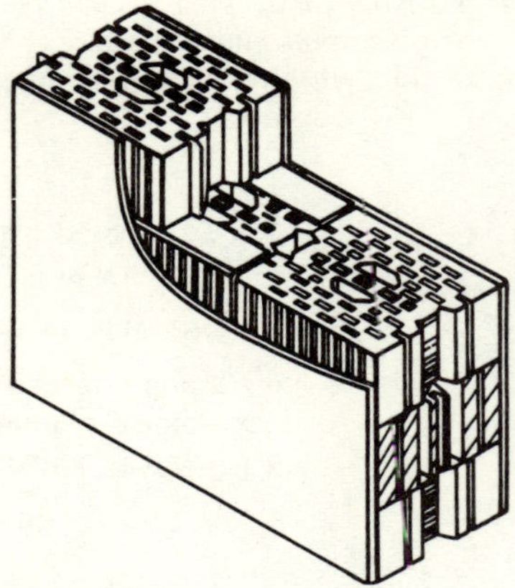

Tested 1957–8 and 1958–9.

	1957–8	1958–9
24 h mean U-value	1.2	1.2
Day-time mean U-value	1.5	1.3
Night-time mean U-value	1.1	1.2

Thermal resistance of blocks = 0.61 (9.8 in. thick).

Panel No. 20

Single-leaf wall of horizontally-perforated clay bricks, B.R.S. type H.2, 13$\frac{1}{4}$ *in. long* × 8$\frac{3}{4}$ *in.* × 2$\frac{3}{4}$ *in. Laid with two strips of mortar in horizontal and vertical joints. Plastered inside. Built September* 1958.

Tested in 1958–9.

24 h mean U-value = 1.8
Resistance of bricks = 0.31

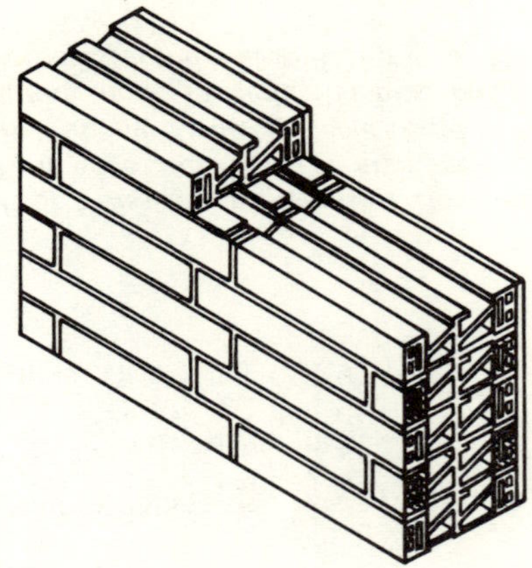

Panel No. 21

Double-leaf wall, outer leaf of Fletton bricks, 3 *in. cavity,* 9 × 18 × 3 *in. foamed-slag concrete blocks, plastered inside. Built in* 1948.

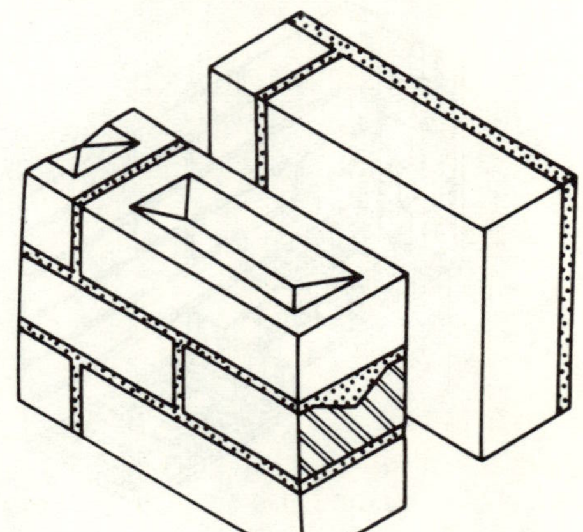

Panel No. 21 (*continued*)

Tested in 1948–9.

Mean $U = 1.6$
Mean k of foamed-slag concrete blocks = 0.43

Panel No. 22

Double-leaf wall, outer of Fletton bricks, inner of $9 \times 18 \times 4\frac{1}{4}$ *in. clinker-concrete blocks plastered inside. Built in August* 1951.

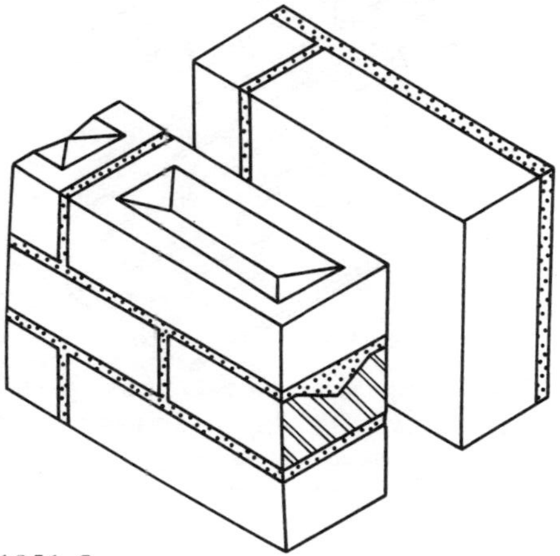

Tested in 1951–2.

Mean $U = 1.6$
Mean k of outer leaf = 1.0
Mean k of inner leaf = 0.75

Mean moisture-content of inner leaf = 12 percent by vol.

Density of concrete blocks = 72.7 lb/ft^3 dried at 105 degC (1.16 g cm^{-3})

Fundamental time-lag for 24 h temperature wave = 5.5 h
Amplitude decrement for 24 h temperature wave = 0.050

Panel No. 23

Double-leaf wall. Outer leaf of sand-faced Fletton bricks, inner leaf of Thermalite aerated concrete blocks, $18 \times 4 \times 9$ *in., dry density* 52 *lb/ft*^{3} (= 0.83 g/cc). *Plastered inside. Built June* 1957.

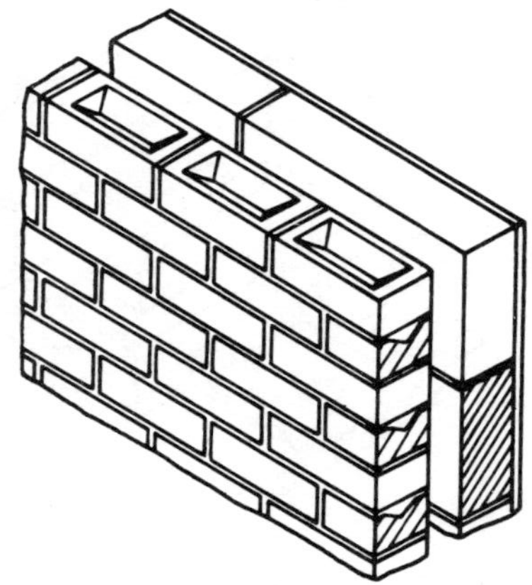

Tested in 1957–8, 1958–9, 1959–60.

	1957–8	1958–9	1959–60
Mean 24 h U-value	1.4	1.4	1.3
Day-time U-value	1.8	1.7	1.5
Night-time U-value	1.3	1.3	1.2
Mean conductivity of brick outer leaf	1.0	1.0	0.94
Mean conductivity of concrete inner leaf	0.39	0.35	0.33

Panel No. 23 (*continued*)

Moisture contents, percent by vol.

Concrete inner leaf	January 1958	6.3
	March 1958	5.2
	February 1960	4.0
	August 1960	3.5
Brick outer leaf	August 1960	4.3

Panel No. 24

11-in. cavity wall. Outer leaf of dark sand-faced Fletton bricks, density 109 *lb/ft*^{3} (1.75 g cm^{-3}). *Inner leaf of concrete blocks* 18×4×9 *in., dry density* 77 *lb/ft*^{3} (1.23 g cm^{-3}) *with aggregate of Aglite expanded clay. Plastered inside. Built August* 1959.

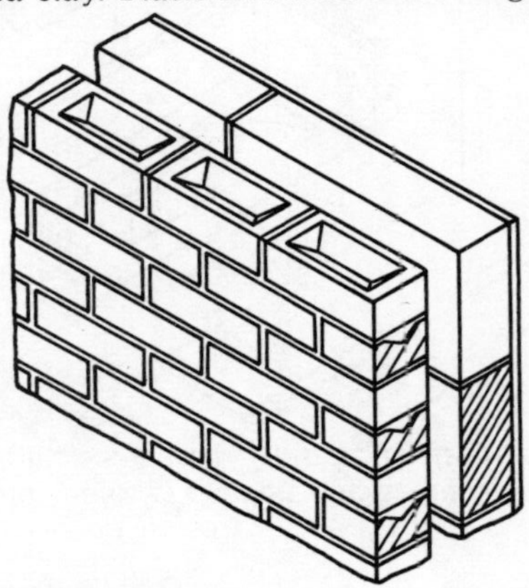

Tested 1959–60 and 1960–1.

	1959–60	1960–1
24 h mean U-value (whole season)	1.5	1.5
Day-time mean U-value (whole season)	1.7	1.8
Night-time mean U-value (whole season)	1.4	1.4

Mean thermal conductivity		
of brick outer leaf	0.97	1.1
of concrete inner leaf	0.54	0.54

Moisture content percent by vol.
in February 1960, concrete blocks = 8.2
in March 1961, concrete blocks = 6.4

Panel No. 25

Double-leaf wall, outer of Fletton bricks, inner of $9\times18\times4\frac{1}{4}$ *in. clinker-concrete cored blocks, with* 30 *percent voids, plastered inside. Built* 3 *October* 1950. *The perforations in these blocks are closed at one end.*

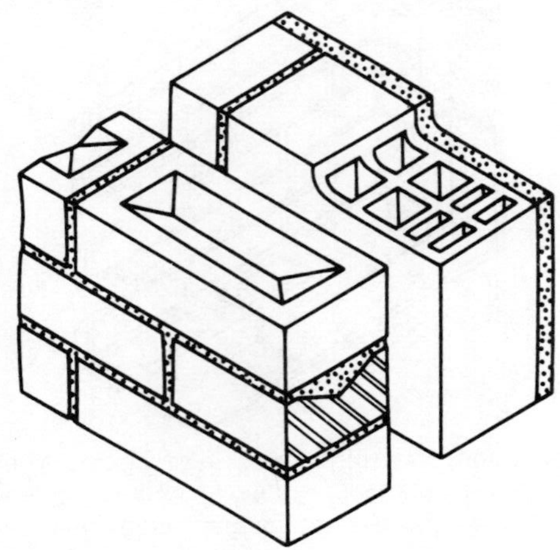

Tested in 1950–1.

Mean $U = 1.5$
Mean k of outer leaf = 1.1
Equivalent k of inner leaf = 0.50

Panel No. 26

Double-leaf wall, outer of Fletton bricks, inner of $9\times18\times4\frac{1}{4}$ *in. clinker-concrete cored blocks, three large holes giving* 34 *percent void, plastered inside. Built* 5 *October* 1950.

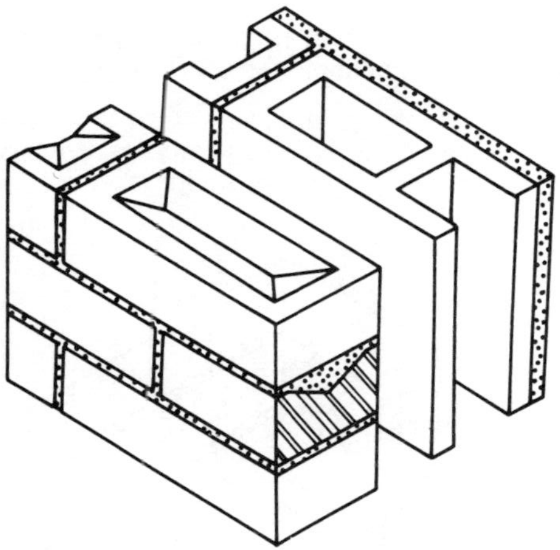

Tested in 1950–1.

Density of concrete = 1137 dried at 105 degC

Mean $U = 1.6$
Mean k of outer leaf = 1.1
Equivalent k of inner leaf = 3.55

Panel No. 27

Cavity wall. Outer leaf of sand-faced Fletton bricks. Inner leaf of Thermalite aerated concrete blocks $18\times4\times9$ *in.,* 52 *lb/ft*^{3}*, with resin-bonded rockwool slabs, density* 5.9 *lb/ft*^{3}*, nailed to cavity face.* $\frac{1}{2}\times1\frac{1}{2}$ *in. wood battens nailed to room face, with inner lining of aluminium-backed plasterboard. Wall built July* 1957.

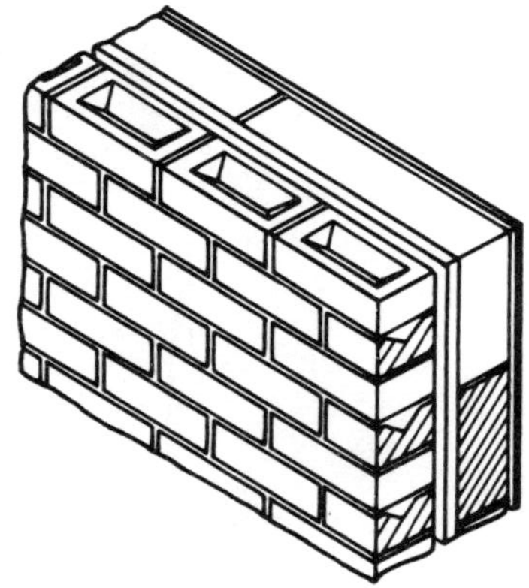

Tested in 1957–8.

24 h mean U-value	0.57
Day-time mean U-value	0.74
Night-time mean U-value	0.57

Mean conductivity of brick outer leaf = 1.1
Mean conductivity of concrete inner leaf = 0.35
Mean conductivity of rockwool slabs = 0.04

Moisture contents, September 1958,
brick = 2.1 percent by vol.
concrete = 5.3 percent by vol.
rockwool = 0.1 percent by vol.

Panel No. 28

Double-leaf wall, outer of Fletton bricks, 2 in. cavity, inner leaf of 2 ft × 8 ft × 4 in. cored gypsum plaster panels, built in September 1953.

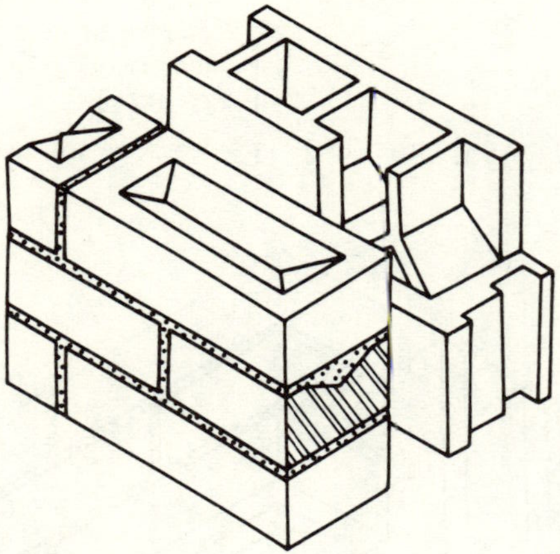

Tested in 1953–4.

Mean $U = 1.3$

Mean k of outer leaf = 0.94

Mean resistance of inner leaf = 0.23

Fundamental time-lag for 24 h temperature wave = 3.6 h

Amplitude decrement for 24 h temperature wave = 0.049

Panel No. 29

Single-leaf wall of gravel no-fines concrete, 1:0:10 *mix, cast* in situ 8 *in. thick, on* 5 *November* 1946. *Rendered outside and plastered inside.*

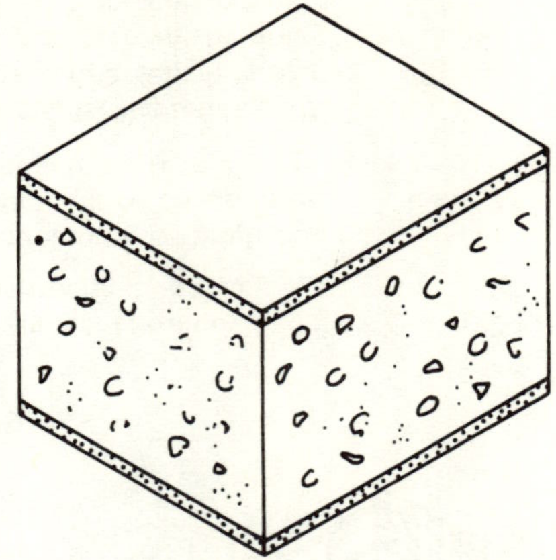

Tested in 1947–8.

Density of concrete = 184.2

Mean $U = 2.4$

Mean k of concrete = 1.0

Panel No. 30

Single-leaf wall of Whinstone no-fines concrete, 1:0:18 *mix, cast* in situ 8 *in. thick on* 24 *February* 1939. *Rendered outside and plastered inside.*

Tested in 1939.

Dry density about 1762

Mean U, April 1939 = 1.9

k of concrete = 0.86

Panel No. 30 (*continued*)

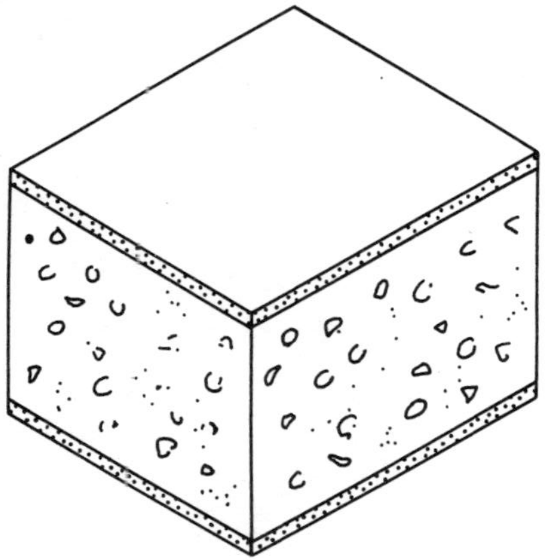

Panel No. 31

Single-leaf wall of dense concrete. Ham River gravel aggregate, 1:2:4 *mix, cast* in situ 8 *in. thick in November* 1946.

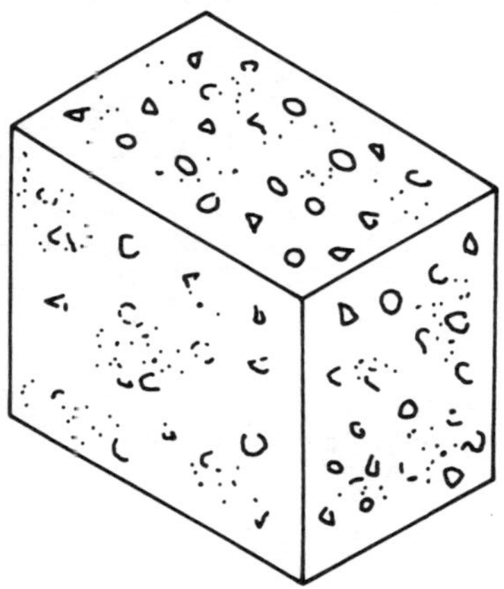

Tested in 1947–8.

Density 2291 (dried at 105 degC)

$U = 2.1$

k of concrete $= 2.1$

Panel No. 32

Single-leaf wall of dense concrete, Ham River gravel aggregate, 1:2:4 *mix, cast* in situ 8 *in. thick, in September* 1948 *(same specification as panel* 31).

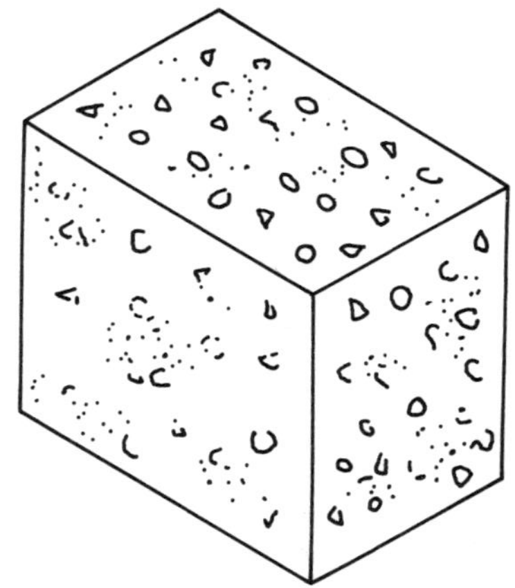

Tested in 1948–9.

Density = 2291 (dried at 105 degC)

$U = 2.3$

k of concrete $= 2.3$

Fundamental time-lag for 24 h period temperature-wave = 4.9 h

Panel No. 33

Single-leaf wall of dense concrete, Accrington brick aggregate, 1:2:4 *mix, cast* in situ 8 *in. thick on* 6 *October* 1948.

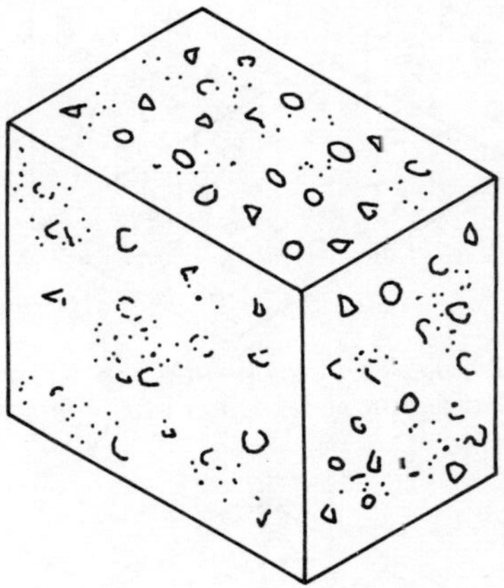

Tested in 1948–9.

Density = 2163 (dried at 105 degC)
$U = 2.7$
k of concrete = 1.1

Fundamental time-lag for 24 h temperature-wave = 5.8 h

Panel No. 34

Single-leaf wall of foamed-slag concrete, $1:2\frac{1}{2}:7\frac{1}{2}$ *mix, cast* in situ, 8 *in. thick. Rendered outside and plastered inside. Cast on* 13 *January* 1947.

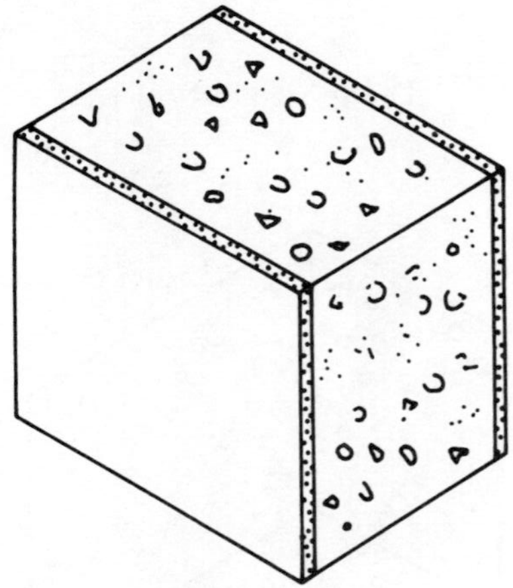

Tested in 1947–8.

Density of concrete not measured, probably about 1169 dry.

Mean thermal transmittance = 1.4
Mean conductivity of concrete = 0.39

Panel No. 35

Single-leaf wall of foamed-slag concrete, $1:2\frac{1}{2}:7\frac{1}{2}$ *mix, cast* in situ, 8 *in. thick. Rendered outside and plastered inside. Cast on* 17 *July* 1951.

Tested in 1951–2 and 1952–3.

Panel No. 35 (*continued*)

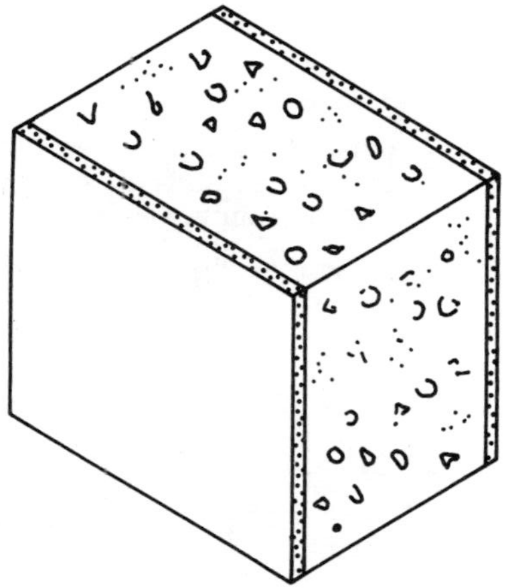

Density of concrete = 1169 (dried at 105 degC)

Measurements in 1951–2
Mean moisture content during test = 16 percent by vol.
Mean thermal transmittance = 1.5
Mean conductivity of concrete = 0.46
Fundamental time lag (24 h period temperature wave) = 7.5 h
Amplitude decrement of temperature wave = 0.093

Measurements in 1952–3
Mean moisture content during test = 11 percent by vol.
Mean thermal transmittance = 1.4
Mean conductivity of concrete = 0.43

Panel No. 36

Single-leaf wall of foamed-slag concrete, 1:3:6 *mix, rendered outside and plastered inside. The concrete was cast* in situ, *the thickness being* 8 *in. Made in* 1950.

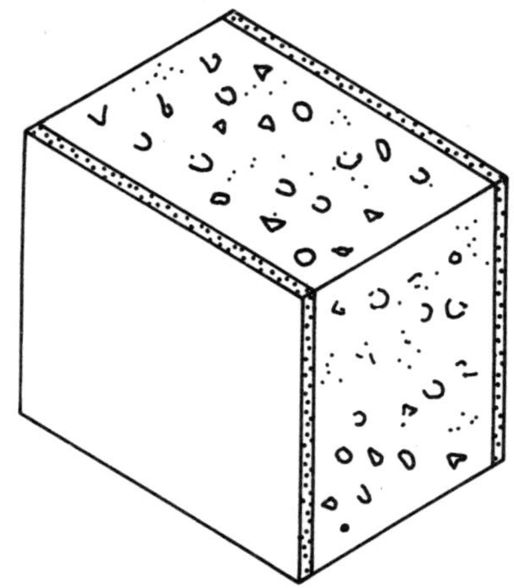

Tested in 1950–1.

Density of concrete 1250 (after drying at 105 degC)

Mean moisture content during test = 19 percent by vol.
Mean thermal transmittance = 1.7
Mean conductivity of concrete = 0.53

Panel No. 37

Single-leaf wall of foamed-slag concrete, 1:3:6 *mix, rendered*

Panel No. 37 (*continued*)

outside and plastered inside. The concrete was cast in situ, *the thickness being* 8 *in. Made in* 1952.

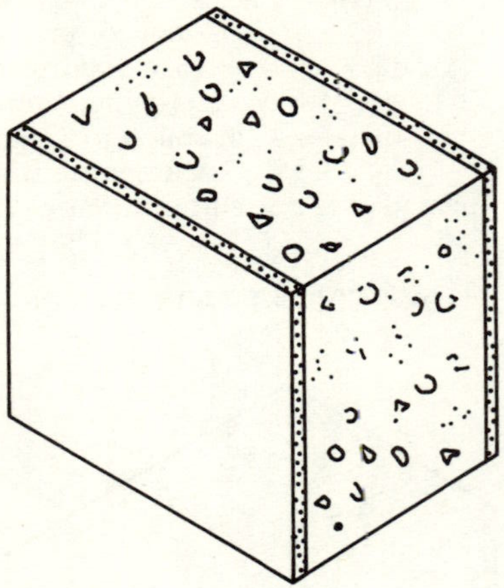

Tested in 1952–3.

Density of concrete 1234 (after drying at 105 degC)

Mean moisture content during test = 16 percent by vol.
Mean thermal transmittance = 1.5
Mean conductivity of concrete = 0.49

Panel No. 38

Single-leaf wall of foamed-slag concrete, 1:2:8 *mix, cast* in situ, 8 *in. thick; designed to give a dry density of* 65 *lb/ft*^{3}. *Cast in* August 1952. *Rendered outside and plastered inside.*

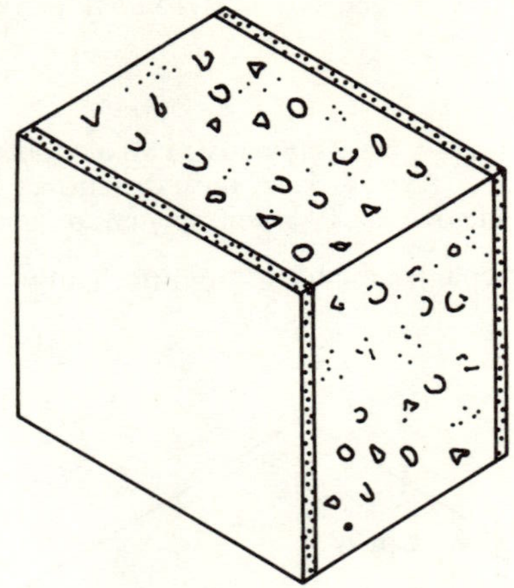

Tested in 1952–3.

Density of concrete = 1062 (dried at 105 degC)

Mean moisture content during test = 17 percent by vol.
Mean U = 1.4
Mean k of concrete = 0.40

Fundamental time-lag for 24 h period temperature-wave = 8.0 h

Amplitude decrement for 24 h period temperature-wave = 0.078

Panel No. 39

Single-leaf wall of aerated concrete blocks, 18 × 6 × 9 *in., rendered outside and plastered inside. Built on* 27 *August* 1951. *Dry*

density of concrete 51.4 *lb/ft*$^{-3}$ (0.82 g cm^{-3}).

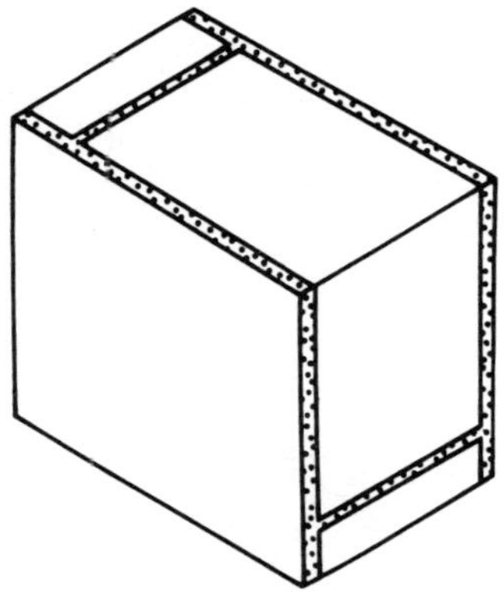

Tested in 1951–2.

Mean U = 1.9
Mean k of blocks = 0.52
Mean moisture content = 16 percent by vol.
Fundamental time-lag for 24 h period temperature-wave = 5.3 h
Amplitude decrement for 24 h period temperature-wave = 0.165

Panel No. 40

Single-leaf wall of cored foamed-slag concrete blocks, 18 × 9 × 6 *in., rendered outside, with plasterboard on* 1 *in. battens inside. Built in* 1946.

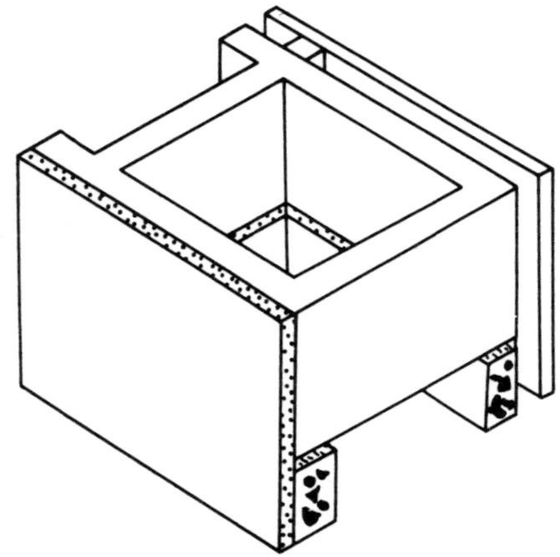

Tested in 1947–8.

Mean thermal transmittance = 1.5
Surface-to-surface resistance = 0.53

Panel No. 41

Single-leaf wall of 18 × 9 × 9 *in. foamed-slag concrete blocks, made on the Besser machine. Rendered and plastered. Built* 25 *September* 1952.

Tested in 1952–3.
Dry density 1442

Panel No. 41 (*continued*)

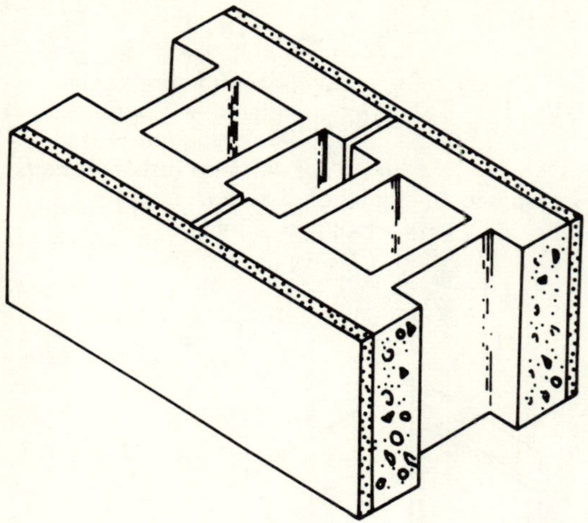

Mean thermal transmittance	= 1.7
Mean moisture content in June 1953	= 7.5 percent by vol.
Fundamental time-lag to 24 h temperature wave	= 4.8 h
Amplitude decrement to 24 h temperature wave	= 0.079

Panel No. 42

Single-leaf wall of 18×9×9 *in. clinker-concrete blocks, made on the Besser machine. Rendered and plastered. Dry density* 71 *lb/ft*^{3}.

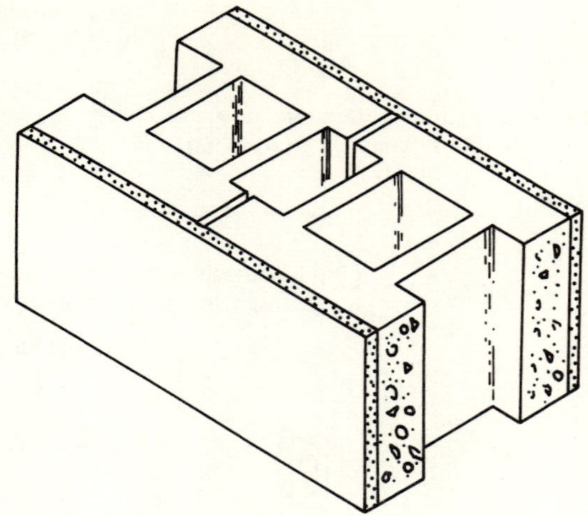

Tested in 1953–4.

Mean thermal transmittance	= 1.8
Moisture content of inner leaf (January 1954)	= 9 percent by vol.
Moisture content of outer leaf (January 1954)	= 18 percent by vol.
Fundamental time-lag to 24 h temperature wave	= 2.2 h
Amplitude decrement to 24 h temperature wave	= 0.097

Panel No. 43

Single-leaf wall of Siporex aerated concrete, in 4 *storey-height slabs* 2 *ft wide* ×8 *in. thick, density* 34 *lb/ft*$^{-3}$ (0.54 g cm^{-3}). *Three*

Panel No. 43 (*continued*)

reinforcing bars in each slab. Plastered inside, outside finish of silicate paint loaded with spar chips, sprayed on. Built August 1960.

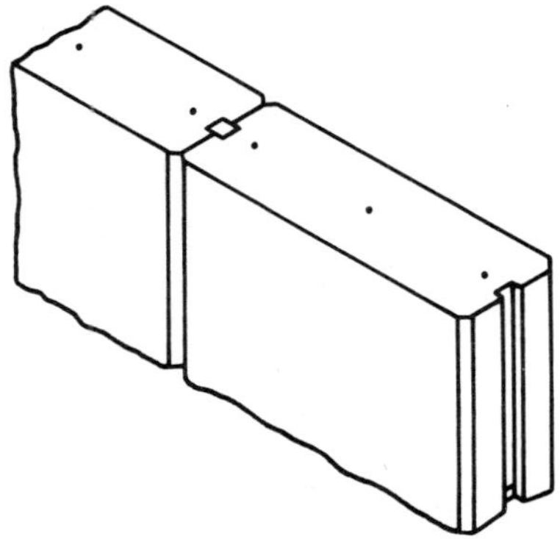

Tested in 1960–1.

Mean thermal transmittances

	24 h	day	night
Whole season	1.2	1.4	1.1
Dec.–Jan.–Feb.	1.2	1.2	1.1

Mean thermal conductivity of concrete

Whole season = 0.32
Dec.–Jan.–Feb. = 0.31

Mean moisture content in March 1961 = 12 percent by vol., when mean conductivity was 2.03 Btu in./ft^2h degF

Panel No. 44

Double-leaf wall, each of 18×9×3 *in., foamed-slag concrete blocks, with* 3 *in. air space. Rendered outside and plastered inside. Built in* 1948.

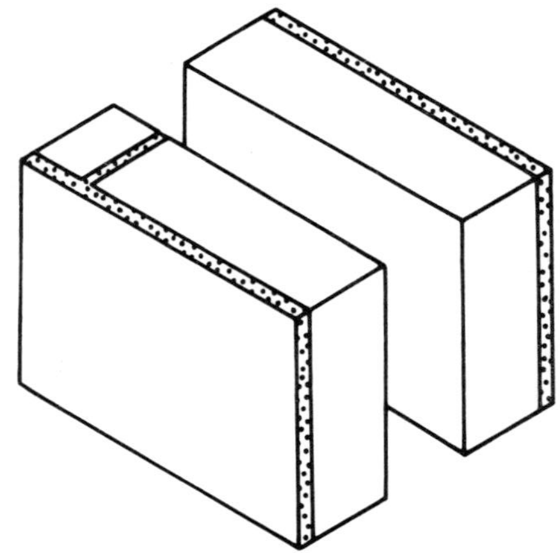

Tested in 1948–9.

Mean $U = 1.6$

Mean k of concrete in March 1949 = 0.46

Panel No. 45

Double-leaf wall of foamed-slag concrete blocks, $18\times9\times4\frac{1}{4}$ *in.,* $1:4\frac{1}{2}:4\frac{1}{2}$ *mix, with* $2\frac{1}{2}$ *in. cavity. Plastered inside when built in June* 1952, *but left unrendered in* 1952–3, *with drainage weep-holes at foot. Rendered outside August* 1953.

Panel No. 45 (*continued*)

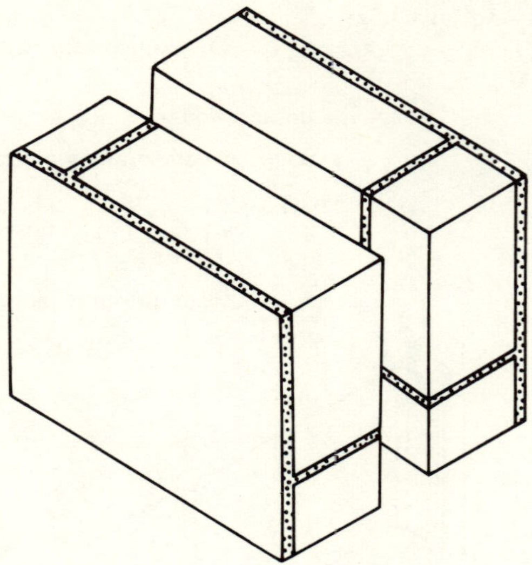

Density of concrete = 1330 (dried at 105 degC)

Measurements in 1952–3 (wall unrendered)

Mean $U = 1.3$

Mean k outer leaf = 0.61 in November 1952

Mean k inner leaf = 0.48 in November 1952

Fundamental time-lag for 24 h period temperature wave = 7.0 h

Amplitude decrement for 24 h period temperature wave = 0.030

Measurements in 1953–4 (wall rendered)

Mean moisture content, outer leaf, in Jan. 1954 = 7 percent by vol.

Mean moisture content, inner leaf, in Jan. 1954 = 6.5 percent by vol.

Mean $U = 1.1$

Measurements in 1954–5

Mean moisture content of outer leaf in April 1955 = 7.5 percent by vol.

Mean $U = 1.1$

Mean moisture content in October 1955, outer leaf 5.5 percent by vol.

Mean moisture content in October 1955, inner leaf 4.6 percent by vol.

Panel No. 46

Double-leaf wall of lightweight, expanded-clay aggregate concrete blocks (*L.E.C.A.*). $18 \times 9 \times 4\frac{1}{4}$ *in., with* $2\frac{1}{2}$ *in. cavity. Rendered outside and plastered inside. Built August* 1953.

Tested in 1953–4.

Density of concrete 1218 dried at 105 degC^{-3}

Mean moisture content of outer leaf in January 1954 = 15 percent by vol.

Mean moisture content of inner leaf in January 1954 = 9 percent by vol.

Mean $U = 1.5$

Conductivity of inner leaf = 0.68 with m/c = 8.5 percent by vol.

Panel No. 46 (*continued*)

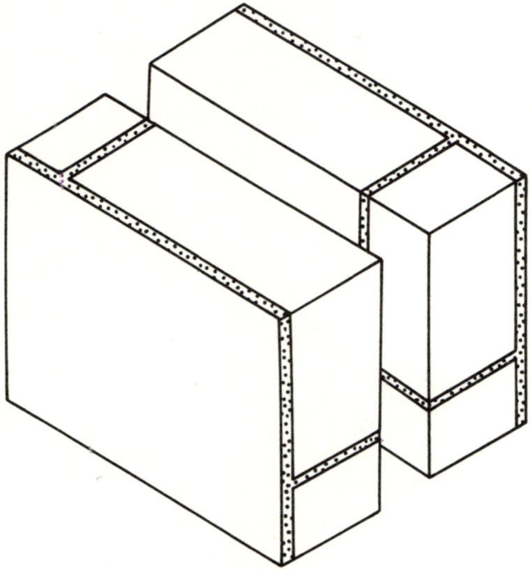

Fundamental time-lag for 24 h period temperature wave = 4.9 h

Amplitude decrement for 24 h period temperature wave = 0.044

Panel No. 47

Double-leaf wall of concrete blocks, 18 × 4 × 9 *in., with* 2 *in. cavity. Rendered and plastered. Blocks made of* 1 : 9 *mix, cement : pulverized fuel-ash pellets. Dry density* 69.5 *lb/ft*^{3} (1.11 g cm^{-3}). *Wall built* 5 *September* 1958.

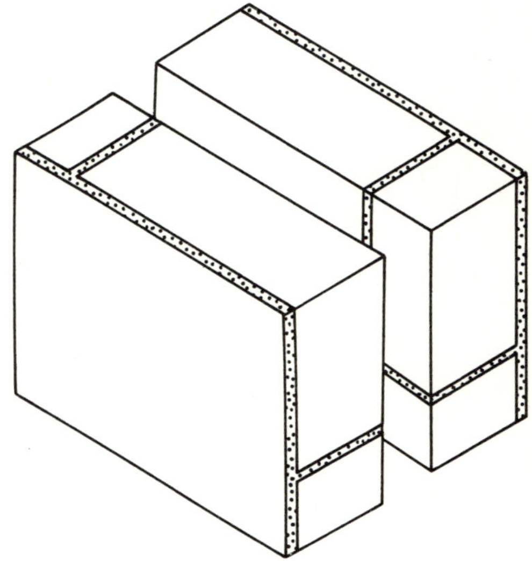

Tested in 1958–9 and 1959–60.

	1958–9	1959–60
24 h mean U-value	1.3	1.2
Day-time mean U-value	1.6	1.5
Night-time mean U-value	1.2	1.2
Mean thermal conductivity in December–February		
Inner leaf	0.48	0.45
Outer leaf	0.58	0.51

Moisture content, percent by vol.
Inner leaf in February 1960 = 5.1
Inner leaf in May 1960 = 5.0
Outer leaf in May 1960 = 7.1

Panel No. 48

Double-leaf wall of 2 in. thick, storey-height, aerated-concrete panels, bolted to cross-shaped dense-concrete posts, giving 2 in. cavity. Rendered outside. Built in 1949.

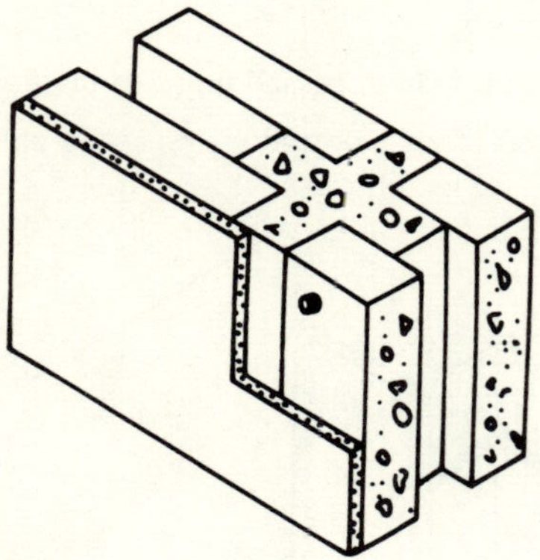

Tested in 1949–50.
Mean $U = 2.4$

Panel No. 49

Double-leaf wall of aerated-concrete slabs, exactly like No. 48 but without the posts. Leaves clamped together with bolts and spacing washers. Rendered outside. Built in 1949.

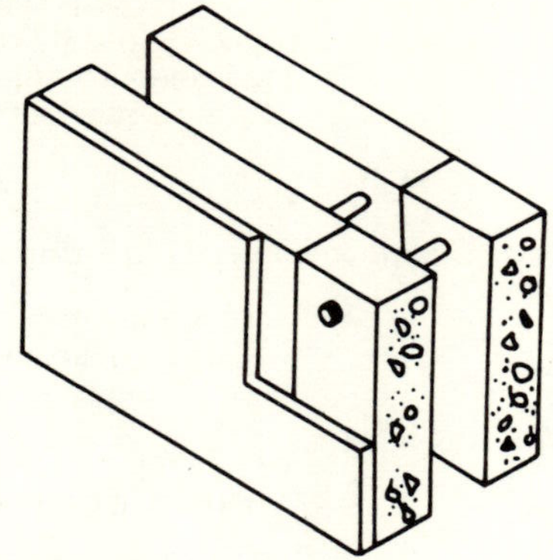

Tested in 1949–50.

Mean $U = 2.1$

Panel No. 50

Frame wall with precast concrete posts, outer cladding of weathered dense concrete slabs, clamped to posts with copper wire ties. Inner

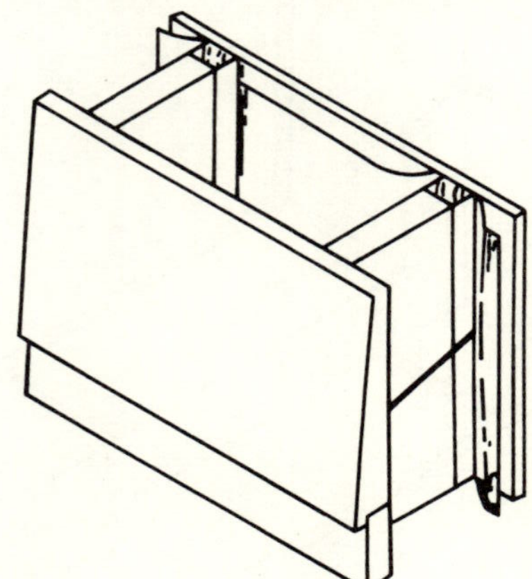

Panel No. 50 (*continued*)

leaf of plasterboard nailed to wood-fillets on the posts, with a membrane of aluminium foil giving a double cavity. Built in 1949.

Tested in 1949–50.

Mean $U = 1.2$

Panel No. 51

Dense concrete storey-height slabs, cored to give two leaves. Half-inch thick fibreboard stuck to inner face. For the test an 8 ft square panel was cast in one piece; built in 1949.

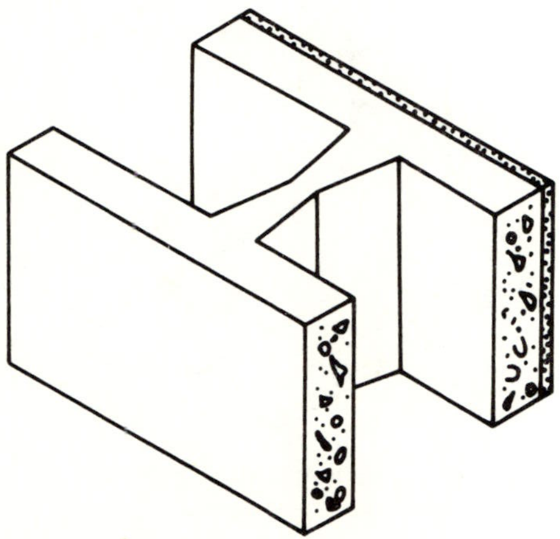

Tested in 1949–50.

Mean $U = 1.5$

Panel No. 52

Double-leaf wall with concrete posts giving a cavity width of $5\frac{1}{2}$ *in. Outer leaf of dense concrete slabs 2 ft high by 4 ft wide,* $1\frac{1}{2}$ *in. thick except at edges where they thicken to 2 in. Inner leaf of foamed-slag concrete blocks* $18 \times 2\frac{1}{2} \times 9$ *in., plastered inside (Orlit wall).*

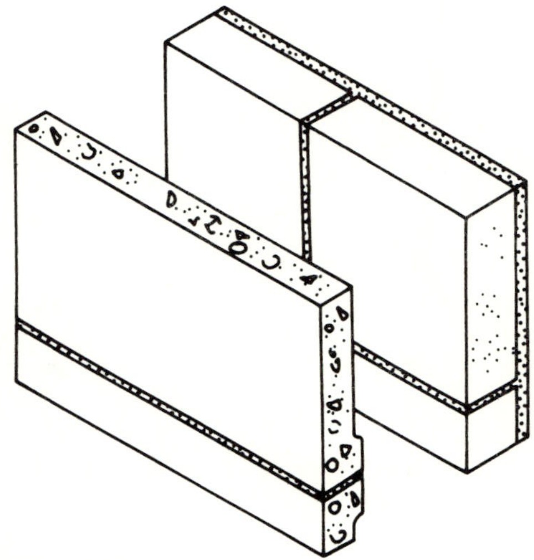

Tested in 1950–1.

Mean thermal transmittance = 2.0

Panel No. 53

Wall of double-leaf dense concrete blocks $18 \times 8 \times 9$ *in., the leaves of each block being joined by three 1 in. square concrete ties. Along*

Panel No. 53 (*continued*)

the top of each inner leaf runs a 1 in. square fillet of nailable concrete, to which is nailed a plasterboard inner lining. Wall built in July 1951. On 28 January 1952 the plain plasterboard lining was replaced by one with aluminium foil facing the cavity.

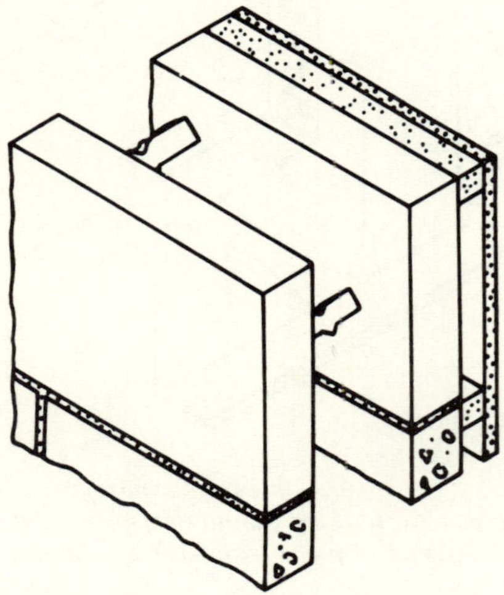

(*a*) Measurements in October 1951–January 1952 (plain plasterboard)

Mean $U = 1.7$

Fundamental time-lag for 24 h period temperature wave = 3.9 h

Amplitude decrement for 24 h period temperature wave = 0.048

(*b*) Measurements in February–March 1952 (foil-backed plasterboard)

Mean $U = 1.3$

Fundamental time-lag for 24 h period temperature wave = 4.1 *h*

Amplitude decrement for 24 h period temperature wave = 0.030

Note: The outer face of these blocks was made with exposed aggregate, giving a rough surface to simulate natural stone.

Panel No. 54

Three-leaf wall, with two 1 in. cavities. Outer leaf of steel sheet with vertical stiffeners, other leaves of plasterboard separated by wood strips, the second one being aluminium-foil backed. Built in 1949.

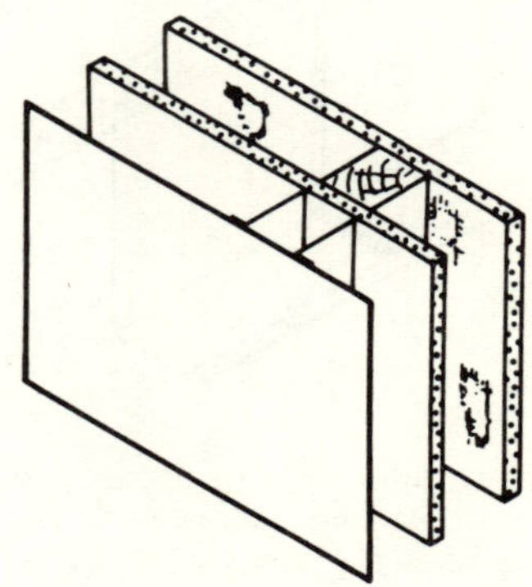

Tested in 1949–50.
Mean $U = 1.8$

Panel No. 55

Prefabricated wall, with wood frames, corrugated aluminium outer skin, hardboard inner skin, insulated with 3 in. glass fibre quilt.

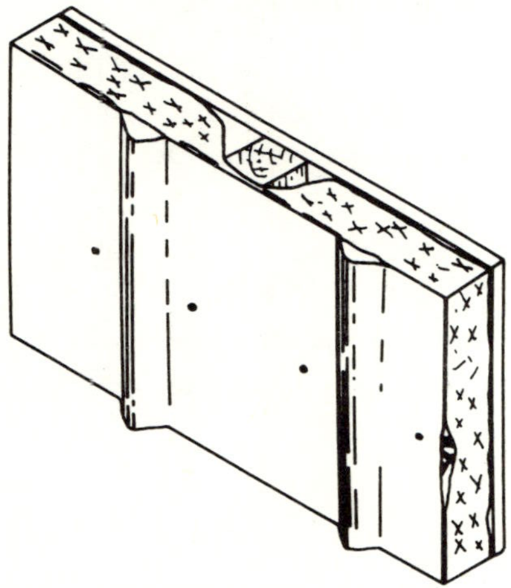

Tested in 1949–50 and 1950–1.

Mean $U = 0.97$

Surface-to-surface resistance = 0.92

Panel No. 56

Prefabricated wall with aluminium framing and outer skin of aluminium with metal stiffeners. The inner skin is of hardboard with aluminium stiffeners, connected to the main frames by wood strips. The panels are insulated with a glass fibre quilt, 1.5 in. thick.

The test wall consisted of two 4 ft wide sections with framing between as shown in the sketch.

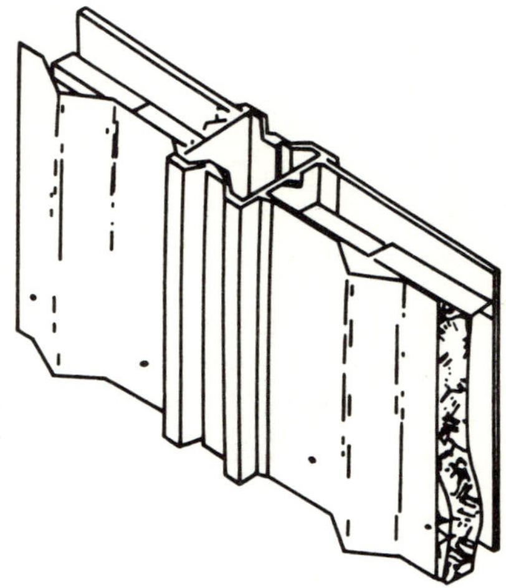

Tested in 1953–4.

Mean thermal transmittance = 2.0

Panel No. 57

Metal wall designed for factories, not load-bearing. Inner leaf of steel trays 2 ft wide, corrugated aluminium outer leaf screwed to horizontal stiffners on steel trays, with 1 in. resin-bonded fibre glass quilt clamped between the two leaves. Built in 1958.

Panel No. 57 (*continued*)

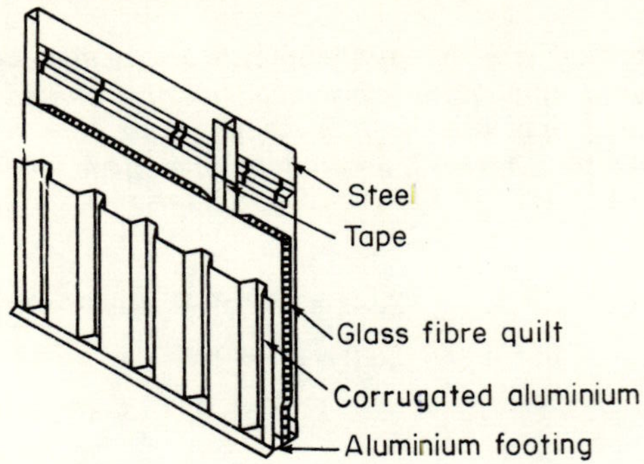

Tested in 1958–9.

Mean $U = 1.2$

Panel No. 58

Wood-framed wall, with light-weight infill panels. Plywood outer

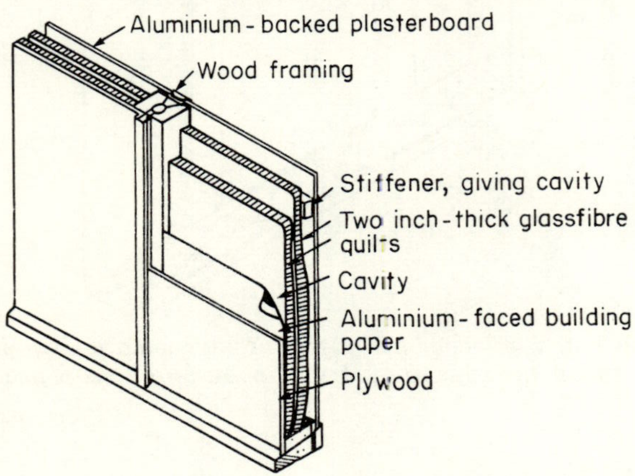

face, lined with aluminium foil, glassfibre quilts 2 in. thick with cavity either side, inner face of aluminium-backed plasterboard. Built in 1957.

Tested in 1957–8.

24 h mean U-value	0.68
Daytime mean U-value	0.68
Night-time mean U-value	0.62

Panel No. 59

Framed wall of 3×2 *in. softwood, clad on outside with* 40 *in. square panels of* 1 *in. thick wood, backed with* 1 *in. thickness of sawdust-cement. Lined internally with plasterboard. Built in August* 1955.

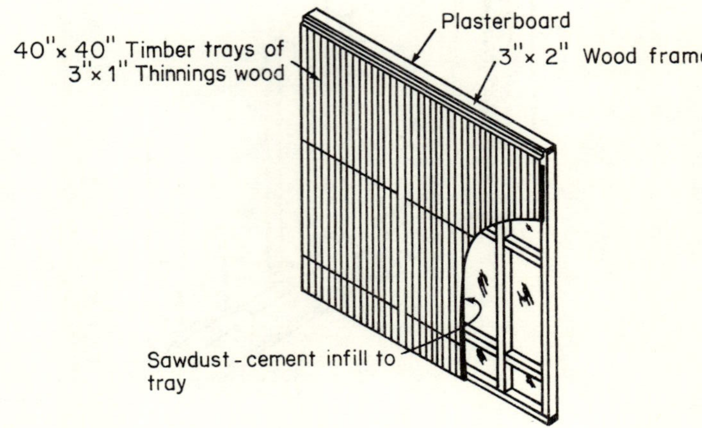

Tested in 1955–6.

Mean $U = 1.5$

Panel No. 60

Single-leaf wall of hollow glass blocks $5\frac{3}{4}$ in. square and 4 in. thick, set in 1:2:9 mortar with joints approximately $\frac{1}{4}$ in. thick. Expanded metal reinforcement in strips $2\frac{1}{2}$ in. wide was included in every third horizontal joint. The blocks were laid with the outside lenticular pattern vertical.

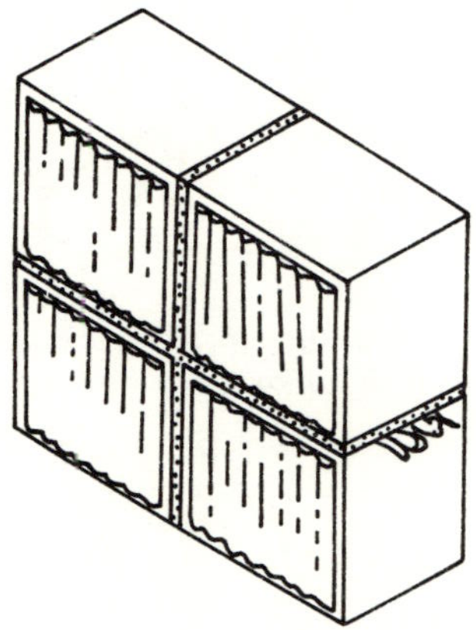

Tested in 1937–8.

Mean thermal transmittance = 2.5

Panel No. 61

Single-leaf wall, 9 in. solid Fletton bricks, plastered inside, with steel-framed window 4 ft high × 3 ft 3 in. wide, set in wood outer frame, sill 3 ft 4 in. above floor. In alternate weeks the window was double-glazed with removable inner lights, giving a cavity $2\frac{1}{4}$ in. wide. Built 1937.

Tested 1937–8.

Panel No. 61 (*continued*)

Thermal transmittances

	Window single-glazed	Window double-glazed
October 1937	2.5	2.1
November 1937	2.8	2.4
December 1937	2.9	2.4
January 1938	2.8	2.5
Whole period	2.7	2.3

Panel No. 62

School wall, with aluminium sub-frame at 3 ft 4 in. centres, bolted to steel main-frame. Single glazing and lower panels fixed with U-section aluminium glazing beads on stainless-steel spring clips. Built in 1955.

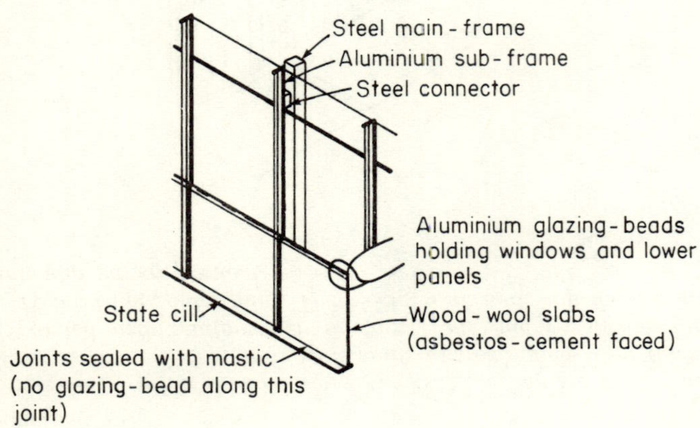

Tested in 1955–6 and 1956–7.

Mean thermal transmittances of 8 ft square panel,

	1955–6	1956–7 Oct.–Mar.	1956–7 Nov.–Feb.
24 h (09–09 GMT)	3.4	3.1	3.4
Day-time (09–1630 GMT)		1.6	
Night-time (1630–09 GMT)		3.5	

Estimated U of wood-wool slab panels = 2.0

Distribution of heat flow

Part of wall	Area as proportion of whole panel	Proportion of total heat-flow (%)
Glass (single-glazed)	0.59	71
Metal framing	0.09	13
Wood-wool slab panels	0.32	16

Panel No. 63

Glass-clad wall on T-section steel sub-frame, with bracket fixings to steel main frame (box-section posts and lattice horizontal members). Single-glazed windows. Separate gypsum plaster inner leaf below window cill, resting on structural floor. Built in 1955.

Panel No. 63 (*continued*)

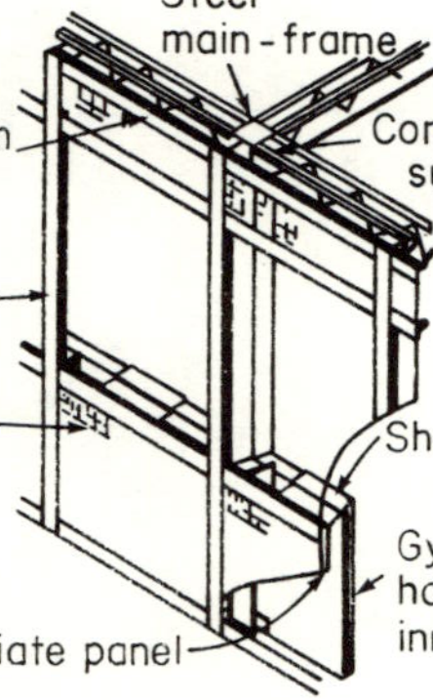

Tested in 1955–6.

Mean $U = 2.2$

Estimated U of panel below window cill $= 1.0$

Distribution of heat flow

Part of wall	Area as proportion of whole panel	Proportion of total heat-flow (%)
Glass, including metal sub-frames exposed to room air	0.66	86
Panel below window	0.34	14

Panel No. 64

This is panel No. 63 modified in August 1956 by the substitution of new glazing units for the old ones. The new windows were sealed double-glazed ones with a cavity about $\frac{3}{16}$ in. wide. The other glass-faced panels consisted essentially of a dished steel tray, painted inside, with a roughcast glass face. They were held in place by the "patent-glazing" technique previously used. The steel perimeter-beam was boxed in.

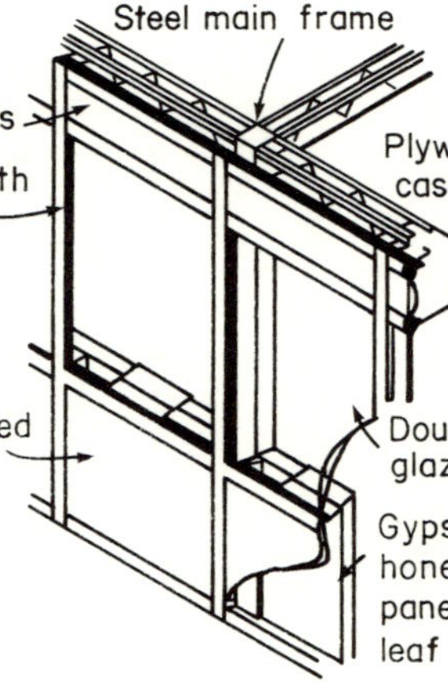

Tested in 1956–7.

Mean U,	
24 h (09–09 GMT)	1.6
Day-time (09–1630 GMT)	0.86
Night-time (1630–09 GMT)	1.8

Estimated mean U of lower panels $= 0.75$

Estimated mean U of window $= 2.8$

Panel No. 64 (*continued*)

Distribution of heat flow

Part of wall	Area as proportion of whole wall	Proportion of total heat-flow (%)
Glass, double with air-space about $\frac{3}{16}$ in.	0.51	71
Window framing	0.05	8
Upper spandrel panel	0.13	9
Lower spandrel panel	0.32	12

Panel No. 65

Wood-framed wall, with outer leaf of $\frac{3}{4}$ in. wood boards, $\frac{3}{4}$ in. air-space and inner leaf of 2 in. thick compressed-strawboard. Centre-hung, wood-framed double windows. Built in 1956.

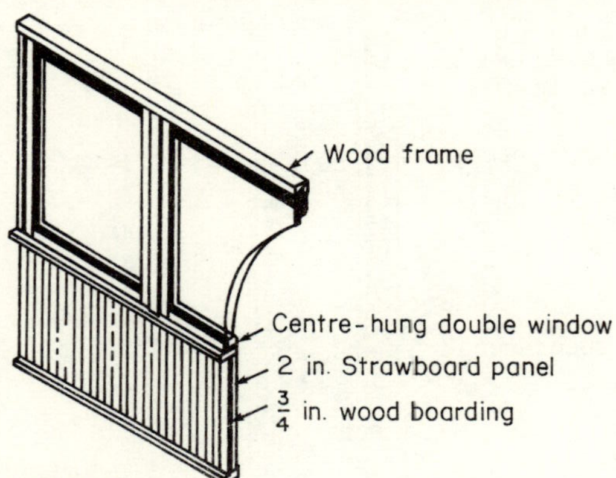

Tested in 1956–7.

Mean U of panel (24 h)	1.5
Mean U of panel (day-time 0915–1630 GMT)	0.74
Mean U of panel (night-time 1630–0915 GMT)	1.8
Estimated U of spandrel panel alone	0.97
Estimated U of double glazing (without framing)	2.8

Distribution of heat flow

Part of wall	Area of proportion of whole panel	Proportion of total heat-flow (%)
Glass	0.45	65
Wood framing	0.25	18
Strawboard panels	0.31	17

Panel No. 66

Wall with aluminium subframe which is fixed to each section of the panel. Each mullion is in two channel-shaped parts fixed to their respective panel section, with a sliding junction to allow for movements. Spandrel panels of cored plastic construction $1\frac{5}{8}$ in. thick, the vertical channels being filled with glass-fibre insulation. Double-glazed windows, with sealed units ($\frac{1}{2}$ in. airspace).

Tested in 1956–7.

Mean thermal transmittances

24 h (09–09 GMT)	2.1
Day-time (09–1630 GMT)	1.0
Night-time (1630–09 GMT)	2.4

Estimated overall U of plastic panel = 1.4

Panel No. 66 (*continued*)

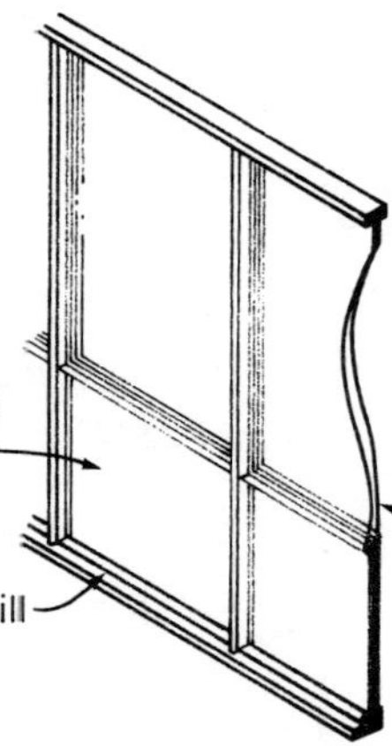

Distribution of heat-flow

Part of wall	Area as proportion of whole panel	Proportion of total heat-flow (%)
Glass (double-glazed)	0.55	56
Metal framing	0.15	30
Plastic panels	0.29	14

Panel No. 67

Factory-assembled panel, wood-framed of softwood, except for the hardwood window-cill. The details of the panel-infill below the windows are shown in the figure. Single-glazed windows. Built in 1955.

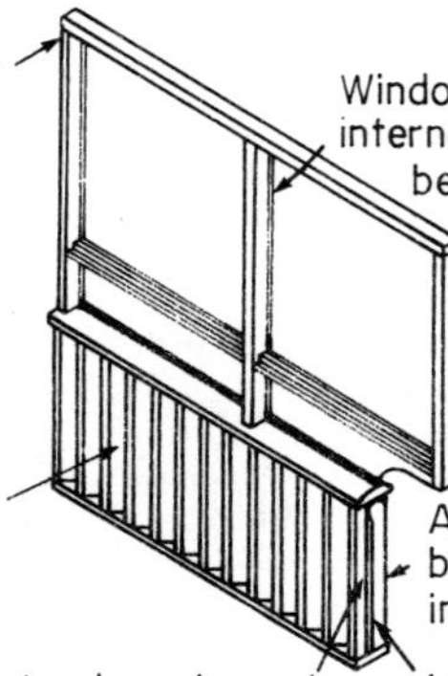

Tested in 1955–6.

Mean $U = 3.0$
Estimated U of spandrel panels $= 0.74$

Distribution of heat flow

Part of wall	Area as proportion of whole wall	Proportion of total heat-flow (%)
Glass, single	0.55	86
Wood window frame	0.13	6
Spandrel panel	0.33	8

Panel No. 68

Herts County Council 8 ft 3 in. grid wall. Main frame of 5 in. square steel lattice columns and steel lattice horizontal members. Columns covered with precast plaster casing. Outer cladding of dense concrete slabs. Inner leaf of wood-wool slabs covered with plasterboard. Single-glazed window with aluminium frame.

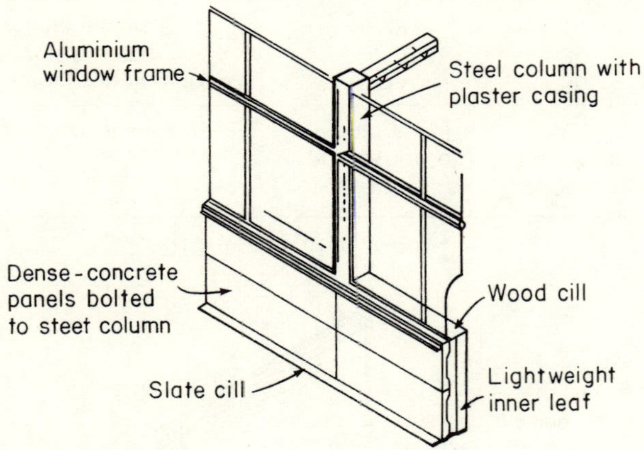

Tested in 1955–6.

24 h mean U-value = 3.5

Mean U-value for spandrel = 1.8

Distribution of heat flow

Part of wall	Area as proportion of whole wall (%)	Proportion of total heat-flow (%)
Glass and aluminium window framing	65	83
Spandrel panel	35	17

Panel No. 69

The same as panel No. 68, but with double windows. A wooden frame was built on to the inside of the metal window frame, and glazed from inside with wooden beads, giving an airspace 2 in. wide.

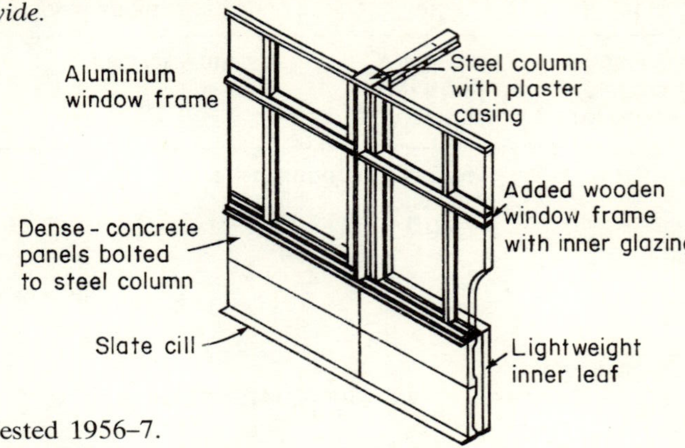

Tested 1956–7.

24 h mean U-value	1.8
Day-time mean U-value	0.85
Night-time mean U-value	2.0
Mean U-value for spandrel	1.8

Distribution of heat-flow

Part of wall	Area as proportion of *total (%)	Proportion of heat-flow (%)
Glass (double-glazed)	49	52
Wood window frame	10	8
Spandrel panel	35	26
Column (above window-cill)	12	14

* Based on projected area of panel.

Panel No. 70

Experimental curtain-wall with wood-framed windows. Built 1955.

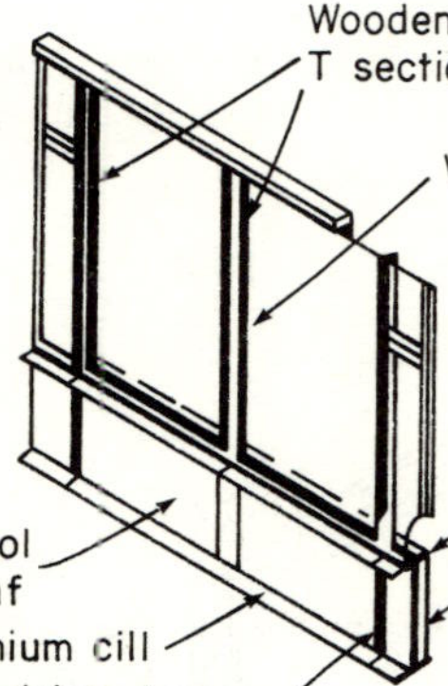

Tested 1955–6.

Mean thermal transmittance = 3.2
Estimated thermal transmittance of spandrel panel = 1.4

Distribution of heat-flow

Part of wall	Area of proportion of whole wall	Proportion of total heat-flow (%)
Glass, including frames	0.55	81
Spandrel panel	0.45	19

Panel No. 71

Wood-framed single-glazed window, with 18 *lights each* 2 *ft* 2 *in.* × 1 *ft* 1 *in. clear* (68 *percent of wall area glazed*). *Built January* 1939.

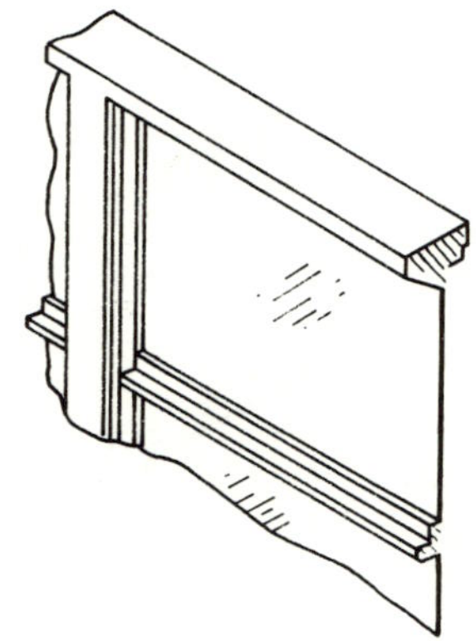

Tested February–March 1939.

Mean thermal transmittance,
6 February to 26 March = 3.3

Panel No. 72

Wood-framed window, exactly similar to panel 71, but double-glazed. Built January 1939.

Tested in February–March 1939.

Mean thermal transmittance,
6 February to 26 March = 1.7

Panel No. 72 (*continued*)

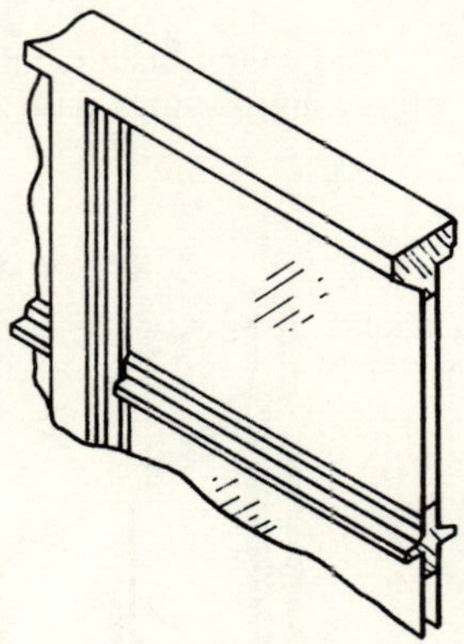

Panel No. 73

Teak frame with four panes, each approx. 3 *ft* 9 *in. square, of 'Insulight' sealed double-glazed units. Each unit made of* $\frac{1}{4}$ *in. plate glass with* $\frac{1}{4}$ *in. air-space, set into frame with mastic bedding and held by teak beads. Wall erected* 17 *June* 1960.

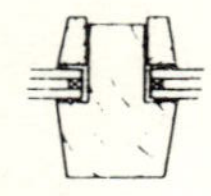

Section of Framing

Tested in 1960–1.

Mean night-time U-value = 2.7

REFERENCES

American Society for Testing Materials (1954) *Tentative Method of Test for Thermal Conductance and Transmittance of Built-up Sections by Means of the Guarded Hot Box*, ASTM no. C236-54T.

Bastings, L. and Benseman, R. F. (1950) *N.Z. J. Sci. and Tech.*, **31,** B, 4 21.

Blackwell, M. J. (1954) Five years continuous recording of total and diffuse solar radiation at Kew Observatory; Met. Res. Paper no. 895 (unpublished).

Cooling, L. F. (1929) *Trans. Brit. Cer. Soc.*, **29,** 39.

Dines, W. H. and Dines, L. H. G. (1927) *Mem. R. Met. Soc.*, **2,** 11.

Dinnie, A., Beard, R., and Richards, R. (1959) *Trans. Brit. Cer. Soc.*, **58,** 113.

Dufton, A. F. and Sheard, H. (1937) *J. Inst. Heat. and Vent. Engineers*, **5,** 82.

Griffiths, E. (1942) *J. Inst. Heat. and Vent. Engineers*, **10,** 112, 106.

Haller, P. (1957) *Schw. Bauz.*, **75,** 49, 3.

Holmquist, N. (1948) *Teknisk Tidskrift*, no. 5 (NRC Canada Translation TT-96, 1950.)

Illig, W. (1952) *Gesundheizung*, **73,** 124.

Meteorological Office (1956a) *Handbook of Meteorological Instruments*—Part 1, HMSO, London.

Meteorological Office (1956b) *Observer's Handbook, No. 554*, HMSO, London.

Penman, H. L. (1940) *Quart. J. R. Met. Soc.*, **66,** 401.

Pratt, A. W. (1969) *Build. Sci.*, **3,** 147.

Reiher, H. (1954) *Veroff, Inst. Tech. Phys. Stuttgart*, **30**.

Ryder, N. V. and McIlraith, A. H. (1946) *N.Z. J. Sci. and Tech.* **27,** B, (5). pp. 383–93.

Roach, W. T. (1955) Measurements of atmospheric radiation and the heat balance of the ground at Kew, MRP, 1936, An unpublished paper for the Meteorological Research Committee, London.

Solvason, K. R. (1959) *ASHRAE Trans.*, **65,** 541.

Tveit, A. (1953) En metode ti maling av varmegjhen nomgangstall for vegger of bjelkelag, Norges Byggforskuing—institutt, Tapport Nr. 8.

CHAPTER 9

Experiments on Heat Transmission Through Walls

Many of the walls tested in the Wall Laboratory at the Building Research Station, Garston, were examples of established construction. In the first few seasons it was the practice to measure in the main walls whose transmittance had previously been calculated from conductivity measurements at the National Physical Laboratory as a check on the effects of moisture and of the outside weather. Soon it became possible to make tests on experimental walls and the work developed into a continuing investigation over several years of heat transfer processes in walls generally. A detailed account of these studies forms the content of this chapter.

1. MEASUREMENT OF MOISTURE CONTENT IN WALLS

Three methods were used for determining the moisture content of test walls in the Wall Laboratory:

(a) Drilling cores of material from the walls, weighing and drying (i.e. direct sampling);

(b) Inserting plaster moisture gauges to give a continuous record of moisture content; and

(c) Measuring the attenuation of high-frequency radio waves caused by the presence of moisture in the walls.

Direct Sampling

Cores, 25 or 75 mm diameter, were drilled out of the wall, weighed immediately, then dried to a constant weight and re-weighed. The drills used are shown in Fig. 1. It was possible to use a hand-brace with the 25 mm drill on exceptionally soft material such as aerated concrete, but for harder material an electric drill was needed, turning the tool at 250 rev/min.

The method is satisfactory only if a core about 100 mm long can be bored out within two or three minutes. Most bricks are so hard that the prolonged drilling necessary generates enough heat to evaporate some of the water in

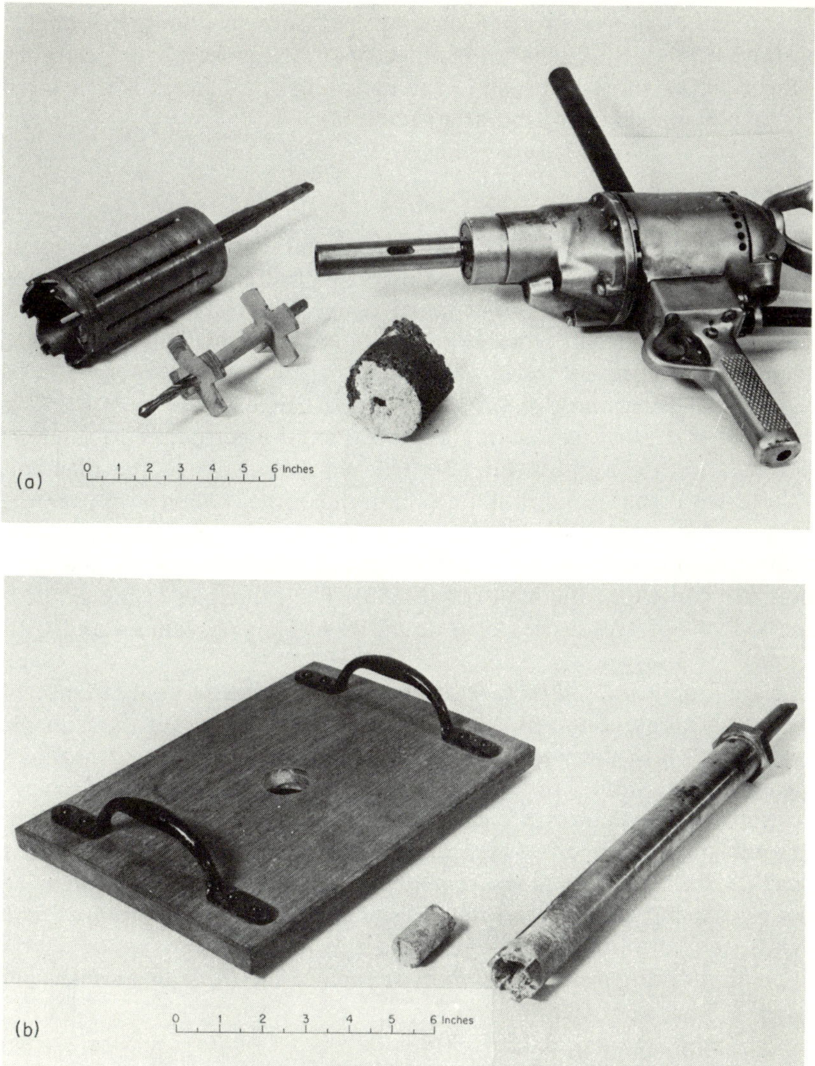

Fig. 1. Drills for removing sample cores from walls (a) Three-inch core-drill with tungsten carbide teeth and removable starting bit, 250 r.p.m. electric drilling machine and core produced with the drill. (b) One-inch core drill with tungsten carbide teeth, jig for locating the drill and core produced with the drill

the core and invalidate the results. Dense concrete cannot be sampled in this way, but satisfactory results may be obtained with all types of lightweight concrete. If it is required to determine the distribution of moisture through the thickness of the wall the core may be cut into suitable lengths before weighing and drying.

The specimens are oven-dried at a temperature of about 105 degC. A drying time of 48 h is adequate for most materials; indeed, it is desirable to keep the time as short as possible for lightweight concretes which tend to absorb carbon dioxide into the atmosphere.

Plaster Moisture Gauge

The electrical resistance of a porous material varies with its moisture content, but it is not feasible to make direct measurements of the resistance of building materials *in-situ.* The most convenient way of employing the principle is to use a small block of a suitable material containing electrodes and inserted in the test sample.

The material particularly suitable for such a gauge is plaster of Paris, cast into a cylindrical block containing a central axial electrode, surrounded by another wound into a helix (Fig. 2). The characteristics of the plaster are such that, with the usual building materials, the balance between the moisture contents of the gauge and of the material in which it is set gives a reasonable relationship between the electrical resistance of the gauge and the moisture content of the material (Fig. 3). The principles of the gauge and the method of use are discussed in detail by Croney, Coleman, and Currer (1951).

The main practical disadvantage of the method is the necessity to insert the gauges into a block of the material under test and calibrate these in place before building into the wall. This takes about three months or more.

Six gauges were built into a block of aerated concrete, as shown in Fig. 4. The resistance of each moisture gauge was measured at least once a day with an a.c. ohm-meter. Since the resistances of the gauges also varies with the temperature simultaneous measurements were made of the block near each gauge with thermo-couples and the resistance read was converted to that at a standard temperature of 20 degC.

The mean monthly moisture content at each of the three positions is given

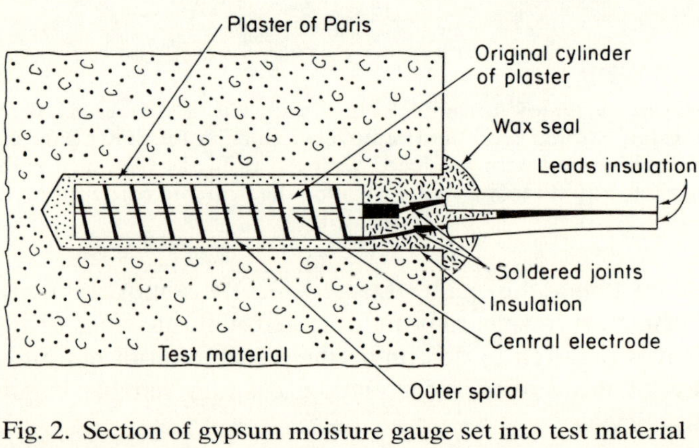

Fig. 2. Section of gypsum moisture gauge set into test material

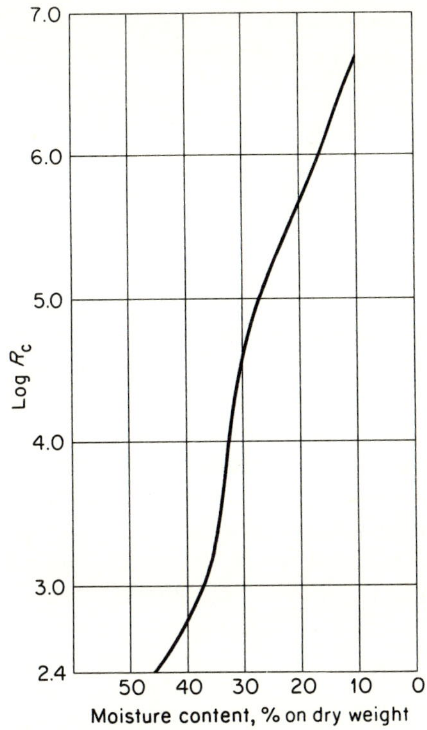

Fig. 3. Calibration of plaster gauges used in aerated concrete wall (mean of six drying curves)

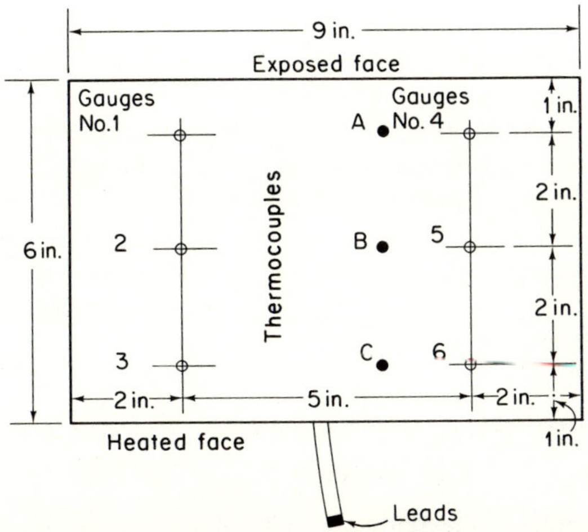

Fig. 4. Positions of moisture gauges and thermojunctions in aerated concrete block

Table 1 Mean moisture content of aerated concrete block (percent of dry weight)

	Oct.	Nov.	Dec.	Jan.	Feb.	Mar.	Apr.	May	June
Outer	22	22	21	20	20	19	18	15	14*
Middle	20	21	20	19	17	17	15	14	14*
Inner	21	21	18	16	15	14	14	14	13*

*Gravimetric readings.

in Table 1, the values being the average of the readings of the pair of gauges to each depth in the block. The mean temperature difference between the inner and outer gauge points, 100 mm apart, was about 7 degC.

The values for June are those obtained when the wall was demolished and parts were weighed and dried to obtain the moisture content directly. There is good agreement with the figures from the resistance gauges.

Although through the winter there was a steady and quite slow reduction in moisture content there were, on occasions, some more rapid changes. These are illustrated in Fig. 5 and were identified with rain driven on to the wall by strong wind.

Attenuation of Radio Waves

This method depends upon the fact that high frequency radio waves are absorbed by water. The amount of absorption is independent of the medium holding the water although it is increased if the water dissolves salts from the medium.

In the apparatus developed at the Building Research Station (Boot and Watson, 1964) the transmitter works at a frequency of 3000 Mc/s. Radiation is directed at the wall through a horn, and that transmitted is picked up by a second horn and the attentuation measured. For a given type of material there is a constant relationship between the attenuation and the moisture content, so that by calibration on a sample specimen the amount of moisture in the wall is obtained directly. In Fig. 6 is shown a 185 mm square hole in the inner leaf of a cavity wall, this being normally filled with a removable section. By removing the section the radio-wave attentuation may be measured to give the moisture content for each leaf separately as well as for the whole wall.

2. VARIATION OF THERMAL CONDUCTIVITY WITH MOISTURE CONTENT

Mention has already been made (see Chapter 2) of the relationship between the thermal conductivity of a porous inorganic material and its moisture content quoted by Jakob (1949). In the discussion on the Wall Laboratory

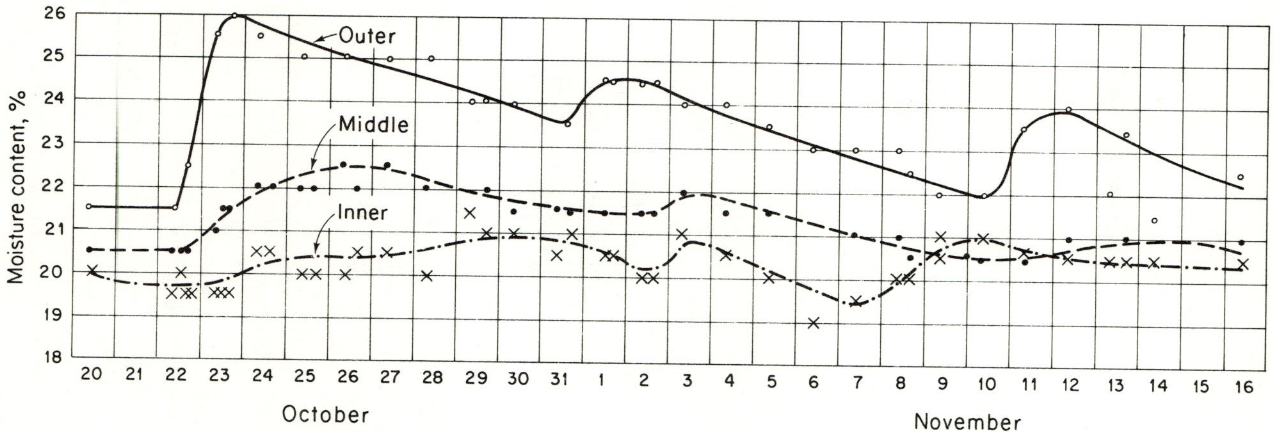

Fig. 5. Readings of moisture gauges in aerated concrete wall

Fig. 6. Receiver horn of radio-wave attenuation apparatus, hung on inside face of wall. *Note*: 9 in. square hole in inner leaf of wall, from which removable secton has been taken to permit independent measurement of moisture contents of two leaves

Table 2 The relation between the moisture content and thermal conductivity of building materials, after Jakob (1949) and Jesperson (1953)*

Moisture content (vol. %)	Conductivities relative to that at 0% m/c			
		Jesperson		
	Jakob	l.w. brick	l.w. concrete	glass wool
0	1.0	1.0	1.0	1.0
1	1.30	1.06	1.25	1.88
2.5	1.55	1.15	1.43	2.03
5	1.75	1.30	1.64	2.17
10	2.10	1.57	2.03	2.43
15	2.35	1.75	2.40	2.70
20	2.55	1.88	2.78	—
25	2.75	2.02	3.16	—

* Note that Jakob's ratios are relative to the conductivity of dry material; Jesperson's are relative to the conductivity of material in 'laboratory-dry' state, which in some cases includes up to 1 percent hygroscopic water.

results below frequent use is made of this, and of similar relationships given by Jesperson (1953).

Jakob does not give the source of his data but it is evident that they are based on results from a variety of materials and that the relationship is an average one. It will be seen from the discussion that there is good agreement between this relationship and the results from the Wall Laboratory experiments for most bricks and concretes.

The data in Table 2 are extracted from curves that he gives, and show that while the relationship for lightweight concretes is similar, up to 15 percent moisture content, to that of Jakob, for glassy fibrous materials it is very different.

It should be noted that whereas Jakob's data are based on the conductivity of dry (refers to 'oven-dry' implying zero moisture content) material, those of Jesperson are based on the conductivity of materials in the 'laboratory-dry' state. He does not state the conditions in the laboratory but a temperature of 20 degC and a relative humidity of 40–50 percent may be assumed. Under these conditions the brick specimens would be virtually dry, but the concretes held, according to the curves, about 1 percent hygroscopic moisture. This low water content has very little effect on the thermal conductivity but should be remembered when using Jesperson's results.

In the discussion below the thermal conductivity of dry materials has been calculated from the results of measurements in the Wall Laboratory and in the guarded hot plate thermal conductivity apparatus at the Building Research Station (BRS) or at the National Physical Laboratory (NPL)

In Figs. 7 and 8 the measured conductivities are plotted against the moisture content and density as tested and there is considerable scatter of

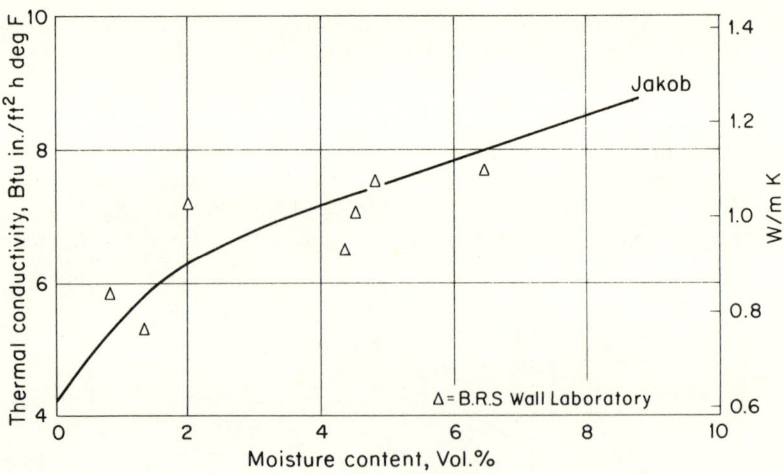

Fig. 7. Relation between moisture content and thermal conductivity of Fletton brickwork

the points. The computed dry values, however, plotted against corresponding densitites, fall in most cases close to a straight line that is characteristic of a particular group of materials.

Brickwork

The experimental points for brickwork, shown plotted in Fig. 7, are seen to be in reasonable agreement with Jakob's curve. The agreement is good enough to suggest that the curve may be used to relate the thermal conductivity of Fletton brickwork with moisture content.

Several thermal conductivity measurements have been made in different laboratories on various types of bricks. Some of these data are given in Fig. 8 showing smooth curves drawn through values of thermal conductivity plotted against the density as tested, including any water present. There are considerable variations in the slopes of the curves, as between one material and another. Only one value is available for each wall tested in the Wall Laboratory (denoted by an open triangle in the graph and a number that distinguishes the particular test wall), but in each case a dry conductivity has been calculated using the Jakob relationship (filled triangles). Two further curves in the Figure show how the conductivity of a brick with a dry density of 1600 kg/m^3 and corresponding thermal conductivity of 0.43 W/mk (Cammerer, unpublished value) would vary with moisture content using Jakob's relationship (points marked J) and a relationship due to Cammerer for bricks only (points marked C). The latter shows a greater rate of increase in conductivity with moisture content but since different types of bricks differ so much in this respect there is probably little justification for choosing one

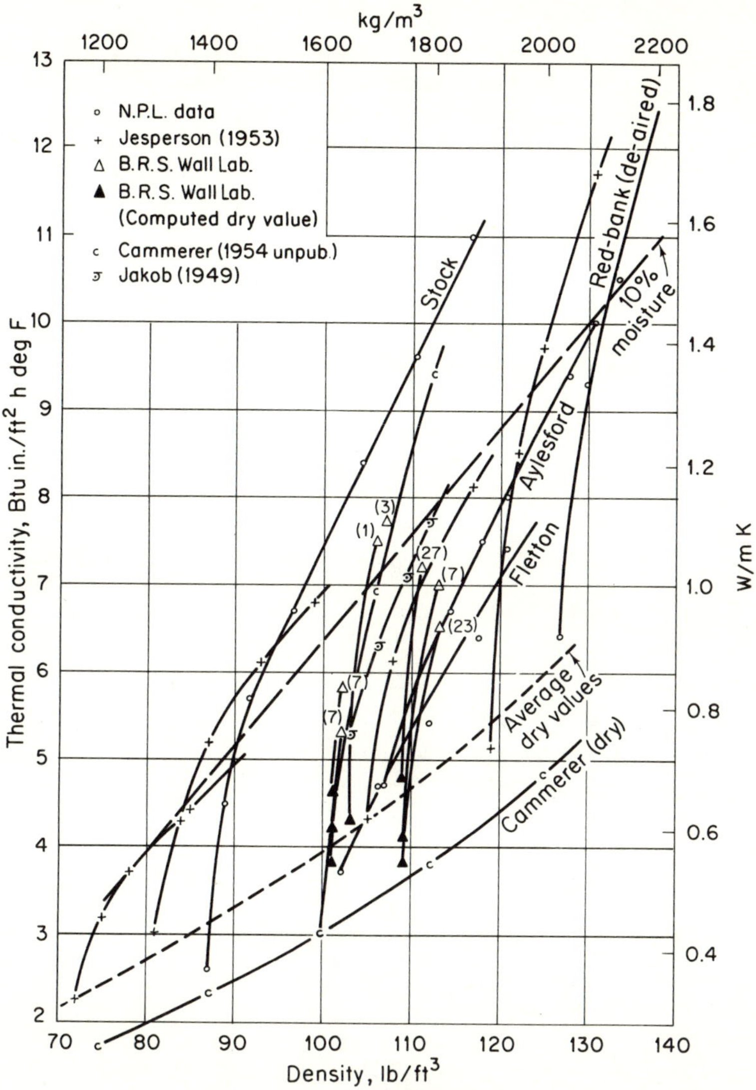

Fig. 8. Relation between thermal conductivity, density, and moisture content of clay bricks

relationship rather than the other for use as a general correction factor for the effect of moisture.

The disagreement between the average dry curve in the figure and that due to Cammerer may also be due to differences between the materials, although discrepancies of 10–15 percent, and more, between the conductivities measured by different laboratories have been recorded. It is uncertain how much of this is due to differences in measurement technique and how much to real differences between the materials tested.

Although it was possible to draw the average dry curve in Fig. 8, no such universal curve could be fitted for moist bricks because of the scatter of the results. A line marked 10 percent moisture, which fits the data for a number of types of bricks reasonably well, has been indicated but the NPL data for Fletton and Aylesford bricks do not conform with this.

Dense Concrete

Five walls of dense concrete, cast *in situ*, were tested in the Wall Laboratory. Two walls, denoted 31 and 32, were of similar mix with Ham River gravel aggregate and dry density 2290 kg/m^3, a third being of the same mix 1:2:4 (proportions refer to cement: fine aggregate: coarse aggregate) but having Accrington brick aggregate, giving a dry density of 2163 kg/m^3 (wall 33). The conductivities of these two walls (31 and 32), and a wall of no-fines concrete with gravel aggregate (wall 29) are compared in Fig. 9 with values obtained at the NPL on 300 mm square specimens in the guarded hot plate apparatus. The Wall Laboratory value for wall 29 agrees well with that of

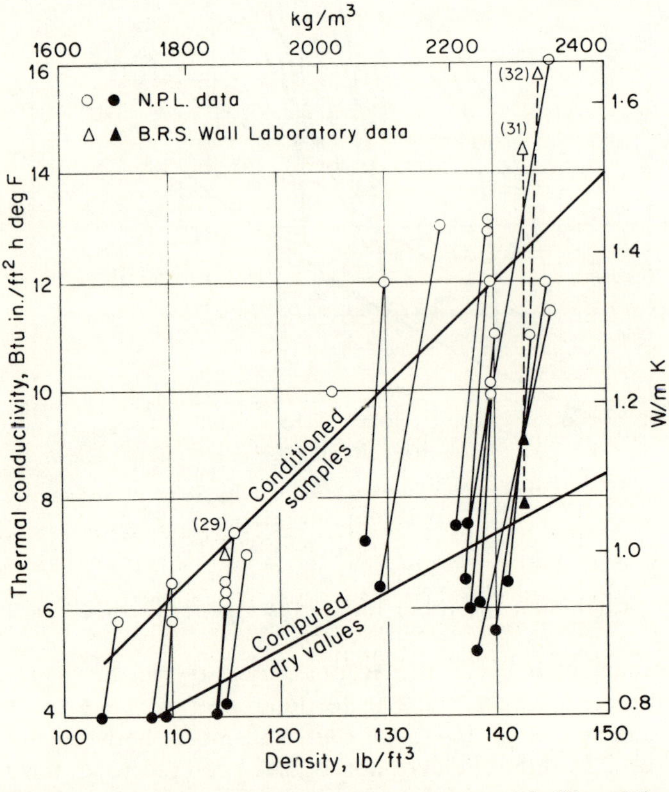

Fig. 9. Themal conductivity of concretes with Ham River gravel aggregate

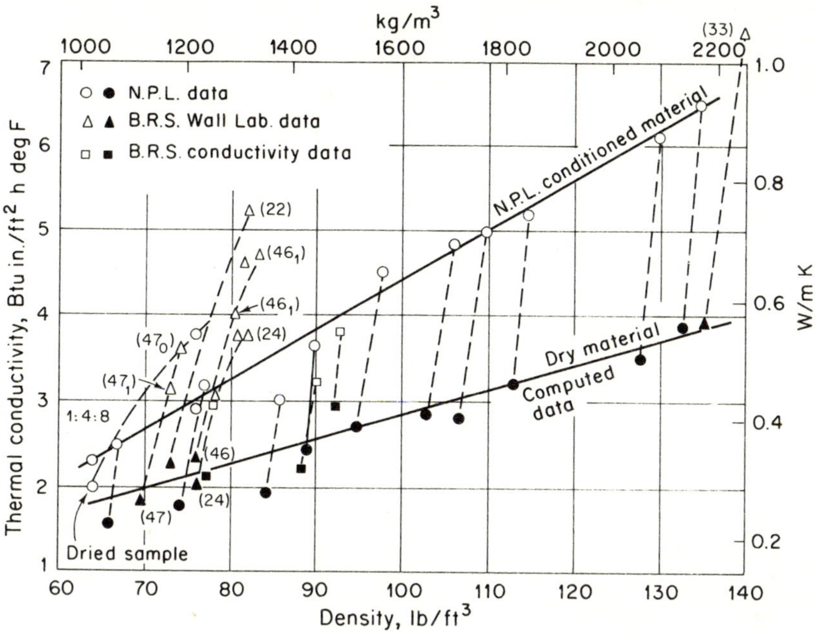

Fig. 10. Crushed brick, expanded clay and clinker concretes—relationship between conductivity, density, and moisture content (filled symbols indicate computed conductivities at zero moisture content)

the NPL but those for the walls marked 31 and 32 are high. The conductivities for these concretes at zero moisture content calculated with Jakob's relationship, were 1.3 and 1.1 W/mK, respectively. These are appreciably higher than the corresponding values calculated for specimens of similar concretes measured by the NPL which averaged about 0.91. However, since the moisture contents of these two walls were not measured, values had to be estimated and it is probable that the concretes were wetter than has been assumed.

Data for burnt-clay aggregate blocks are plotted in Fig. 10 and marked 24, 46, and 47. These measured results when extrapolated to zero moisture content with the Jakob relationship give values in the range 0.50–0.56 W/mK. These data obtained with naturally exposed walls in the Wall Laboratory are seen in the Figure to agree closely with those obtained with the standard guarded hot-plate method.

Foamed-slag Concrete, Cast *in situ*

Five walls of foamed-slag concrete were included in the many different wall constructions tested in the Wall Laboratory. The curves in Fig. 11 have been

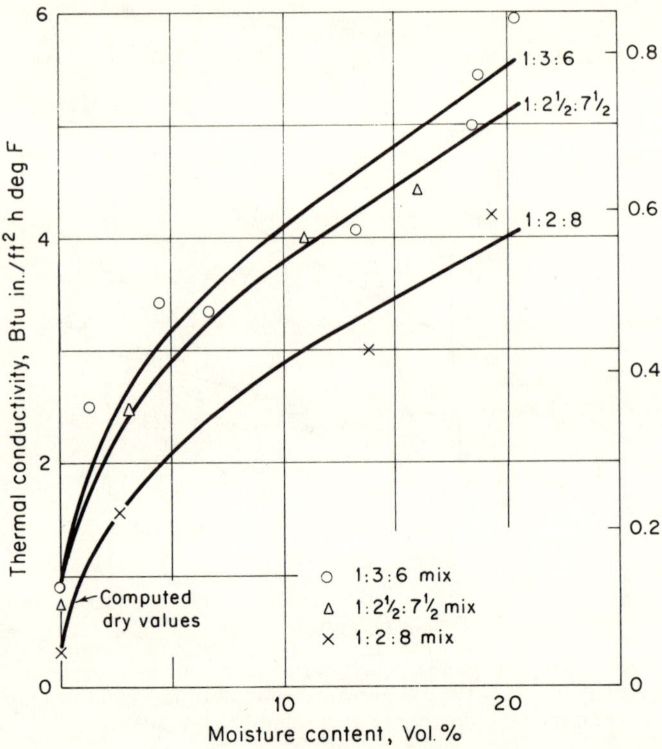

Fig. 11. Variation in conductivity of foamed-slag concrete with moisture content

drawn from the results of these experiments and on 300 mm square specimens in the guarded hot-plate apparatus at the BRS and NPL. For example, Wall Laboratory measurements on a wall of $1:2\frac{1}{2}:7\frac{1}{2}$ mix gave, during the first year of test, a thermal conductivity of 0.46 W/mK when the average moisture content was 16 percent. At this level Jakob's ratio is 2.39 and gives an estimated dry conductivity of 0.19. The mean conductivity of the same test wall in the following year was 0.43 with a moisture content of 11 percent (Jako ratio = 2.17) giving a dry conductivity of 0.20. A third determination, in the standard conductivity apparatus on specimens of the same mix made when the wall was cast, gave a conductivity of 0.32 with a moisture content of 3 percent. The estimated dry value is 0.20. The mean value was taken to be 0.20 W/mK and from this the curve in Fig. 11 was drawn. The curves for the other two mixes were similarly constructed and it will be seen that they fit the data well.

Fig. 12 shows the variation of thermal conductivity with density, the figures for the latter being the actual densities during the test. It has been drawn to show how the conductivity varies with both the density of the dry

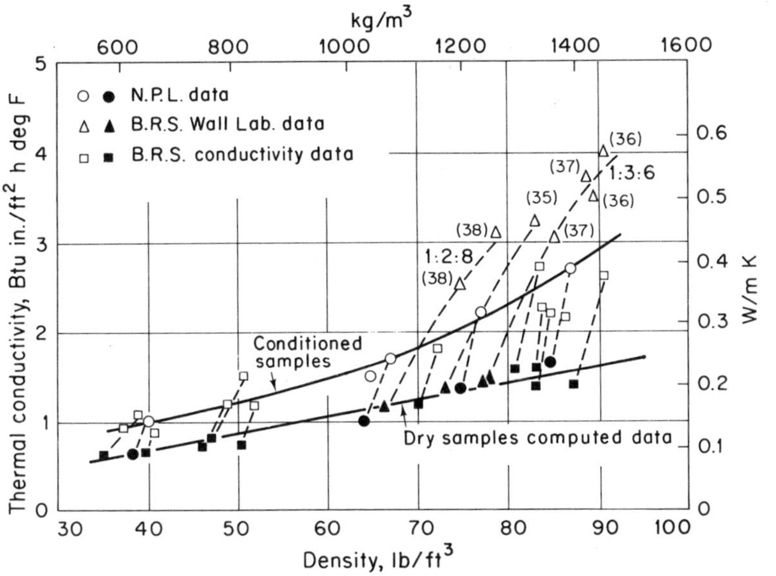

Fig. 12. Thermal conductivity of foamed-slag concretes (filled symbols indicate computed conductivities at zero moisture content)

material and the moisture content. The rate of variation in the conductivity with water content is seen to be twice or more the rate of change with dry density. The conductivities at the higher moisture contents are derived from Wall Laboratory data, the others being obtained with the guarded hot-plate apparatus.

Although the results from the Wall Laboratory measurements cannot be expected to be as accurate as those determined with the guarded hot-plate, it is clear from these diagrams that the results are consistent with one another.

Aerated Concrete

Data for various aerated concretes are given in Fig. 13. It will be seen that the several results obtained for material with a dry density of 825 kg/m³ fit one curve—the relationship given by Jesperson (1953) for lightweight concretes. It fits the data for this class of material more closely than Jakob's relationship which, it may be recalled, appears to be an average one based on results from a variety of materials. Jesperson gives the mean relationships between conductivities in wet and dry states separately for different classes of materials. Again, comparing the two sources of results, Jakob's data refer to dry (oven-dry) material, whereas Jesperson's relationships are based on the conductivity of materials in the 'laboratory-dry' state. The laboratory conditions that determine this state are not given, but an air temperature of 20 degC and a relative humidity of 40–50 percent may be assumed. In this

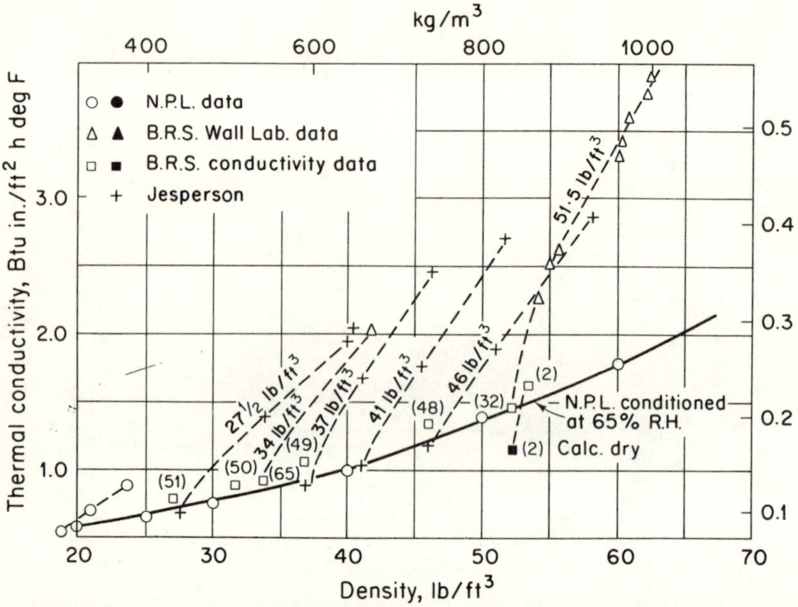

Fig. 13. Thermal conductivity of aerated concretes—variation with density and moisture content

atmosphere concretes hold, according to Jesperson's curves, about 1 percent hygroscopic moisture.

Summary of Dry Conductivity Data

Certain of the diagrams in the discussion above show values of thermal conductivity for material in the dry state calculated using the relationships of Jakob or Jesperson or, in some cases, actual measured values. These results are brought together in Fig. 14 with average curves drawn through the points for each class of material—aerated concretes, foamed-slag concretes, fired-clay bricks, burnt-clay and clinker concretes, gravel concretes—and showing Jakob's curve for comparison. As already admitted, there are appreciable differences between specimens of the same nomimal constitution, while in addition there may be appreciable discrepancies between the values of thermal conductiity measured by individual laboratories. For instance, Jakob states that individual values may deviate by ±15 percent from those given by his average curve. Therefore the curves in Fig. 14 are not precisely defined but the pattern is similar to that given by Cadiergues and Genevay (1952) in a comprehensive summary of the thermal conductivity of materials.

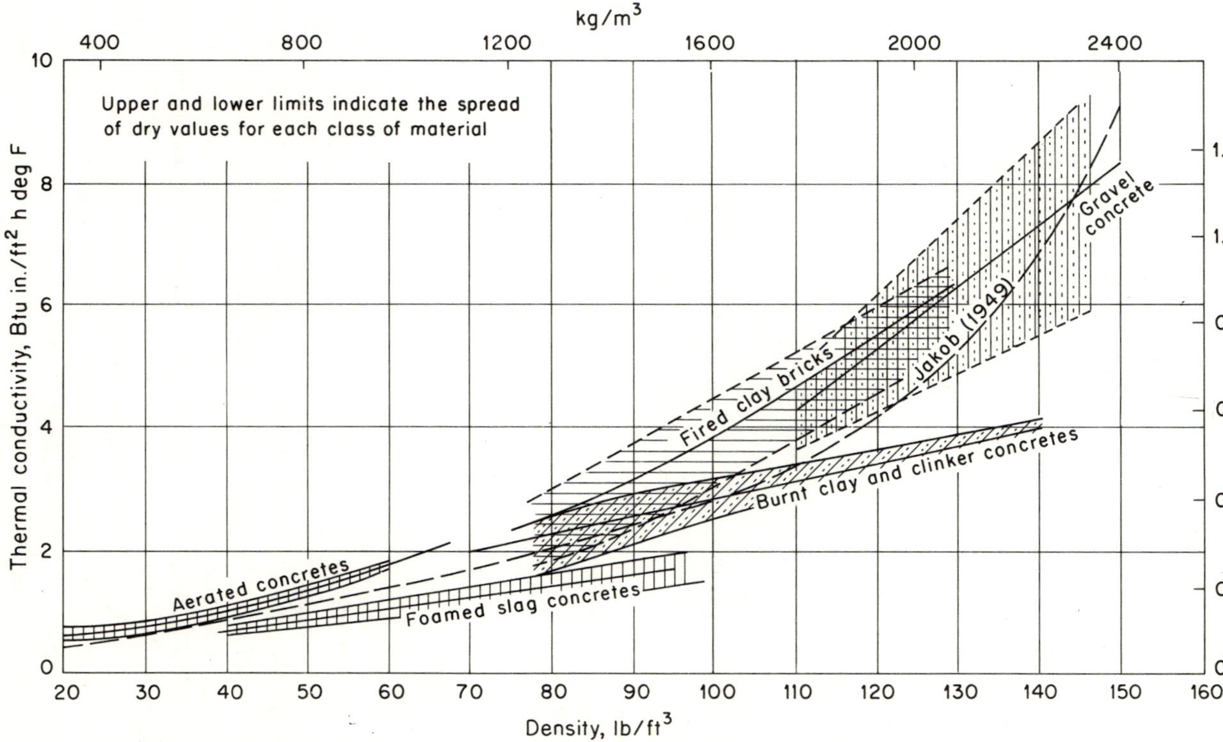

Fig. 14. Relation between density of materials and their dry thermal conductivity. Jakob's curve shown for comparison

3. PERFORATED BRICKS

Perforated bricks with a small ratio of void to solid have been used for a considerable time, those with a larger void ratio having been developed later in British practice. Many have been used in lightweight partition walls and the same type has been used as the inner leaf of a cavity wall (see Chapter 8, Panel 11). These blocks had six large horizontal holes and were 114 mm thick; they were found to have a mean thermal resistance of 0.26 m^2K/W, about twice that of a leaf of Fletton brickwork. A somewhat similar block was used in both leaves of Panel 12.

Panels 14–18 and 20 were tested as part of a programme of development by the Building Research Station of large load-bearing perforated clay bricks. Panel 19 was built with a standard Swiss perforated brick for comparison with the BRS patterns.

The leading dimensions and thermal properties of each type of block are summarized in Table 3. The types V2, V4, V5, and V6 were experimental products, using the same clay in each case, a mixture of 85 percent Keuper Marl and 15 percent London Clay. The commercially made bricks, types V6/2 and H2, were extruded from a much denser clay. The material of the Swiss B25 bricks was like that used in the BRS bricks.

The B25 brick is a standard unit in Switzerland and a similar pattern is also used in Finland. It is used extensively in load-bearing walls up to four storeys high. In Switzerland it is used with a thick rendering, but the panel in the Wall Laboratory was given the normal rendering used in Britain with two coats of 1 : 1 : 6 mix to a total thickness of 19 mm. All the BRS V bricks and the horizontally perforated H2 were designed to be used as facing bricks unrendered.

The V4 design (Panel 15) was developed from the V2 (Panel 14), the outer transverse webs being omitted with the object of reducing the conduction heat loss. In fact, this modification made no detectable difference.

The V5 brick (Panel 16) was designed to have the weight of a single common brick, while having the volume of two; the void ratio at 57 percent is therefore appreciably greater than that of preceding designs. In spite of this, and the use of thin staggered transverse webs, the surface-to-surface resistance was little better than that of the V2 and V4 bricks. In proportion to thickness of wall the design was not quite as good, in fact, as is shown in column 9 of Table 3, where the effectiveness of each panel is expressed as an 'equivalent conductivity' or conductance per 25 mm thickness of wall.

The two other designs of vertically perforated bricks tested, the V6 and V6/2, were much less satisfactory, the equivalent conductivities being respectively 31 and 79 percent higher than that of the V5 design. The increased heat loss was due in part to the introduction of straight-through transverse webs, and in the V6/2 was made worse by the use of a much denser clay with high thermal conductivity.

The low thermal resistance of the H2 bricks with horizontal perforations

Table 3 Thermal properties of perforated bricks

Panel no.	Type of brick	Void (%)	Sizes (mm)* Length	Width	Height	Year of test	Thermal resistance (m^2K/W)	Equivalent conductivity (W/mK)	Thermal transmittance (W/m^2K)
14	V2	41	208	208	102	1957–58	0.493	0.423	1.5
15	V4	44	208	208	102	1957–58	0.493	0.423	1.5
16	V5	57	222	222	70	1958–59	0.484	0.456	1.5
16	V5	57	222	222	70	1959–60	0.490	0.443	1.5
17	V6	50	226	226	71	1959–60	0.389	0.581	1.7
18	V6/2	40	218	218	66	1960–61	0.276	0.790	2.1
19	B25	37	300	249	135	1957–58	0.613	0.407	1.2
19	B25	37	300	249	135	1958–59	0.669	0.372	1.2
13	LBC 6 in.		216	152	140	1948–49	0.310	0.492	2.1
20	H2		349	222	70	1958–59	0.315	0.705	2.0

*Width of brick is dimension through wall.

was probably also largely caused by the high conductivity of the dense clay with which they were made. This is indicated by comparison with the values given in Table 3 for the LBC 6 in. bricks which were made from a relatively porous clay.

4. SURFACE RESISTANCE

The transfer of heat between a surface and its surroundings depends upon several factors, as shown by the surface heat balance equations (1.4) in Chapter 1. For practical purposes it is expressed most conveniently as a resistance. At the outside surface of a north-facing wall of moderate to sheltered exposure the recommended value for the resistance, R_{so}, is 0.055 m^2K/W. This is based upon mean winter values obtained by measurement on many test walls, although the results varied considerably. Since the outside surface resistance is such a small fraction of the air-to-air resistance of a wall—about 5 percent or less—the temperature difference between the surface and the air will be correspondingly small, say 1 degC. This is especially so for a north-facing wall where forced convection accounts for the greater part of the heat exchange. Any error in temperature measurement will tend to magnify the effect of this small temperature difference in the calculation of R_{so} and thereby contribute markedly to the spread of experimental results.

From experiments on heat transfer in forced convection between a plane surface and air moving over it at velocity v (measured in m/s) Jürges (1924) gives the following results for the transfer coefficient

for smooth surfaces, e.g. glass

$$h_c = 5.7 + 3.9v \quad \text{W/m}^2\text{K}$$

for rough surfaces, e.g. brick

$$h_c = 6.2 + 4.2v \quad \text{W/m}^2\text{K}$$

These expressions were derived from experiments with boundary conditions carefully controlled. Although it was not possible, nor indeed required, to control the temperatures and air flow in the experiments referred to above on test walls in natural exposure, nevertheless the results obtained support the use of Jürges's formulae for calculating the convection heat transfer coefficient h_{co} at a naturally exposed surface.

The value of the inside surface resistance for walls recommended by the IHVE *Guide* is 0.12 m^2K/W. This is supported by the results of experiments on many walls in the Wall Laboratory.

At the inside surface the test walls in these experiments faced a sealed cubic enclosure exchanging heat by natural convection with air, temperature θ_i, thermostatically controlled at 18.3 degC, and by radiation with five interior plane surfaces, Assuming that these five surfaces have the same temperature, θ_ρ, the heat flux, q_{si}, at the surface of the test wall, temperature θ_{si}, is

$$q_{si} = h_{ci}(\theta_i - \theta_{si}) + Eh_{ri}(\theta_\rho - \theta_{si})$$

or, by slight rearrangement

$$q_{si} = h_{ci}(\theta_i - \theta_{si}) + Eh_{ri}\{(\theta_i - \theta_{si}) - (\theta_i - \theta_\rho)\}$$

Dividing q_{si} by $(\theta_i - \theta_{si})$ gives the surface heat transfer coefficient

$$h_{si} = h_{ci} + Eh_{ri}\left\{1 - \frac{\theta_i - \theta_\rho}{\theta_i - \theta_{si}}\right\}$$

The value of E, the effective emissivity for radiation heat exchange, is not known precisely owing to uncertainty about the emissivity of the walls at the mean room temperature. Emissivity data tabulated by Fishenden and Saunders (1950) suggests that 0.95 might reasonably be assumed; with this figure the value calculated for Eh_{ri} is 4.9 W/m^{2}K. Values of h_{ri}, the radiation coefficient, may be obtained from Fig. 2, Chapter 1.

The convection transfer coefficient, h_{ci}, varies according to whether the air flow in free convection over the wall surface is laminar or turbulent. Following McAdams (1954), in the region of streamline flow of warm air at mean room temperature over the cooler surface of the wall (height greater than 600 mm)

$$h_{ci} = 1.78(\theta_i - \theta_{si})^{1/4} \quad \text{W/m}^2\text{K}$$

In the turbulent region

$$h_{ci} = 1.66(\theta_i - \theta_{si})^{1/3} \quad \text{W/m}^2\text{K}$$

The separate regions are defined by the value of the Nusselt number (Nu): for streamline flow McAdams recommends Nu between 6 and 100, and between 100 and 1000 where the flow is turbulent. In the Wall Laboratory it was difficult to establish the type of flow, although observations of air and surface temperature suggest that it was probably turbulent. Assuming therefore $\theta_i = \theta_\rho = 18.3$ degC, values of the surface resistance R_{si} $(=1/h_{si})$ have been calculated for selected values of θ_{si} (Table 4).

Measurements on forty four test walls gave a mean inside surface resistance of 0.137 m^{2}K/W with 80 percent of the values falling between 0.123 and 0.158. The value recommended in the IHVE *Guide* is 0.123 based probably on early work in the Wall Laboratory at the BRS.

Table 4 Calculated values of R_{si}

θ_{si}	12.8	14.4	15.6	16.7 degC	
R_{si}	0.136	0.139	0.143	0.148 m^{2}K/W	laminar flow
R_{si}	0.122	0.127	0.132	0.139 m^{2}K/W	turbulent flow

5. CAVITY RESISTANCE

The cavity wall was little known before 1912. It began with various origins but of these the heat insulating property of the airspace and protection from water penetration are the most important attributes. It was used increasingly in the small-house building drive between the wars and more so since 1945. In the early period both leaves were of brick but this has now given way almost entirely to lightweight concrete blocks for the inner leaf.

For practical purposes calculations of heat transmission through walls assume the thermal resistance of a cavity to be 0.18 m^2K/W, corresponding to 1.0 degF ft^2h/Btu following on the work of Rowley and Algren (1929). They showed experimentally under controlled laboratory conditions that at a given temperature the resistance of a vertical cavity bounded by fairly smooth surfaces (hardboard, plasterboard, and similar materials) was constant at all cavity thicknesses above about 20 mm. At a mean cavity temperature of −6.7 degC the resistance of a 25 mm airspace was 0.17 m^2K/W, decreasing to 0.15 at 10 degC (see Fig. 2, Chapter 3).

Many of the walls investigated in the BRS Wall Laboratory were of cavity construction (see Table 1, Chapter 8) and included a variety of building products. The mean values of cavity resistance for a whole winter divided broadly into two groups, one with a resistance averaging below 0.16 m^2K/W, the other with a resistance of over 0.18. The first group consisted almost wholly of walls constructed of brick or concrete blocks, the cavities being 2.4 m high and 2.4 m horizontally. The average thermal resistance of the group was 0.14 m^2K/W, with individual values ranging from 0.11 to 0.17. The walls in the second group with the higher resistance values had smoother cavity surfaces. A probable explanation of the increased airspace resistance of the latter group is that the smooth surfaces of the materials forming the cavities created less turbulence in the airflow than the rough suface of brick and lightweight concrete.

In their measurements of the thermal resistance of cavities, Rowley and Algren used plates 225 × 225 mm for airspaces up to 17.5 mm and 900 m square for air spaces 12.5–37.5 mm to examine the effect of varying the thickness of the cavity. In order to determine whether an increase in the size of the heated panel would affect the result, a special wall was built in the Wall Laboratory. It consisted simply of two leaves of 4.8 mm hardboard enclosing a cavity that could be altered in thickness from 12.5 to 200 mm. Table 5 gives the mean values of the observations for each series of tests, while Fig. 15 shows the variation in thermal resistance, calculated from the measurements, with cavity width.

It will be seen that there is little change in the thermal resistance as the thickness of the cavity is increased above 25 mm. This agrees with the result obtained by Rowley and Algren, although the actual thermal resistance determined from data for the larger panel is greater than their value. The reason for this is not known.

Table 5 Heat flow measurements on a cavity wall

	Cavity thickness (mm)					
	12.7	19	25.4	50.8	101	203
Heat flow (measured) (W/m^2)	32.8	34.3	32.2	31.5	23.8	19.4
Temperature drop across cavity (degC)	1.8	6.1	6.2	6.3	5.0	4.0
Theral resistance of cavity (m^2K/W)	0.123	0.178	0.192	0.201	0.210	0.206
Mean air temperature in cavity (degC)	11.1	9.6	10.1	10.1	11.8	13.2
Computed conduction transfer (W/m^2K)	1.97	1.31	0.988	0.494	0.256	0.125
Computed convection transfer (W/m^2K)	0.017	0.392	0.596	0.886	0.897	0.863
Computed radiation transfer (W/m^2K)	4.32	4.25	4.27	4.28	4.34	4.42
Computed total transfer (W/m^2K)	6.31	5.95	5.85	5.66	5.49	5.41
Measured total transfer (W/m^2K)	7.12	5.62	5.20	4.97	4.76	4.85

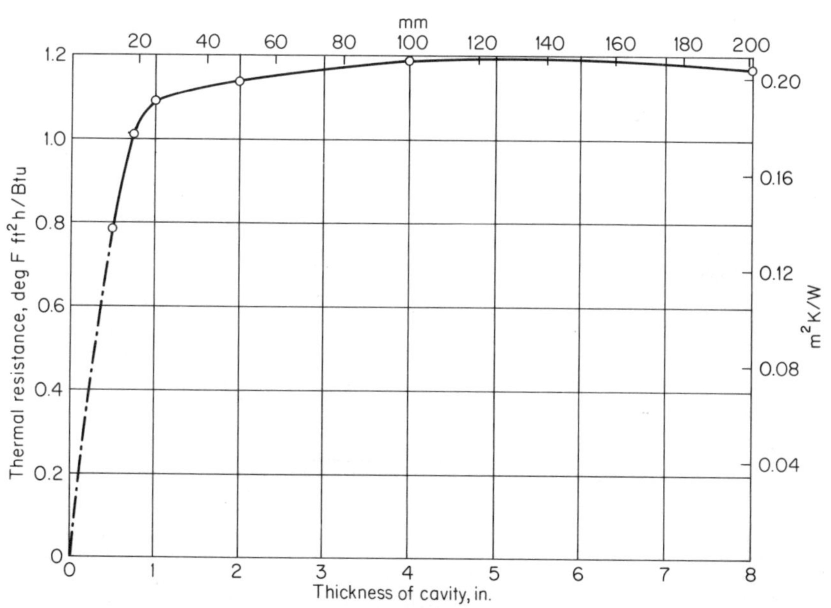

Fig. 15. Thermal resistance of a cavity in a wall. From measurements made in the BRS Wall Laboratory on an experimental wall with hardboard leaves

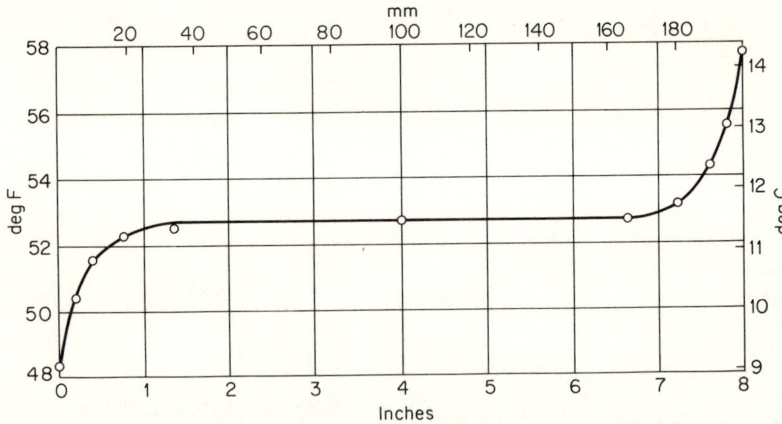

Fig. 16. Temperature profile across a 200 mm thick vertical cavity

With the hardboard-faced cavity at its maximum thickness of 200 mm, measurements were made of the temperature of the faces and of the air at various points in a line across the cavity at its mid position. The result is shown in Fig. 16. Considerable short-period fluctuations were recorded in the temperature of the air within about 20 mm of the walls but nearer the middle of the cavity the readings were steady.

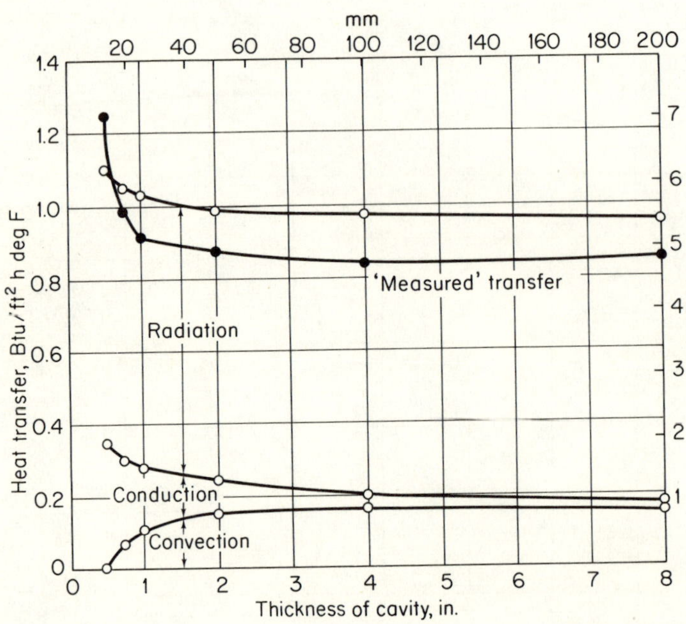

Fig. 17. Heat transfer across a wall cavity. Comparison of the measured transfer and that computed from surface temperature measurements on an experimental wall with hardboard leaves

Heat Transfer across a Cavity

To illustrate the contribution due to the various processes in the transfer of heat across a cavity, the figures in Table 5 have been calculated with Wall Laboratory data, following Fisheden and Saunders (1950) (see Appendix 1) and are illustrated in Fig. 17.

Direct conduction through the air is responsible for about 30 percent of the total heat transfer at the smallest cavity thickness (12.7 mm) but decreases to a negligible proportion as the thickness increases. Convection contributes little in a 12.7 mm airspace but increases to about 20 percent of the total amount at 50.8 mm. The radiation heat transfer is constant in amount over the whole range of cavity thickness, so that its proportional contribution increases slightly up to 50.8 mm as the combined amount due to conduction and convection decreases. This explains the large increase in the heat insulating value of a cavity obtained by using a low emissivity surface.

Ventilated Cavity

It is stated by Sheard (1938) that when the BRS Wall Laboratory was built one of the principal investigations in mind was to determine the increase in heat transmission due to ventilating the airspace in a cavity wall.

Accordingly experiments were made on a cavity wall of Fletton brickwork with forced and natural ventilation. The results are given in Table 6. The volume flow rate of 0.07 m^3/s corresponds to a ventilation rate of about 900 air changes per hour for the 50.8 mm airspace. This is considerably in excess of cavity ventilation rates likely to occur in normal building practice. It is instructive however to compare the measured result for this extreme condition with the theoretical value calculated with the set of steady state heat

Table 6 Coefficients of heat transmission for cavity walls

Description of wall	Thermal transmittance (W/m^2K) Average value for whole period of tests
280 mm Fletton brick cavity wall, extreme ventilation of cavity, approx rate of air flow 0.07 m^3/s	2.6
280 mm Fletton brick cavity wall, moderate natural ventilation of cavity	2.0
280 mm Fletton brick cavity wall, cavity unventilated	1.8

balance equations (1.4) in Chapter 1, using the following test data

$l_1 = l_2 = 114$ mm
$k_1 = 0.84$ W/mK
$k_2 = 0.94$ W/mK
$n = 900$ air changes/h
$c = 1340$ J/m^{3}K
$l = 50$ mm
$L = 2.4$ m
$\theta_o = 4.8$ degC
$\theta_i = 18.3$ degC
$\theta_{s.o} = 2$ degC
$h_{si} = 7.9$ W/m^{2}K
$h_{ci} = 5.7 + 4.1v$ W/m^{2}K
$v = nL/3600 = 0.6$ m/s
$h_{co} = 13$ W/m^{2}K

The calculation gives the temperature distribution through the wall shown in Fig. 18. By definition the thermal transmittance is

$$U = q/(\theta_i - \theta_o)$$
$$= \frac{k_1}{l_1}\left(\frac{\theta_{si} - \theta_1}{\theta_i - \theta_o}\right)$$
$$= 2.7 \text{ W/m}^2\text{K}$$

For similar conditions Sheard obtained in the Wall Laboratory a U-value of 2.6. In view of the practical difficulties of the experiment the agreement between the experimental and calculated values is as good as may be expected.

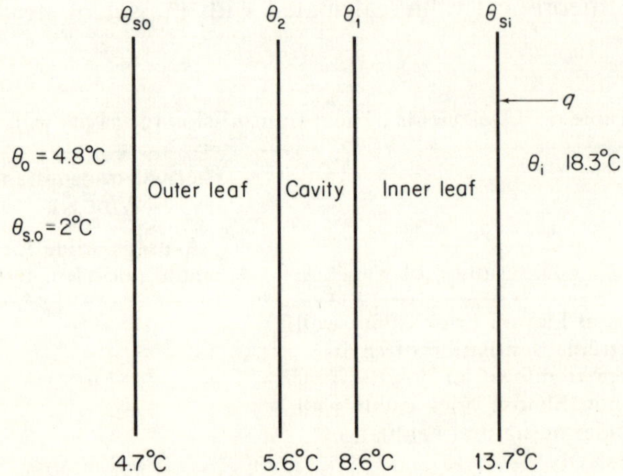

Fig. 18. Temperature distribution due to cavity ventilation in a hollow wall

For the 280 mm Fletton brick cavity wall, cavity unventilated, the equations (1.4) in Chapter 1 give a calculated thermal transmittance of 1.8 W/m^2K which agrees with the average value obtained by measurement (Table 6).

6. HEAT TRANSMISSION THROUGH SINGLE- AND DOUBLE-GLAZED WINDOWS

Windows form a substantial proportion of the envelope of most buildings. Because of their high thermal conductance, and the diathermancy of glass, windows may have a strong influence on the thermal behaviour of buildings. From time to time therefore walls with windows were included among the many test panels investigated for their thermal properties in the Wall Laboratory. Although not insolated directly, a considerable amount of diffuse radiation entered the windows in these north-facing panels, scattered and reflected by sky and clouds, and by the ground and surrounding buildings. In October and March, particularly, the amount was often sufficient to maintain the test room at the design temperature of 18.3 degC or more during the day without the addition of any electrical heating.

The heat balance for the constant temperature room with a glazed test panel is

$$Q + ItAg = UA(\theta_i - \theta_o) + H \tag{6.1}$$

where Q is the electrical heat input to the room, $ItAg$ is the solar heat entering the room through an area Ag of glass, A is the area of the test panel, θ_i and θ_o are the inside and outside air temperatures, and H is the heat loss from the room through surfaces other than the test panel.

By definition the thermal transmittance (U-value) of the panel is the unit rate at which it transfers heat in steady flow between the air on opposite sides. For test panels that include glazing the U-value may be determined most directly as the mean rate of heat flow during the night provided (see Section 7 below) the thermal capacity of the panel is sufficiently small that the flow may be assumed steady over a typical night period. The transient temperature response curves in Fig. 5, Chapter 7, indicate this is acceptable for single and double glazing though becoming less so as the thermal capacity of the conducting medium increases.

During the course of the winter measurements were made of the thermal transmittance of the windows in some of the walls, using heat-meter plates on the warm side of the windows to determine the mean heat flow through the glass during the night. The mean thermal transmittances obtained were:

Single-glazed window	$U = 5.4$ W/m^2K
Double-glazed window 4.8 mm space	$U = 4.4$ W/m^2K
Double-glazed window, 12.7 mm space	$U = 3.0$ W/m^2K
Double-glazed window, 50.8 mm space	$U = 2.8$ W/m^2K

The values confirm that an increase in cavity thickness over 12.7 m gives no appreciable improvement in the thermal insulation. They agree also with the generally accepted transmittance values for single and double glazing given in the IHVE *Guide Book A* (1970).

Of the various coefficients defining heat transmission in structures, h_{so}, the coefficient of heat transfer at the outside surface, is the most variable. This is due very largely to the weather with air movement and radiation heat exchange the principal causal factors. The relationship is complex, though it may be expressed empirically with a regression eqation

$$h_{so} = a_o + a_1 v + a_2 \left(\frac{T_o^4 - T_{s.o}^4}{\theta_{so} - \theta_o} \right)$$

Even so, the mean outdoor temperature for long-wave radiation heat exchange, $\theta_{s.o}$, is normally not known and the above expression therefore may not be suitable for general use.

Measured values of h_{so} and the relevant wind speed v from experiments in the Wall Laboratory on heat transmission through glass are shown in Fig. 19. Two linear relationships are drawn through the plot of points, one representing the regression equation $h_{so} = 5.9 + 4.6v$, and the other $h_c = 5.7 + 3.9v$ due to Jürges (1924) correlating the results of several experiments to determine the effect of air flow on the convection heat transfer coefficient for a smooth surface. For values of v ($\leqslant 4.9$ m/s) specified by Jürges, the two relationships are very similar, and demonstrate that his simple formula may be used in practical calculations for the numerical prediction of heat flow through the outside surface of a glass window.

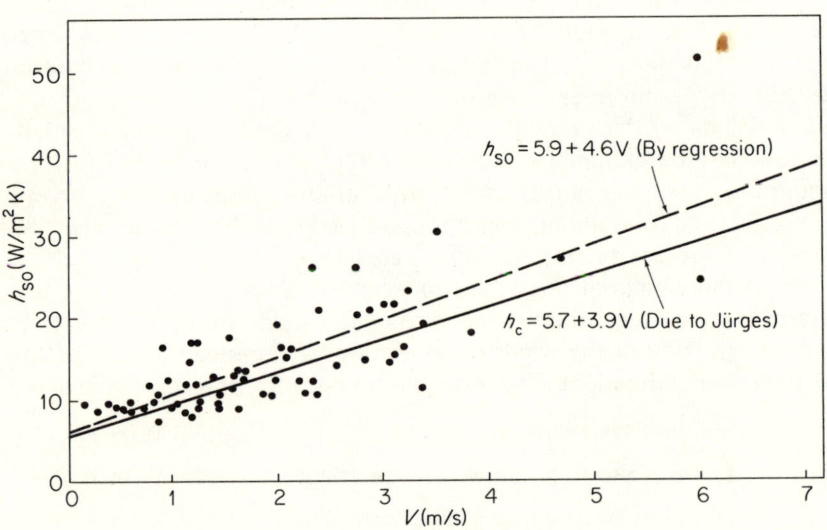

Fig. 19. Variation of outside surface coefficient h_{so}, with wind speed at night

Radiation Exchange between Windows and the Sky Vault

The net amount of radiation interchanged between two surfaces, dA_1 and dA_2, separated by a distance r in a non-absorbing medium is

$$dq_{1.2} = I\, dA_1 \cos\theta_1\, dA_2 \cos\theta_2 / r^2 \tag{6.2}$$

where $\cos\theta_1$ and $\cos\theta_2$ convert the actual areas to apparent areas as viewed by each surface, and I is the difference between the radiation intensities in the normal direction at the two surfaces. Referring to Fig. 20, let dA_1 be the window with its outside surface viewing a quarter sphere representing the sky vault. Then dA_2 denotes an elementary area of sky with co-ordinates (r, θ, ϕ). Assuming that sky radiation is independent of the azimuth angle ϕ, then by (6.2) the flux of diffuse radiation at the window surface is

$$\frac{dq_{1.2}}{dA_1} = I \cos\theta\, d\omega_1$$

where $d\omega_1$ is the small solid angle subtended by dA_1 at dA_2. From Fig. 20

$$d\omega_1 = \frac{\pi r \sin\theta}{r^2} r\, d\theta$$

$$= \pi \sin\theta\, d\theta$$

so that

$$\frac{dq_{1.2}}{dA_1} = \pi I \int_0^{\pi/2} \sin\theta \cos\theta\, d\theta$$

$$= \frac{\pi I}{2} \int_0^{\pi/2} \sin 2\theta\, d\theta$$

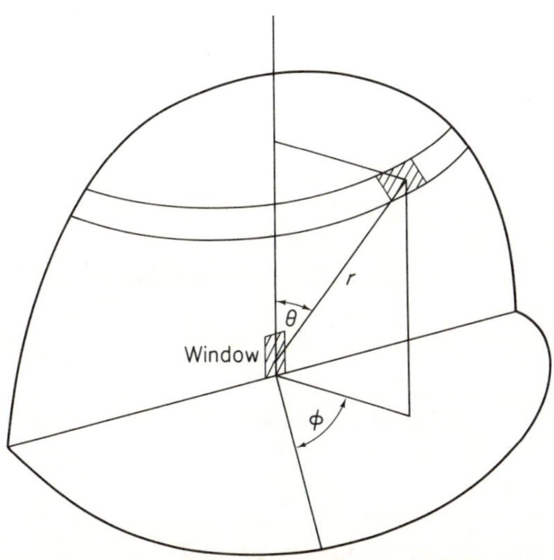

Fig. 20. Representation of a vertical window exposed to diffuse radiation from the sky

At the angle of incidence θ a fraction $\nu(\theta)$ of the incident radiation is reflected from the glass; the total fraction of diffuse radiation reflected therefore is

$$R = \frac{\dfrac{\pi I}{2}\displaystyle\int_0^{\pi/2} \nu(\theta) \sin 2\theta \, d\theta}{\dfrac{\pi I}{2}\displaystyle\int_0^{\pi/2} \sin 2\theta \, d\theta}$$

$$= \int_0^{\pi/2} \nu(\theta) \sin 2\theta \, d\theta$$

Radiation from the sky is polarized due to scattering by particles about the same size as the wavelength of the radiation; (θ) may be represented therefore as the sum of two components, one for each of the two mutually perpendicular planes into which the scattered radiation may be resolved. These components may be calculated separately by Fresnel's equations relating the specular reflection of each from a single surface to the angle of incidence, θ

$$R_1 = \frac{\sin^2(\theta - \theta')}{\sin^2(\theta + \theta')}$$

$$R_2 = \frac{\tan^2(\theta - \theta')}{\tan^2(\theta + \theta')}$$

where

θ' = angle of refraction,
R_1 = fraction reflected of component parallel to plane of incidence, and
R_2 = faction reflected of component perpendicular to plane of incidence.

In addition to surface reflection a further loss in the energy transmitted occurs due to absorption in the glass. This may be calculated by assuming a decay relationship of the usual exponential form with the absorbing property of a single sheet characterized by a quantity KL, the product of the extinction coefficient, K, and the thickness of the glass, L.

From these considerations Parmalee (1945) calculated the fractions of sky radiation reflected, absorbed, and hence transmitted by single and double glass for different KL-values. For double glass with $KL = 0.054$ for each sheet, he calculated a transmission factor of 0.63.

A topic of wide practical interest, included in these experiments, is the effect of curtains in reducing the heat loss through windows in winter. A reduction of 19 percent was obtained with lightweight curtains and 27 percent with heavy ones. Both test panels had substantial wood frames occupying about 30 percent of the area; with lighter-weight sections it is expected that the improvement would be greater.

Apart from the reduction in heat loss there is an additional advantage in using curtains, for the occupants of the room are thereby screened from the cold glass and the radiant temperature of their surroundings is increased. The measurements were made on a north-facing window, but it is expected that the results would apply to any direction in winter. However, it should be remembered that, even in mid-winter, there is a useful solar heat gain available on many mornings through windows facing east or south-east.

7. THERMAL TRANSMITTANCE DEFINING HEAT TRANSFER IN VARIABLE FLOW

The flow of heat through the external structure of a building is rarely steady. Explicit expressions for the flow consist of a steady state term and a time dependent quantity determined by the boundary conditions. The former is the U-value of the structure, for all practical purposes it is assumed to be constant. The variable state quantity diminishes with time increasing at a rate determined by the thermal capacity of the conducting medium, light-weight structures reaching steady state more quickly than heavier structures. This is illustrated by the transient response curves of temperature in Fig. 4, Chapter 1, calculated with equation (2.1.42). They compare the thermal behaviour of high and low thermal capacity panels exposed to a constant temperature on one side, representing the indoor surface, and on the other facing an outdoor temperature to which a single step-function change has been applied and maintained for a prolonged period. Practical conditions may be simulated by reversing, through the same amount after time t_1, the direction of the ambient temperature change at the outside surface and, after a further time t_2, repeating this pattern in a sequence so that $t_1 + t_2$ is the period of a complete cycle. If t_1 and t_2 represent day and night periods, respectively, then the corresponding values of temperature and heat flow at the inside surface are given by equations (2.1.31) through (2.1.39) in Chapter 1. Alternatively the result may be expressed as a mean value of thermal transmittance for each period, calculated with equations (2.1.40) and (2.1.41). Calculated values of these quantities are given in Table 7.

Table 7 Calculated mean values of day and night thermal transmittances

	Thermal transmittance (W/m^{2}K)		
Wall description	Day time mean $U(t_1)$	Night time mean $U(t_2)$	24 h mean $U(t_1 + t_2)$
229 mm brickwork (heavy)	3.44	2.26	2.58
280 mm fibreboard (light)	1.51	1.48	1.49

The calculations assume the following data

$$\theta_i = 18.3 \text{ degC}; \qquad \theta_{e.1} = 4.4 \text{ degC}; \qquad \theta_{e.2} = -1.1 \text{ degC}$$

$$t_1 = 8 \text{ h}; \qquad t_2 = 16 \text{ h}$$

Brickwork:

$$l/k = 0.21 \text{ m}^2\text{K/W}; \qquad K = 0.07 \times 10^{-5} \text{ m}^2/\text{s}; \qquad l = 229 \text{ mm}$$

Fibreboard:

$$l/k = 0.49 \text{ m}^2\text{K/W}; \qquad K = 0.015 \times 10^{-5} \text{ m}^2/\text{s}; \qquad l = 25.4 \text{ mm}$$

It is seen that the large thermal capacity wall of 229 mm brickwork has a considerable difference between the mean day and night U-values, the calculated day-time transmittance being about 33 percent higher than the steady state value $U(t_1 + t_2)$. The longer night period diminishes the influence of heat storage in the wall, giving a value for $U(t_2)$ that is 11 percent below the 24 h mean.

With the light wall of 25.4 mm fibreboard the effect of thermal capacity is barely perceptible, as indicated by the respective transmittance values which are almost identical.

The experimental results shown in Table 8 of transmittance measurements in the Wall Laboratory are consistent with the above theoretical considerations. The walls are identified by their prefix number among those illustrated in Chapter 8. All but wall no. 58 are heavy masonry structures.

Table 8 Thermal transmittances measured in the Wall Laboratory 1957–8. Whole winter, October to March inclusive (number of observations in brackets)

	Thermal transmittance (W/m²K)		
Wall description	24 h 09.15–09.15 GMT	Day time 09.15–16.30 GMT	Night time 16.30–09.15 GMT
19. Swiss B25 perforted bricks rendered and plastered	1.2 (168)	1.5 (122)	1.1 (122)
15. BRS V4 perforated bricks, plastered	1.5 (168)	1.9 (123)	1.4 (123)
14. BRS V2 perforated bricks, plastered	1.4 (162)	1.8 (121)	1.3 (120)
23. Brick—cavity—aerated concrete, plastered	1.4 (164)	1.8 (123)	1.3 (121)
27. Well-insulated heavy wall	0.57 (163)	0.74 (118)	0.57 (115)
58. Well-insulated light wall	0.68 (160)	0.68 (113)	0.62 (108)

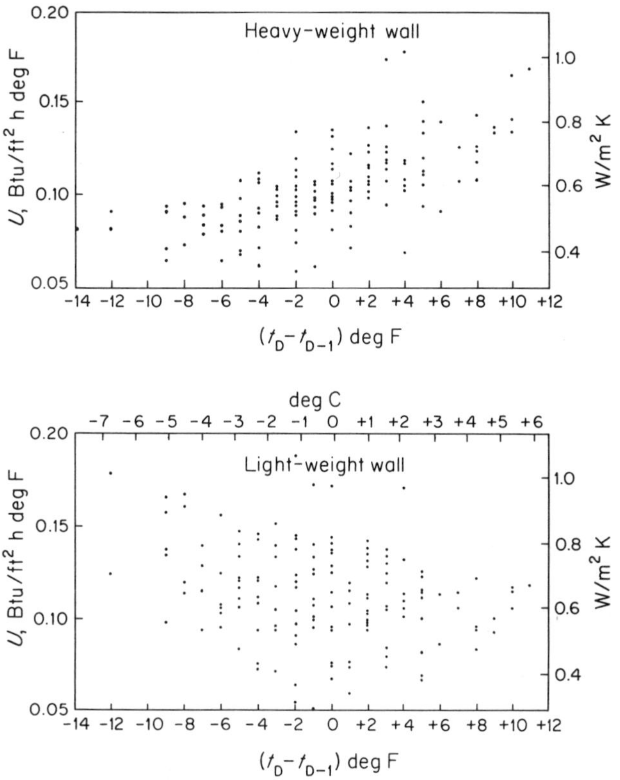

Fig. 21. Variation in thermal transmittance with day-to-day change in mean air temperature

The effect of thermal capacity on the thermal transmittance may extend beyond a 24 h cycle and create differences between mean values of $U(t_1+t_2)$ for successive days. This is shown by the dot diagrams of experimental data in Fig. 21. The measured thermal transmittance for each 24 h period is plotted against the difference between the mean outdoor air temperature for that period and the corresponding temperature for the preceding 24 h. The transmittance of the lightweight wall is seen to be almost independent of day-to-day temperature differences. The heavyweight wall behaves differently, the transmittance increasing when the mean air temperature rises, and decreasing when the temperature falls below that of the previous day.

The calculations above, using the equations specified in Chapter 1, involve extensive arithmetic in evaluating the quantity under the summation sign. This may be greatly simplified, without signficant loss of accuracy depending upon the thermal capacity of the wall and the length of the time interval considered, by writing

$$\sum_{n=1}^{\infty} 2a_n H(\alpha_n) \simeq \frac{1+(B_{si}+B_{so})/2+B_{si}B_{so}/6}{B_{si}+B_{so}+B_{si}B_{so}} \tag{7.1}$$

Table 9 Calculated mean values of day and night thermal transmittance: comparison of exact and approximate values

Wall description		Thermal transmittance (W/m^2K) Day-time mean $U(t_1)$	Night-time mean $U(t_2)$
229 mm brickwork	Exact	3.44	2.26
(heavy)	Approx	3.51	2.23
280 mm fibreboard	Exact	1.51	1.48
(light)	Approx	1.51	1.48

as shown in Chapter 1. The values of thermal transmittance given in Table 7 have been recalculated using the approximation (7.1) and the result is shown in Table 9.

It will be seen that the exact and approximate methods give identical results for the lightweight wall. With the heavier wall of 229 mm brickwork the maximum error due to the approximation is no more than about 2 percent and occurs with the day-time value. The error is half this amount for the longer night period.

The magnitude of the effect of the dynamic term, and hence that of the error due to the approximation (7.1), depends upon a ratio of temperature differences: $(\theta_{e.1}-\theta_{e.2})/(\theta_i-\theta_{e,1})$ for the day time, and a similar factor for the night period. For values therefore of the ambient temperatures other than those used in Table 9 the result may be different. With north-facing walls, however, the mean winter values of temperature are such that the approximation is sufficiently accurate for practical purposes.

Although the result (7.1) applies only to a solid wall of uniform material it is probable that an approximate expression of similar form may be obtained for a more complex structure.

8. THERMAL DIFFUSIVITIES OF SINGLE-LAYER WALLS

Walls of different construction, though designed to have the same value of thermal transmittance—a coefficient of steady heat flow—may differ in their response to fluctuations in the ambient temperature such as occur constantly in natural exposure. The physical property of a material that determines this behaviour in non-steady heat flow is the thermal diffusivity, K. It is calculated easily with the defining formula $K = k/\rho c$, and may be determined experimentally by measuring a temperature versus time relationship in the passage of a temperature change through a test specimen. The thermal diffusivities of some single-layer walls were determined in the BRS Wall Laboratory by applying a periodic temperature variation to the outside surface and measuring the time-lag between this and the resulting temperature change at the inside face. As a check on the accuracy of the method the

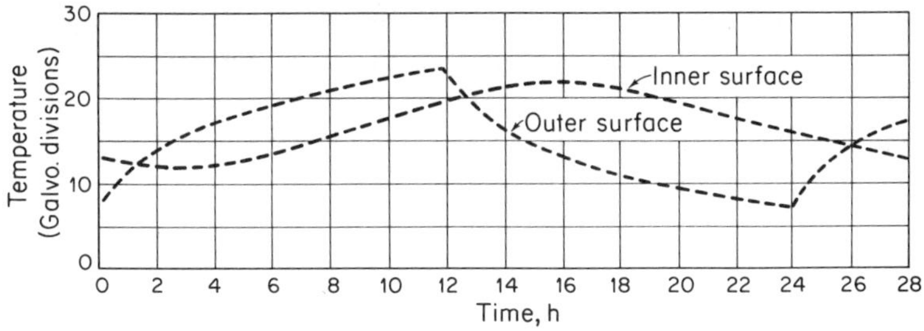

Fig. 22. Part of a typical temperature record

theoretical ratio of the amplitudes of these temperature oscillations was calculated using the diffusivity values determined experimentally and compared with the amplitude attenuations observed.

The periodic temperature variation was arranged by intermittent switching of an electric heater fixed to the outside surface of the wall. The regular train of heat pulses was controlled to provide alternate heating and cooling cycles of equal duration with a period of 24 h. A typical record of steady periodic temperature is shown in Fig. 22 where it will be seen that the temperature pattern at the inside surface is very nearly sinusoidal. The succession of rise and decay curves that formed the regular variation at the outside surface was examined by Fourier analysis to determine the first harmonic. The component was then compared with the temperature wave at the inside surface, assumed to be sinusoidal, to give the time-lag at the switching frequency (Parsons and Burnand, 1951).

In Chapter 1, equation (2.2.1) gives the steady cyclic temperature of the inside surface of the wall following a sinusoidal variation in the ambient temperature at the outside surface. By putting $1/B_{so} = 0$ in this result it follows that the steady periodic temperature at the inside surface of a homogeneous wall where the ambient temperature is constant, transmitting the circular function sin $2\pi t/T$, periodicity T, applied to the temperature θ_{so} of the outside surface as in the experiments described above, is

$$\theta_{si}(t) = \frac{2\sqrt{(2\pi/KT)}\exp\{-l\sqrt{(\pi/KT)}\}\,\theta_{si}\sin\{(2\pi t/T) - l\sqrt{(\pi/KT)} + \psi\}}{\sqrt{\{(h_{si}/k)^2 + (2h_{si}/k)\sqrt{(\pi/KT)} + 2\pi/KT\}}} \tag{8.1}$$

with

$$\tan\psi = \{1 + (2k/h_{si})\sqrt{(\pi/KT)}\}^{-1} \tag{8.2}$$

From (8.1) the phase lag, symbol λ is

$$\lambda = l\sqrt{(\pi/KT)} - \tan^{-1}[1/\{1 + (2k/h_{si})\sqrt{(\pi/KT)}\}] \tag{8.3}$$

equivalent to a time-lag, δ, given by

$$\delta = \lambda T/2\pi \tag{8.4}$$

The phase lag for a similar temperature function at a distance l into a semi-infinite solid is $l\surd(\pi/KT)$. Comparing this result with (8.1) indicates that ψ may be regarded as a correction to the semi-infinite solid relationship that allows for the finite thickness of the wall. As a first step therefore in obtaining the diffusivity from the experimental data a value for K may be estimated using the approximate result

$$K = \frac{T}{4\pi}\left(\frac{l}{\delta}\right)^2 \tag{8.5}$$

The exact expression for the diffusivity is given by equations (8.3) and (8.4) as

$$K = \frac{Tl^2}{4\pi[\delta + (T/2\pi)\tan^{-1}\{1 + 2(k/h_{si})\surd(\pi/KT)\}^{-1}]^2} \tag{8.6}$$

Using the approximate value of K already obtained a closer approximation may then be calculated with (8.6) and the process repeated until the value of the diffusivity becomes stable.

The period of the applied symmetrical variation of temperature in these experiments for the determination of thermal diffusivity need not be 24 h. With this particular valuc of T however, the experiment has the added interest of measuring directly the dynamic response of the wall to a diurnal variation in the ambient temperature, a property of fundamental importance in the interpretation of the thermal behaviour of buildings.

The calculation of K by the method described above may be lengthy. Values can be obtained more simply, though possibly with slightly less accuracy, from a chart prepared by writing equation (8.4) as

$$\delta = \sqrt{\left(\frac{T}{2\pi}\right)}\sqrt{\left(\frac{Q}{C}\right)} - \frac{T}{2\pi}\tan^{-1}\left\{\frac{1}{1 + \frac{2}{h_{si}}\sqrt{\left(\frac{2\pi}{T}\right)}C\sqrt{\left(\frac{Q}{C}\right)}}\right\} \tag{8.7}$$

where Q is the quantity of heat contained by the slab in the steady state for unit temperature difference betwcen opposite faces and equal therefore to $\rho cl/2$, and C is the thermal conductance, k/l. Choosing C as the parameter and plotting Q/C against δ, a family of curves may be constructed to give the diffusivity directly from the observed time-lag and known conductance of the wall. Such a chart is shown in Fig. 23. It has been constructed for a sinusoidal variation in the temperature of the outside surface, period 24 h, and with the inside surface heat transfer coefficient, h_{si}, equal to 1.4 Btu/ft^{2}h degF ($=7.9$ W/m^{2}K) which is the value generally used for calculating heat conduction in buildings. The corresponding chart for the amplitude ratio has been similarly prepared and is shown in Fig. 24.

Time-lag measurements and calculated values of the thermal diffusivity for some homogeneous walls are given in Table 10 (Pratt and Lacy, 1966). The thermal conductances and surface transfer coefficients of the walls needed for the calculations were obtained from mean values of the heat flow

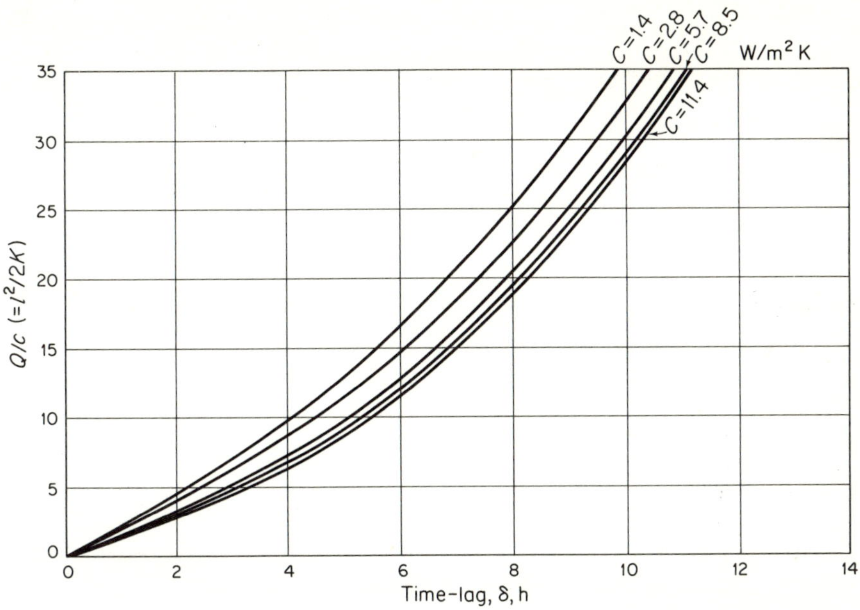

Fig. 23. Periodic heat flow through a uniform slab Q/C versus time-lag, δ

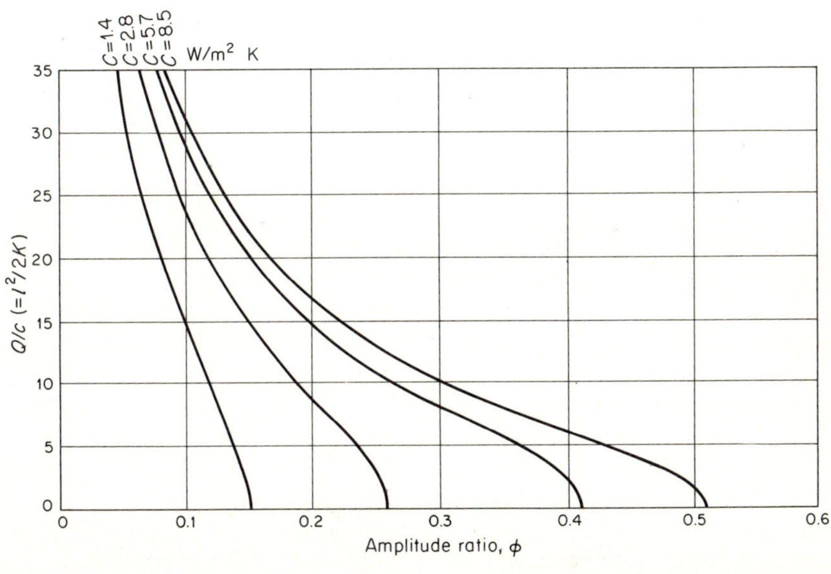

Fig. 24. Periodic heat flow through a uniform slab Q/C versus amplitude ratio ϕ

Table 10 Results of measurements of periodic heat flow through homogeneous walls

Wall construction	Measured			Calculated
	k (W/mK)	l (mm)	δ (h)	K (m^2/s) $\times 10^{-5}$
1. Solid cast brick aggretate concrete	1.19	203	5.8	0.0467
2. Solid cast gravel concrete	1.95	203	4.9	0.0668
3. Fletton brickwork	0.962	229	5.8	0.0557
4. Aerated concrete blocks, rendered, plastered	0.548	178	5.3	0.0359
5. Foamed-slag concrete 1 : $2\frac{1}{2}$: $7\frac{1}{2}$* rendered, plastered	0.464	229	7.5	0.0333
6. Foamed-slag concrete 1:2:8 rendered, plastered	0.410	229	8.0	0.0297

* Mix proportions refer to cement: fine aggregate: coarse aggregate.

and surface temperatures measured over a continuous period of several days.

The results indicate an increase in the thermal diffusivity with conductivity as illustrated by the curve fitted to these particular points in Fig. 25 and supported by data reported by Billington (1949, 1952) for other materials covering a similar range of thermal values. A method due to Ångström was used by Billington (1949) for measuring the diffusivities of laboratory

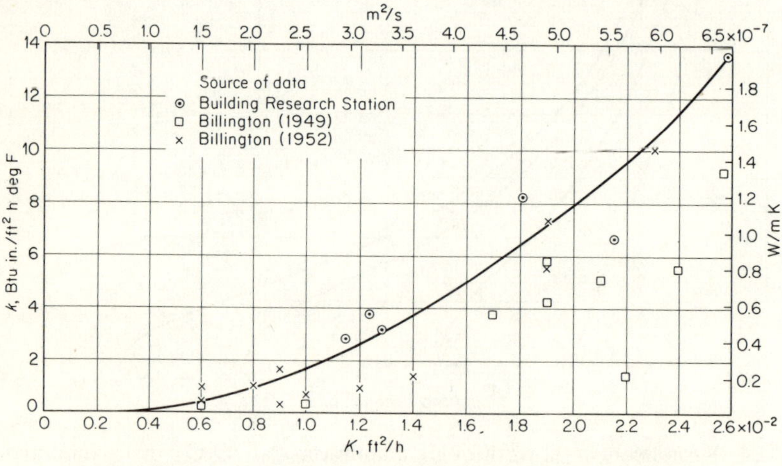

Fig. 25. Variation of thermal conductivity with diffusivity

specimens of some poorly conducting building materials and gave results that agree reasonably well with those obtained in the Wall Laboratory.

The thermal diffusivities of various other materials, including metals, have been determined by Strong, Bundy, and Bovenkerk (1960) from measurements of the temperature propagation through a semi-infinite solid due to a suddenly activated plane heat source.

Because of the brief heating period dynamic methods have been used in an attempt to determine the moisture dependence of thermal conductivity (see Chapter 2). This technique was extended by Hatton (1960) to a similar examination of thermal diffusivity using specimens of damp fibreboard and baked cork slab; otherwise very few practical results on the thermal diffusivity of damp material are available.

The values of thermal transmittance and other properties measured in the Wall Laboratory on specimen panels exposed to the weather are more representative of practical performance than values obtained by guarded hot-box methods using a controlled temperature environment. The test panels in the Wall Laboratory faced due north and had a moderate to sheltered exposure. Attempts to relate the test data with simple expressions that evaluate the thermal transmittance for other orientations and degrees of exposure unfortunately proved unsuccessful, owing to the inadequate range of variation in the weather factors recorded at the site. A new laboratory specially designed to overcome this difficulty followed the previous building when it was taken out of commission. The new laboratory was sited to have more severe exposure than its predecessor and constructed to be rotatable in azimuth through a complete circle thus enabling the test panels to be exposed in any direction as required. No results of experiments in the rotatable heat transmission wall laboratory have been published.

APPENDIX 1

Computation of the Conduction, Radiation, and Convection Heat Transfer across a Cavity

Symbols used

θ_1 = temperature of inner leaf
θ_2 = temperature of outer leaf
θ_a = temperature of air = $(\theta_1 + \theta_2)/2$
θ = temperature difference between surfaces = $(\theta_1 - \theta_2)$
l = cavity thickness
H_k = conduction heat transfer
H_r = radiation heat transfer
H_c = convection heat transfer
E_1 = emissivity of inner surface
E_2 = emissivity of outer surface
Nu = Nusselt number = hl/k

Gr = Grashof number = $ag\theta l^3 \rho^2/\mu^2$
Pr = Prandtl number = $c\mu/k$
h = heat transfer coefficient (conduction + convection)
k = conductivity of air
ρ = density of air
g = acceleration due to gravity
a = coefficient of cubic expansion of air
μ = viscosity of air
c = specific heat of air at constant pressure

Taking the cavity thickness of 25 mm as an example

$$\theta_1 = 13.2 \text{ degC}$$

$$\theta_2 = 7.0 \text{ degC}$$

$$k = 0.025 \text{ W/mK}$$

$$l = 25 \text{ mm}$$

therefore heat transfer by conduction

$$\begin{aligned} H_k &= k(\theta_1 - \theta_2)/l \\ &= 6.2 \text{ W/m}^2 \\ &= 1 \text{ W/m}^2\text{K} \end{aligned}$$

The heat transfer by radiation between two parallel surfaces is given by the expression

$$H_r = 54.6 \times 10^{-10} \frac{E_1 E_2}{E_1 + E_2 - E_1 E_2} \{(\theta_1 + 273)^4 - (\theta_2 + 273)^4\} \text{ W/m}^2$$

Taking the emissivities of both surfaces to be 0.90 gives for the 25 mm cavity a radiation heat transfer of 26.6 W/m^2, or 4.29 W/m^{2}K. The convection heat transfer across the cavity is obtained by use of the relationships between the Grashof, Prandtl, and Nusselt numbers. The product (Gr · Pr) = $agc\theta l^3 \rho^2/k\mu$ is calculated with values for the properties of air given in Fishenden and Saunders (1950, Table XVIII) and the Nusselt number then obtained from Fig. 9, Chapter 3, relating Nu to $\log_{10}$ (Gr · Pr). This gives, for the present calculation, Nu = 1.60.

The sum of the conduction and convection heat transfer

$$\begin{aligned} H &= Nu \cdot H_k \\ &= 1.60 \times 6.2 \\ &= 9.92 \text{ W/m}^2 \end{aligned}$$

i.e.

$$h = 1.60 \text{ W/m}^2\text{K}$$

Therefore convection heat transfer = 1.6 − 1 = 0.6 W/m^{2}K.

REFERENCES

Billington, N. S. (1949) *J. Sci. Instrum.*, **26,** 20.

Billington, N. S. (1952) *Thermal Properties of Buildings*; Cleaver-Hume Press, Limited.

Boot, A. A. and Watson, A. (1964) Proc. Int. Conf. on Materials, Philadelphia, *ASTM,* **19.**

Cadiergues, R. and Genevay, J. (1952) *Annales de l'Institut Technique du Batiment et des Travaux Publics,* **5,** 52–53, 471.

Croney, D., Coleman, J. D., and Currer, E. W. H. (1951) *Brit. J. Appl. Phys,* **2,** 85.

Fishenden, M. and Saunders, O. A. (1950) *An Introduction to Heat Transfer, Clarendon Pess, Oxford.*

Hatton, A. P. (1960) *J. Mech. Eng. Sci,* **2,** 1, 45.

Jakob, M. (1949) *Heat Transfer,* Vol. 1, Chapman and Hall, London.

Jesperson, W. B. (1953) *J. Inst. Heat Vent. Eng.*, **21,** 113.

Jürges, W. (1924) *Beiheft zum Gesundheits-Ingenieur,* **1,** 19.

McAdams, W. H. (1954) *Heat Transmission,* McGraw-Hill, New York.

Parmalee, G. V. (1945) *ASHVE Journal,* Section Oct.–Nov.

Parsons, J. J. and Burnand, G. (1951) *J. Inst. Heat. Vent. Eng.*, **18,** 187, 478.

Pratt, A. W. and Lacy, R. E. (1966) *Int. J. Heat Mass Transfer,* **9,** 345.

Rowley, F. B. and Algren, A. W. (1929) *Trans. American Soc. Heating and Ventilating Engineers,* **35.**

Sheard, H. (1938) *J. Inst. Heat. Vent. Eng.*, **6,** 67, 360.

Strong, H. M., Bundy, F. P., and Bovenkerk, H. P. (1960) *J. Appl. Phys,* **31,** 1.

Index of Names

Index of Subjects